Parameter Estimation of Permanent Magnet Synchronous Machines

IEEE Press Series on Control Systems Theory and Applications

Series editor: Maria Domenica Di Benedetto

The series publishes monographs, edited volumes, and textbooks which are geared for control scientists and engineers, as well as those working in various areas of applied mathematics such as optimization, game theory, and operations.

Zi Qiang Zhu, Kan Liu, Dawei Liang

Parameter Estimation of Permanent Magnet Synchronous Machines

ISBN: 9781394280421

Reihaneh Kardehi Moghaddam, Mostafa Rabbani

Methods of Developing Sliding Mode Controllers: Design and Matlab Simulation

ISBN: 9781394314485

Hejia Gao, Wei He, Changyin Sun

Dynamic Modeling and Neural Network-Based Intelligent Control of Flexible Systems

ISBN: 9781394255276

Zi Qiang Zhu, Xi Meng Wu

Sensorless Control of Permanent Magnet Synchronous Machine Drives

ISBN: 9781394194353

Jianying Zheng, Liang Xu, Qinglei Hu, Lihua Xie

Control over Communication Networks: Modeling, Analysis, and Design of Networked Control Systems and Multi-Agent Systems

ISBN: 9781119885795

Levent Guvenc, Bilin Aksun-Guvenc, Sheng Zhu, Sukru Yaren Gelbal

Autonomous Road Vehicle Path Planning and Tracking Control

ISBN: 9781119747949

Leena Vachhani, Pranjal Vyas, Arunkumar G. K.

Embedded Control for Mobile Robotic Applications

ISBN: 9781119812388

Feng Liu, Zhaojian Wang, Changhong Zhao, Peng Yang

Merging Optimization and Control in Power Systems: Physical and Cyber Restrictions in Distributed Frequency Control and Beyond

ISBN: 9781119827924

Jongrae Kim

Dynamic System Modeling and Analysis with MATLAB and Python: For Control Engineers

ISBN: 9781119801627

Milad Farsi, Jun Liu

Model-Based Reinforcement Learning: From Data to Actions with a Python-based Toolbox

ISBN: 9781119808572

Jianying Zheng, Liang Xu, Qinglei Hu, Lihua Xie

Control over Communication Networks: Modeling, Analysis, and Design of Networked Control Systems and Multi Agent Systems over Imperfect Communication Channels

ISBN: 9781119885795

Akira Shimada

Disturbance Observer for Advanced Motion Control with MATLAB/ Simulink

ISBN: 9781394178100

Hasan Komurcugil, Sertac Bayhan, Ramon Guzman, Mariusz Malinowski, Haitham Abu-Rub

Advanced Control of Power Converters: Techniques and Matlab/ Simulink Implementation

ISBN: 9781119854401

Silvia Mastellone, Alex van Delft

The Impact of Automatic Control Research on Industrial Innovation: Enabling a Sustainable Future

ISBN: 9781119983613

Simone D'Amico, Matthew B. Willis

Distributed Space Systems: Dynamics, Navigation, and Control of Multi-Satellite Missions

ISBN: 9781119808954

Parameter Estimation of Permanent Magnet Synchronous Machines

Zi Qiang Zhu
University of Sheffield, UK

Kan Liu
Hunan University, China

Dawei Liang
University of Sheffield, UK

IEEE Press
WILEY

Published by John Wiley & Sons, Inc., Hoboken, New Jersey.

Published simultaneously in Canada.

Library of Congress Cataloging-in-Publication Data:

Names: Zhu, Ziqiang, Ph. D., author. | Liu, Kan (Dean), author. | Liang, Dawei (Research associate), author.
Title: Parameter estimation of permanent magnet synchronous machines / Zi Qiang Zhu, Kan Liu, Dawei Liang.
Description: Hoboken, New Jersey : Wiley, [2025] | Series: IEEE Press series on control systems theory and applications | Includes bibliographical references and index.
Identifiers: LCCN 2025009543 | ISBN 9781394280421 (hardback) | ISBN 9781394280445 (epdf) | ISBN 9781394280438 (epub) | ISBN 9781394280452 (obook)
Subjects: LCSH: Parameter estimation. | Permanent magnet motors.
Classification: LCC QA276.8 .Z48 2025 | DDC 621.46—dc23/eng/20250326
LC record available at https://lccn.loc.gov/2025009543

Cover Design: Wiley
Cover Images: © Bosca78/Getty Images, © Pobytov/Getty Images

Contents

Authors *xiii*
Preface *xv*
List of Abbreviations *xvii*
List of Symbols *xix*

1 General Introduction *1*
1.1 Introduction *1*
1.2 Permanent Magnet Machines *2*
1.3 Basic Equations and Machine Parameters *5*
1.3.1 Fundamental Mathematical Model for PMSMs *5*
1.3.2 Mathematical Model Considering Magnetic Saturation, Thermal Effect, and Iron Loss *7*
1.3.2.1 Influence of Magnetic Saturation *7*
1.3.2.2 Influence of Temperature *9*
1.3.2.3 Influence of Iron Loss *11*
1.4 Drives and Control Strategies *11*
1.4.1 Drive System of PMSM *11*
1.4.2 Space Vector Pulse Width Modulation *12*
1.5 Outline of Parameter Estimation Techniques *14*
1.5.1 Offline Parameter Estimation *14*
1.5.2 Online Parameter Estimation *16*
1.6 Scope of This Book *17*
References *18*

2 Critical Issues with Online Parameter Estimation *23*
2.1 Rank-Deficient Problem *23*
2.1.1 Rank-Deficient Issue *23*
2.1.2 Experimental Analysis and Results *24*
2.2 Non-linearity of VSI *29*
2.2.1 Modelling of VSI Non-linearity *30*
2.2.1.1 VSI Modelling in *abc* Stationary Reference Frame *30*
2.2.1.2 VSI Modelling in *dq* Rotating Reference Frame *32*
2.2.2 VSI Non-linearity Estimation and Compensation *34*
2.2.2.1 Estimation of VSI Non-linearity *34*

2.2.2.2 Online Compensation of VSI Non-linearity *37*
2.2.3 Influences of VSI Non-linearity on Parameter Estimation *39*
2.3 Ill-Conditioned Problem *46*
2.4 Summary *48*
References *48*

3 Online Estimation of Rotor Flux Linkage with the Aid of Thermocouples in Stator Windings *51*
3.1 Introduction *51*
3.2 Online Estimation of Rotor Flux Linkage with the Aid of Thermocouples in Stator Windings *52*
3.2.1 Online Estimation of Rotor Flux Linkage *52*
3.2.2 Thermal Condition Monitoring of Rotor PM *55*
3.3 Summary *57*
References *57*

4 Online Parameter Estimation Based on Current Injections *59*
4.1 Introduction *59*
4.2 Multi-parameter Estimation Based on Current Injection and Error Analysis *60*
4.2.1 Designed Parameter Estimation Scheme *60*
4.2.2 Error Analysis *63*
4.2.3 Experimental Results *65*
4.3 Winding Resistance and Rotor Flux Linkage Estimation Based on Current Injection under Constant Torque/Speed Control *68*
4.3.1 Designed Parameter Estimation Scheme *68*
4.3.2 Error Analysis and Experimental Validation *69*
4.4 Summary *72*
References *73*

5 Online Parameter Estimation Based on Position Offset Injection *75*
5.1 Introduction *75*
5.2 Phasor Analysis of Rotor Position Offset in PMSMs *76*
5.3 Position Offset-based Estimation with $i_d = 0$ Under Constant Torque/Speed Control *77*
5.3.1 Designed Estimation Method *78*
5.3.1.1 Estimation of Rotor PM Flux Linkage ψ_{pm} and Stator Winding Resistance R_s *79*
5.3.1.2 Estimation of q-axis Inductance L_q *81*
5.3.1.3 Estimation of d-axis Inductance L_d *82*
5.3.2 Experimental and FEA Results *84*
5.3.2.1 Experimental Results of Estimated ψ_{pm} and R_s *84*
5.3.2.2 Experimental Results of Estimated L_d and L_q *88*
5.4 Position Offset-based Estimation with $i_d = 0$ Under Variable Speed Control *89*
5.4.1 Designed Estimation Method *90*
5.4.2 Experimental and FEA Results *92*
5.5 Position Offset-based Estimation with $i_d \neq 0$ Under Constant Torque/Speed Control *94*
5.5.1 Designed Estimation Method *94*
5.5.1.1 Estimation of Rotor PM Flux Linkage ψ_{pm} and Inductances' Saliency ΔL *95*

5.5.1.2 Estimation of dq-axis Inductances 95
5.5.2 Experimental and FEA Results 96
5.6 Position Offset-based Estimation with $i_d = 0$ and $i_d \neq 0$ Under Constant and Variable Speed Control 99
5.6.1 Designed Estimation Method 99
5.6.2 Experimental and FEA Results 101
5.7 Analysis of Amplitude of Position Offset Injection 103
5.8 Summary 108
References 109

6 Online Parameter Estimation Under Variable Speed Control 111
6.1 Introduction 111
6.2 Estimation of Stator Resistance, Inductances, and Rotor PM Flux Linkage 112
6.2.1 Identifiability Analysis and Influence of VSI Non-linearity 113
6.2.2 Improved Estimation Scheme 113
6.2.3 Experimental Results 115
6.3 Estimation of dq-axis Flux Linkage Maps with Uncertainties of Circuit Resistance and Inverter Non-linearity 119
6.3.1 Modelling of Cost Functions for Identification of Flux Linkages 119
6.3.2 Minimization of Cost Functions 121
6.3.3 Experimental Results 122
6.3.3.1 Data Recording and FE-predicted Flux Linkage Maps 122
6.3.3.2 Identification of dq-axis Flux Linkage Maps 122
6.3.3.3 Influence of Uncertain Circuit Resistance 123
6.3.3.4 Derivation of dq-axis Inductances 125
6.3.3.5 MTPA Current Trajectory Based on dq-axis Flux Linkage Maps 127
6.4 Summary 128
References 129

7 Estimation of Magnetic Saturation and Cross-coupling Based on High-frequency Signal Injection 131
7.1 Introduction 131
7.2 Magnetic Saturation Modelling and Time Delay Effect in HF Signal Injection 133
7.2.1 Fundamental Mathematical Model 133
7.2.2 Time Delay Effect in HF Signal Injection Methods 134
7.3 HF Rotating Voltage Injection Method 134
7.4 HF Pulsating Voltage Injection Method 137
7.4.1 With the Aid of Position Estimator 137
7.4.2 Without the Aid of Position Estimator 139
7.5 Combined HF Rotating and Pulsating Voltage Injection Method 140
7.6 Experimental Results 142
7.6.1 Evaluation of Estimation Performance 142
7.6.2 Comparison with FE Results 144
7.6.3 Selection of Injected Voltage 145
7.7 Summary 146
References 147

8 Offline and Multi-step Parameter Estimation Methods *149*
8.1 Introduction *149*
8.2 Parameter Estimation at Standstill by Square Voltage Injection *150*
8.3 Parameter Estimation at Standstill by HF Current Injection *153*
8.3.1 Estimation Based on Current Variations *153*
8.3.2 Estimation Based on Zero-crossing Detection *156*
8.4 Multi-step Parameter Estimation *158*
8.4.1 Two-step Estimation Method *159*
8.4.2 Three-step Estimation Method *161*
8.5 Summary *163*
References *163*

9 Estimation of Mechanical Parameters *165*
9.1 Introduction *165*
9.2 Mechanical Parameter Estimation Methods *166*
9.2.1 Acceleration/Deceleration-Based Method *166*
9.2.2 MRAS Observer–Based Method *167*
9.2.3 Fundamental Motion Equation–Based Estimation Method *168*
9.2.4 Critical Issue of Torque Prediction *169*
9.3 Experimental Results *170*
9.3.1 Comparison Between Schemes I and III *171*
9.3.2 Comparison Between Schemes II and III *173*
9.3.3 Influence of Variation in Rotor PM Flux Linkage *173*
9.3.4 Sensitivity Analysis of Sinusoidal Perturbation Signal *175*
9.4 Design of PI Speed Regulator *177*
9.5 Summary *180*
References *181*

10 Modern Control and Optimization Theory–Based Parameter Estimation Algorithms *183*
10.1 Introduction *183*
10.2 Designed Parameter Estimation Scheme *183*
10.3 Recursive Least-Squares *185*
10.3.1 Basic Principle *185*
10.3.2 Application to Parameter Estimation for PMSMs *186*
10.4 Kalman Filter *187*
10.4.1 Basic Principle *187*
10.4.1.1 Conventional Kalman Filter *187*
10.4.1.2 Extended Kalman Filters *189*
10.4.2 Application to Parameter Estimation for PMSMs *190*
10.5 Model Reference Adaptive System *190*
10.5.1 Basic Principle *190*
10.5.2 Application to Parameter Estimation for PMSMs *191*
10.6 Adaline Neural Network *192*
10.6.1 Basic Principle *192*
10.6.2 Application to Parameter Estimation for PMSMs *193*

10.7 Gradient-Based Methods *195*
10.7.1 Basic Principle *196*
10.7.1.1 Steepest Descent Method *196*
10.7.1.2 Newton's Method *196*
10.7.1.3 Gauss–Newton Method *197*
10.7.2 Application to Parameter Estimation for PMSMs *198*
10.8 Particle Swarm Optimization *198*
10.8.1 Basic Principle *198*
10.8.2 Application to Parameter Estimation for PMSMs *200*
10.9 Genetic Algorithm *201*
10.9.1 Basic Principle *201*
10.9.2 Application to Parameter Estimation for PMSMs *203*
10.10 Summary *203*
References *205*

11 Applications of Parameter Estimation *209*
11.1 Introduction *209*
11.2 Improvement in Control Performance *210*
11.2.1 Design of PI Regulators for FOC *210*
11.2.1.1 Parameter-Based PI Regulator Design *210*
11.2.1.2 Parameter Estimation *212*
11.2.1.3 Designed PI Regulators and Experimental Performance *214*
11.2.2 Determination of MTPA Current Trajectory *215*
11.3 Improvement in Sensorless Control *216*
11.3.1 Improvement in Sensorless Control Performance *218*
11.3.1.1 Extended Back-EMF-Based Method *218*
11.3.1.2 High-frequency Signal Injection–Based Method *220*
11.3.2 Application of Parameter Estimation Under Sensorless Control *224*
11.3.2.1 Position-offset-Based Online Parameter Estimation Under Sensorless Control *224*
11.3.2.2 Experimental Results *228*
11.4 Precise Torque Estimation *229*
11.4.1 HF Square Wave Voltage Injection Considering Cross-coupling Effect *230*
11.4.2 Torque Estimation Based on Estimated HF Inductances *231*
11.5 Thermal Condition Monitoring *232*
11.6 Fault Diagnosis *234*
11.6.1 Inter-turn Short-circuit Fault *234*
11.6.2 Demagnetization Fault *235*
11.7 Summary *237*
References *238*

Appendix A: Finite Element Calculation of Winding Inductances *243*
Appendix B: Specifications of Prototype Machines and Experimental Platforms *251*
Index *255*

Authors

Professor Zi Qiang (Z.Q.) Zhu received his BEng and MSc degrees from Zhejiang University, Hangzhou, China, in 1982 and 1984, respectively, and his PhD degree from the University of Sheffield, Sheffield, UK, in 1991, all in electrical engineering.

After working as a Lecturer/Assistant Lecturer at Zhejiang University from 1984 to 1988, he has been with the University of Sheffield since 1988, initially as a Visiting Research Fellow sponsored by the British Council (1988–1989), then as a Research Associate working with Philips (1989–1992), a Senior Research Scientist/Officer – an established university post (1992–2000), Professor (2000), Royal Academy of Engineering/Siemens Research Chair (2014–2023), and Head of the Electrical Machines and Drives Research Group (2008). As the Founding Director, he has helped several industries in establishing their research centres, most notably, Siemens Wind Power Research Centre at Sheffield (2009) and Midea Electrical Machines and Control Research Centres at Shanghai and Sheffield (2010).

His research interests include design and control of permanent magnet brushless machines and drives for applications ranging from electrified transportation (electric vehicles, fast trains, and more electric aircrafts) through domestic appliances to wind power generation, on which he has published >1500 papers including >600 IEEE and IET journal papers, as well as >200 patents.

He is a Fellow of the Royal Academy of Engineering, UK; the Institute of Electrical and Electronics Engineers (IEEE), USA; the Institution of Engineering and Technology (IET), UK; the Chinese Society for Electrical Engineering (CSEE); the China Electrotechnical Society (CES); and Editor-in-Chief of IET Electric Power Applications. He is the recipient of the 2024 Global Energy Prize, the 2021 IEEE Nikola Tesla Award, and the 2019 IEEE Industry Applications Society Outstanding Achievement Award.

Professor Kan Liu received his BEng and PhD degrees in automation from Hunan University, Changsha, China, in 2005 and 2011, respectively, and his PhD degree in electrical engineering from the University of Sheffield, Sheffield, UK, in 2013.

From 2013 to 2016, he was a Postdoctoral Research Associate in the Electrical Machines and Drives Group at the University of Sheffield. From 2016 to 2017, he was a Lecturer with the Control Systems Group, Loughborough University, UK. He joined Hunan University as a Professor of electromechanical engineering in 2017. He is currently the Director of the Engineering Research Center of Ministry of Education on Automotive Electronics and Control Technology, China, Chief Scientist of National Key Research and Development Program, China (2022), and Senior Member of IEEE.

His research interests include control, design, parameter identification, and state observation of permanent magnet synchronous machines and high-power density SiC inverters/converters for electric vehicles and locomotives. He has published 14 patents and >100 papers, including >40 IEEE transactions papers, and 3 Best Paper Awards of International Conference on Electrical Machines and Systems.

Dr Dawei Liang received his BEng degree from Harbin Institute of Technology, Harbin, China, in 2014; his MSc degree from Karlsruhe Institute of Technology, Karlsruhe, Germany, in 2018; and his PhD degree from the University of Sheffield, Sheffield, UK, in 2022, all in electrical engineering. Since 2022, he has been a Postdoctoral Research Associate with the University of Sheffield, Sheffield, UK. His research interests include design and control of permanent magnet machines.

Preface

Parameter estimation is an essential requirement for high-performance permanent magnet synchronous machine (PMSM) drives. Indeed, many commercial industrial drives for PMSMs have embodied the features of the parameter estimation function, which many consider as a black box for online and/or offline parameter estimation. However, due to commercial sensitivity, most implemented techniques have been withheld from the public, and the estimation performance can vary significantly from one company to another. Nevertheless, how to develop reliable generic methods and reduce the errors in parameter estimation, particularly under wide speed and load variations and sensorless control, remains challenging.

Since 2005, I have been continuously working on various parameter estimation techniques for PMSMs together with my PhD students, particularly the co-authors of this book, Dr Kan Liu (2008–2016) and Dr Dawei Liang (2018–present) who obtained their PhD degrees and was/is a post-doctoral researcher at the University of Sheffield working on this important topic. We successfully addressed the issues of rank deficiency and inverter nonlinearity and have developed many new online electrical and mechanical parameter estimation techniques for PMSMs, including those with the aid of thermocouples in the stator windings, based on current/voltage injection, and based on position-offset injection, under constant or variable speed and load for sensored or sensorless controlled PMSMs for various applications, accounting for magnetic saturation, cross-coupling, inverter nonlinearity, temperature effects, etc. These parameter estimation techniques as well as those developed globally are comprehensively described in this book with examples of both experimental and simulation results.

Over the last few years, it has been interesting to see an increasing number of publications on modern control theory–based parameter estimation techniques for PMSMs. Therefore, such techniques, including recursive least squares, Kalman filter, model reference adaptive system, Adaline neural network, gradient-based methods, particle swarm optimization, and genetic algorithm, are systematically described with examples in this book.

It is anticipated that the new parameter estimation techniques of PMSMs will continue to be developed and more commonly employed in general industrial drives. They will be more extensively used for a wide variety of applications, including electric vehicles, wind power generators, aerospace, industrial drives, automation systems, robots, and domestic appliances; for improving control performance and sensorless control; and condition monitoring/fault diagnosis. Thus, applications of parameter estimation of PMSMs are also included in a separate chapter in this book.

I dedicate my thanks to our sponsors, including the UK government, particularly the UK EPSRC Prosperity Partnership ("A New Partnership in Offshore Wind" under Grant No. EP/R004900/1), Siemens, CRRC, and Midea. I also thank Mr Ian Parmenter for his proofreading of the book.

We sincerely hope this book will be useful to industrial engineers and researchers, university professionals, post-doctoral researchers, and students alike.

Professor Zi Qiang (Z.Q.) Zhu
FREng, FIEEE, FIET, FCSEE, FCES
August 2024, Sheffield, UK

List of Abbreviations

ac	Alternating current
AI	Artificial intelligence
AlNiCo	Aluminium–nickel–cobalt
ANN	Adaline neural networks
Back-EMF	Back electromotive force
dc	Direct current
DTC	Direct torque control
EA	Evolutionary algorithms
EMF	Electromotive force
EV	Electric vehicle
EKF	Extended Kalman filter
FFT	Fast Fourier transform
FE	Finite element
FEA	Finite element analysis
FOC	Field-oriented control
GA	Genetic algorithms
HF	High frequency
ICA	Immune clonal algorithm
ICQGA	Immune clonal–based quantum genetic algorithm
IPM	Interior permanent magnet
IPMSM	Interior permanent magnet synchronous machine
ITSC	Inter-turn short circuit
KF	Kalman filter
LMS	Least mean square
LPF	Low-pass filter
MMF	Magnetomotive forces
MPC	Model predictive control
MTPA	Maximum torque per ampere
MRAS	Model reference adaptive system
NdFeB	Neodymium–iron–boron
PI	Proportional integral
PM	Permanent magnet
PMSM	Permanent magnet synchronous machine
POPE	Position-offset parameter estimation

PSO	Particle swarm optimization
PWM	Pulse width modulation
QGA	Quantum genetic algorithm
RLS	Recursive least squares
SNR	Signal-to-noise ratio
SPM	Surface permanent magnet
SPMSM	Surface permanent magnet synchronous machine
SVPWM	Space vector pulse width modulation
SWRPO	Square wave rotor position offset
THD	Total harmonic distortion
VSI	Voltage source inverter

List of Symbols

Symbol	Description	Unit
$\boldsymbol{A}$	State transition matrix	
$\boldsymbol{B}$	Control-input model	
B_r	Remanence	T
$B_{r,s}$, $B_{r,0}$	Remanence at actual and reference temperatures	T
$BH_{\max}$	Maximum energy product	kJ/m^3
c_1, c_2	Cognitive and social acceleration constants	
cond	Condition number	
Dd, Dq	dq-axis gain functions of VSI nonlinearity	
$\overline{Dd}$, $\overline{Dq}$	Average values of dq-axis gain functions of VSI nonlinearity	
$E_{\text{ex},d}$, $E_{\text{ex},q}$	dq-axis extended back-EMF	V
F	Friction coefficient	
f_h	Frequency of HF signal	Hz
g	First-order gradient	
H_C	Coercivity	kA/m
H_{cl}	Intrinsic coercivity	kA/m
$\boldsymbol{I}$	Identity matrix	
$i_{a,b,c}$	Three-phase currents	A
i_{as},i_{bs}, i_{cs}	Three-phase currents	A
i_α, i_β	$\alpha\beta$-axis currents	A
$i_{\alpha h}$, $i_{\beta h}$	$\alpha\beta$-axis HF currents	A
i_d, i_q	dq-axis currents	A
i_{dh}, i_{qh}	dq-axis HF currents	A
$I_{dh}^{\hat{r}}$, $I_{qh}^{\hat{r}}$	Amplitudes of dq-axis HF currents	A

Symbol	Description	Unit
$\overline{i_d}$, $\overline{i_q}$	Average values of *dq*-axis currents	A
I_{LSB}	Resolution of sampled current	A
I_{max}	Maximum current	A
I_p, I_n	Amplitudes of positive and negative sequence currents	A
Δi_i, Δi_j	Current perturbations in *i*th and *j*th coils	A
ΔI_d	Variables of injected *d*-axis current	A
J	Rotor moment of inertia	kg m^2
k	Index of discrete sampling instant	
k_{a1}, k_{a2}	PI constants	
K_d, K_q, K_u, K_v, K_m, K_w	Gains of cost functions	
K_k	Gain vector of RLS estimator	
L_{ii}	Self-inductance	H
L_{ij}	Mutual inductance	H
L_m	Phase self-inductance	H
$L_{s\sigma}$	Phase leakage inductance	H
L_{s2}	Second-order harmonic of self-inductance	H
L_d, L_q	*dq*-axis inductances	H
L_{dh}, L_{qh}	*dq*-axis HF inductances	H
L_{dq}, L_{qd}	*dq*-axis mutual inductances	H
L_{dqh}, L_{qdh}	*dq*-axis HF mutual inductances	H
L_d^{ap}, L_q^{ap}	*dq*-axis apparent inductances	H
L_d^{inc}, L_q^{inc}	*dq*-axis incremental inductances	H
L_{dd}^{inc}, L_{qq}^{inc}	*dq*-axis incremental self-inductances	H
L_{dd}^{ap}, L_{qq}^{ap}	*dq*-axis apparent self-inductances	H
L_{dq}^{ap}, L_{qd}^{ap}	*dq*-axis apparent mutual inductances	H
L_{dq}^{inc}, L_{qd}^{inc}	*dq*-axis incremental mutual inductances	H
M_{s0}	Average mutual inductance	H
M_{s2}	Second-order harmonic of mutual inductance	H
M	Overall time error	S
N	Sampling number	
p	Pole pair number	
P, P_{1-5}	Cost function	

Symbol	Description	Unit
$\boldsymbol{P}_k$	Covariance matrix	
$\boldsymbol{Q}$	Model uncertainty matrix	
$\boldsymbol{R}$	Measurement noise covariance	
$\hat{r}$	Estimated synchronous rotating reference frame	
r_{ce}	On-state resistance of IGBT	Ω
r_d	On-state resistance of freewheeling diode	Ω
R_c	Iron loss resistance	Ω
R_s	Stator winding resistance	Ω
R_{err}	Error of estimated stator winding resistance	Ω
$R_{s,s}$, $R_{s,0}$	Winding resistance at actual and reference temperatures	Ω
S_{a}, S_{b}, S_{c}	Switching pattern of three-phase legs	
$\mathbf{T}_{abc-\alpha\beta0}$	Clarke transformation	
t_0, t_1, t_2	Active durations of space vectors $\boldsymbol{V}_1$, $\boldsymbol{V}_2$, and $\boldsymbol{V}_0$	s
T_a	Actual duration of effective time	s
T_a^*	Commanded duration of effective time	s
T_{com}	Compensation time	s
T_d	Dead time of power device	s
$T_{\mathrm{on}}/T_{\mathrm{off}}$	Switch on/off time delay	s
T_{s}	Sampling time	s
T_e	Electromagnetic torque	N m
T_{est}	Estimated torque	N m
T_L	Load torque	N m
$\boldsymbol{u}$	Control vector in KF	
u_α, u_β	$\alpha\beta$-axis voltages	V
u_d, u_q	dq-axis voltages	V
u_d^*, u_q^*	dq-axis reference voltages	V
u_{dh}, u_{qh}	dq-axis HF voltages	V
u_{dh}^r, u_{qh}^r	dq-axis HF reference voltages	V
u_s^*	Voltage vector reference	V
$u_{\mathrm{s,max}}^*$	Maximum output phase voltage	V
u_{sig}	HF voltage signal	V
u_N	System noise	
$\boldsymbol{v}$	Observation/measurement noise	
v	Operator norm	
V_{1-7}	Voltage vectors	V
V_{as}^*, V_{bs}^*, V_{cs}^*,	Reference three-phase voltages	V
V'_{as}, V'_{bs}, V'_{cs}	Distorted three-phase voltage caused by VSI nonlinearity	V

Symbol	Description	Unit
V_{ce}	On-state voltage drop of IGBT	V
V_{ce0}	Threshold voltage drop of IGBT	V
V_d	On-state voltage drop of freewheeling diode	V
V_{d0}	Threshold voltage drop of freewheeling diode	V
ΔV_d	Variable of injected d-axis voltage	V
V_{dc}	dc bus voltage	V
V'_d, V'_q	dq-axis distorted voltages due to VSI nonlinearity	V
V_{dead}	Distorted voltage term due to VSI nonlinearity	V
V_h	Amplitude of injected HF voltage	V
V_{offset}	Non-zero dc offset	V
V_{thr}	Threshold voltage	V
W	Total stored energy	J
W_i	Weight in ANN estimator	
$\boldsymbol{w}$	Zero-mean white Gaussian system noise	
w	Inertia weight in PSO estimator	
X_1, X_2	Characteristic roots	
x_p, x_g	Particle best and overall best locations in PSO estimator	
Δx_k	Step variation	
$\boldsymbol{y}_k$, $\hat{\boldsymbol{y}}_k$	Measured and predicted output variables	
$\boldsymbol{Y}$	Output matrix of RLS estimator	
α_{cu}	Temperature coefficient of copper	%/°C
α_{pm}	Temperature coefficient of magnet	%/°C
a_k	Step length of Newton algorithm	
γ	Current advancing angle	Deg.
δ_0	Finite positive constant	
$\boldsymbol{\varepsilon}_k$	Least-square error	
ξ	Optimal damping ratio	
ζ	Tuning factor	
η	Convergence factor	
ϑ_s	Stator actual temperature	°C
ϑ_{pm}	PM actual temperature	°C
ϑ_r	Reference temperature	°C
θ_h	Angle of injected HF signal	Deg.
θ_r	Rotor position	Deg.
θ_u	Unit electrical angle	Deg.
$\Delta\theta$	Position offset	Deg.
$\Delta\pm\theta_p$	Injected rotor position offset	Deg.

Symbol	Description	Unit
$\Delta\theta_p$, $\Delta\theta_n$	Positive and negative position offsets	Deg.
$\boldsymbol{\Theta}$	Parameter vector in RLS estimator	
λ	Coupling factor	
λ_{rls}	Forgetting factor of RLS estimator	
μ_r	Relative permeability	
$\widehat{O}_k$	Predicted output in ANN estimator	
ρ	Differential operator	
ρ_{ele}	Electrical resistivity	Ω m
$\boldsymbol{\Phi}$	Input matrix of RLS estimator	
φ_{delay}	Phase delay between filtered HF current and demodulation signal	Deg.
φ	Power factor angle	Deg.
$\psi_{a,b,c}$	Three-phase flux linkages	Wb
ψ_α, ψ_β	$\alpha\beta$-axis flux linkages	Wb
ψ_d, ψ_q	dq-axis flux linkages	Wb
ψ_{pm}	PM flux linkage	Wb
$\psi_{pm,s}$, $\psi_{pm,0}$	PM flux linkages at actual and reference temperatures	Wb
$\psi_{pm,\mathrm{error}}$	PM flux linkage error	Wb
ω_h	HF angular velocity	Rad/s
ω_r	Electrical angular velocity	Rad/s
$\overline{\omega_r}$	Average electrical angular velocity	Rad/s
$\Delta\omega_r$	Variables of rotor angular velocity	Rad/s

1

General Introduction

1.1 Introduction

According to statistics from the US Energy Information Administration, electrical machines accounted for 43%–46% of the global electricity consumption in 2020 [1] and played an important role in industries and economics. Electrical machines have different types, e.g., direct current (dc) machines, induction machines, reluctance machines, and permanent magnet (PM) machines. Due to their high torque density and efficiency, permanent magnet synchronous machines (PMSMs) have been widely used in various applications, e.g., aerospace, electric vehicles (EVs), industrial drives, servo drives including robots and automation systems, domestic appliances, and wind power generation.

Accurate parameter estimation of PMSMs, including electrical and mechanical parameters, i.e., stator resistance, inductances, rotor flux linkage, and system inertia, is essential for determining the machine characteristics [2–4], improving general control performance [5–8], sensorless control [9–13], and thermal condition monitoring [14–18] for preventing potential irreversible demagnetization of PMs and damage of winding insulation and further improving torque density, as well as fault diagnosis, etc. For example, in EV applications, accurate electrical parameters are required for determining the machine saliency, the reluctance torque, the base speed, the maximum operating speed, the optimal current trajectories under maximum torque per ampere, and maximum power per voltage strategies. For the most common control methods, i.e., field-oriented control (FOC), direct torque control (DTC), as well as model predictive control (MPC), the accurate PMSM parameters are critical to ensuring system stability and improving efficiency and dynamic response. The mechanical parameters, e.g., moment of inertia and viscous friction coefficients, vary significantly with mechanical loads and are important in the design of speed-loop controllers.

Traditionally, the parameters are determined by static open-circuit and short-circuit tests and/or finite element methods. They have significant disadvantages: (a) the machine dimensions and material parameters should be pre-known, (b) there is a rank deficiency issue in most online parameter estimation techniques, (c) parameters are obtained under different operation conditions with conflicts, and (d) variations in speed and load cannot be tracked accurately, which will result in significant errors accounting for magnetic saturation and cross-coupling effect, etc.

In the last few decades, the techniques of parameter estimation have been extensively developed and the parameter estimation can be implemented both offline and online. Generally, offline parameter estimation is essential both in the machine and controller design and has been widely used and investigated [19], in which various computational methods can be employed, i.e., finite element analysis (FEA) and numerical, observer-, and artificial intelligence (AI)-based methods.

Parameter Estimation of Permanent Magnet Synchronous Machines, First Edition. Zi Qiang Zhu, Kan Liu, and Dawei Liang.

However, changes in operation conditions, temperature, mechanical stress, and other environmental factors may cause parameter variations in electrical machines, particularly PMSMs. These variations could result in deteriorated performance, reduced efficiency, and potential damage to the machine over time. To address these challenges, online parameter estimation techniques aim to acquire electrical, mechanical, or thermal parameters under real-time conditions using data gathered during the machine operation [19]. These techniques allow for the continuous monitoring and adjustment of the machine parameters, ensuring that the control system can adapt to changing conditions to maintain optimal performance, and monitoring that the winding and PM temperature cannot exceed the maximum allowable values. Indeed, many commercial industrial drives for PMSMs have embodied the features of parameter estimation functions to ensure efficient and reliable operation.

This book provides a comprehensive resource on basic and state-of-the-art online and offline parameter estimation techniques for PMSMs, including various new online parameter estimation techniques developed at the University of Sheffield and those developed globally, as well as modern control theory–based parameter estimation techniques, with examples of both experimental and simulation results. It addresses the issues of rank deficiency and inverter non-linearity and reports various new online electrical and mechanical parameter estimation techniques for PMSMs, including those with the aid of thermocouples in stator windings, based on current/voltage injection and position offset injection, under constant or variable speed and load for sensored or sensorless controlled PMSMs, accounting for magnetic saturation, cross-coupling, inverter non-linearity, temperature effects, etc. Various applications of parameter estimation techniques of PMSMs for electric vehicles, wind power generators, aerospace, industrial drives, automation systems, robots, and domestic appliances are also reported for improving control performance and sensorless operation, condition monitoring/fault diagnosis, etc.

This chapter briefly describes PMSMs and drives and introduces mathematical models, machine parameters, and parameter estimation techniques accounting for the influence of magnetic saturation and temperature on parameters.

1.2 Permanent Magnet Machines

A typical electrical drive system for PMSMs mainly consists of an electrical machine, a power converter, a load, a rotor position sensor, and a control unit with an employed control strategy. A schematic diagram of a PMSM drive system is shown in Figure 1.1 and will be described in more detail in Section 1.4.

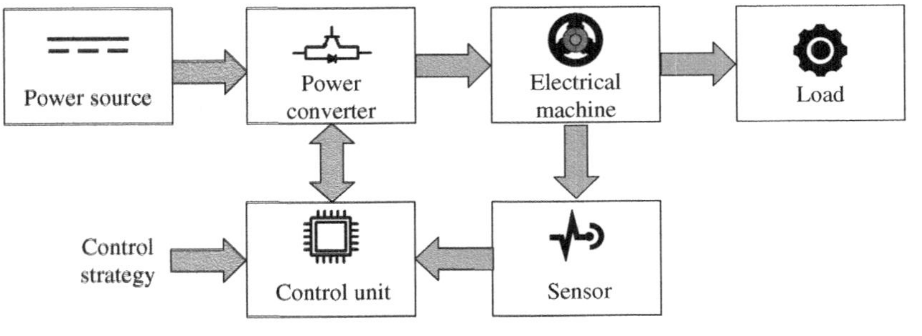

Figure 1.1 Schematic diagram of PMSM drive system.

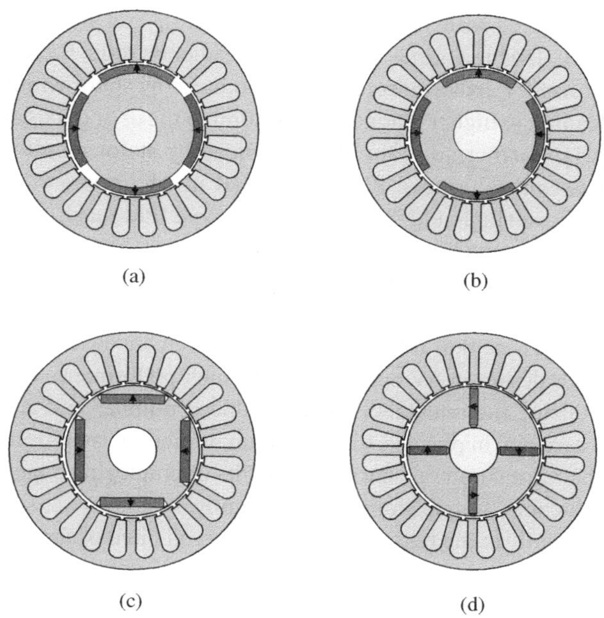

Figure 1.2 Machine configurations of radial-flux internal rotor PMSMs: (a) surface-mounted PM, (b) surface-inset PM, (c) I-type interior PM (radially magnetized), and (d) spoke-type interior PM (circumferentially magnetized).

In most PMSMs, PMs are usually located on the rotor for excitation. According to the directions of the magnetic flux in the air gap, the majority of PMSMs can be categorized into radial and axial flux machines, of which radial-flux rotor PM machines are the most common in terms of electromagnetic performance, manufacturability, cost, etc. The radial-flux rotor PM machines have either an internal or external rotor, while the PMs could be assembled either on the surface or in the interior of the rotor, as shown in Figure 1.2.

In surface-mounted PM (SPM) machines, as shown in Figure 1.2a, the PMs are mounted on the rotor surface and are adjacent to the air gap. There is negligible magnetic saliency and reluctance torque, and PM torque is the major torque component. Since the relative permeability of PMs is almost the same as that of air, the *dq*-axis inductances in SPM machines are almost equal and lower than those of the interior PM (IPM) machines. Furthermore, the field produced by stator windings in SPM machines is low, and hence, the winding inductance is also low. However, SPMs need a rotor sleeve/binding for retaining, particularly in high-speed applications [20]. The sleeve/binding materials can be glass fibre, carbon fibre, or non-magnetic stainless steel. However, in external rotor SPM machines, the rotor back iron can be utilized for retaining PMs. External rotor SPM machines are preferred in high-power low-speed direct-drive wind power generators or hub motors for EVs. Besides, the PM, which can be inset into the rotor surface, as shown in Figure 1.2b, is laterally adjacent to the rotor core. In this case, the magnet pole arc to pole pitch ratio needs to be smaller than one, but PMs can be easily fixed between two adjacent interpole irons. Since the *q*-axis inductance is larger than the *d*-axis inductance, a reluctance torque component exists in inset SPM machines.

In IPM machines, as shown in Figure 1.2c and d, PMs are radially or circumferentially magnetized, respectively. Radially magnetized IPM rotors are more commonly used, e.g., in EV applications, but circumferentially magnetized IPM rotors, i.e., spoke-type IPM rotors, are very popular in domestic appliances since the PM flux focusing effect may be utilized when the pole number is high and cheap ferrite magnets can be used for reducing the cost. Since PMs are embedded in the rotor core, extra rotor PM retaining is not required, but PM leakage fluxes are more significant, and hence, airspace flux barriers are often required next to the iron ribs to reduce the PM leakage flux.

Compared with SPM machines, since the q-axis inductance is usually larger than the d-axis inductance, IPM machines exhibit a high saliency ratio, thereby both PM and reluctance torques can be utilized. Furthermore, the field produced by stator windings in the IPM machines is much higher, leading to higher magnetic saturation in the d- or q-axis and higher cross-coupling between dq-axes [21]. Furthermore, the flux produced by stator windings in radially magnetized IPM machines passes through two PMs, while that in circumferentially magnetized spoke-type machines only passes a single PM. Thus, the spoke-type IPM machines exhibit a higher d-axis inductance.

Besides the I-type IPM machines, there are various other PM configurations as shown in Figure 1.3, including single-/double-layer V-type rotors for utilizing PM flux focusing, double-layer V-type, double-layer I-type, and delta-type rotors for increasing the reluctance torque with higher saliency ratio. These IPM rotor machines have been widely employed in EVs [22, 23]. The main aims for IPM rotor design in EV applications are high performance and low cost, i.e., increased output torque by utilizing the reluctance torque, reduced number and amount of PMs, and ease of manufacturing. Therefore, in practice, the maximum number of layers is usually less than three, with two layers being the most common, since the benefits of employing a layer number higher than three for increasing reluctance torque diminish significantly.

To utilize the reluctance torque, the coil pitch of the stator windings should be as close as possible to the pole pitch of the rotor PMs, and thus, the end windings of different phases will be overlapped, resulting in overlapping windings (Figure 1.4a). Furthermore, to improve the back electromotive force (EMF) waveform and reduce its total harmonic distortion (THD), the rotor PM segments are often step skewed, while distributed overlapping stator windings are often employed. However, overlapping windings have longer end windings, which result in a longer machine axial length and higher copper loss. Non-overlapping stator windings or tooth coils can be utilized to reduce the end winding length and the end winding copper loss in the PM machines, which usually have a fractional slot number per pole per phase, i.e., fractional slot PM machines. In fractional slot PM machines, cogging torque or torque ripple is low, but the reluctance torque will be significantly reduced or even become

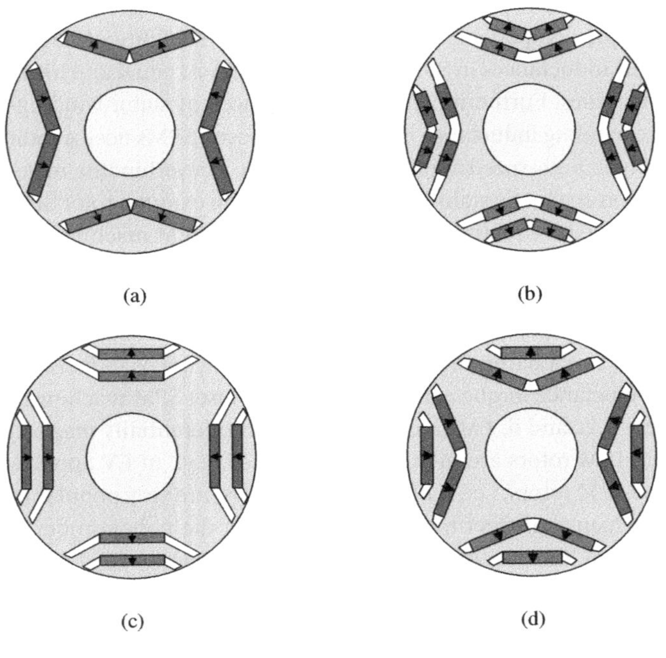

Figure 1.3 Variants of IPM rotor configurations: (a) single V-type, (b) double V-type, (c) double I-type, and (d) delta type.

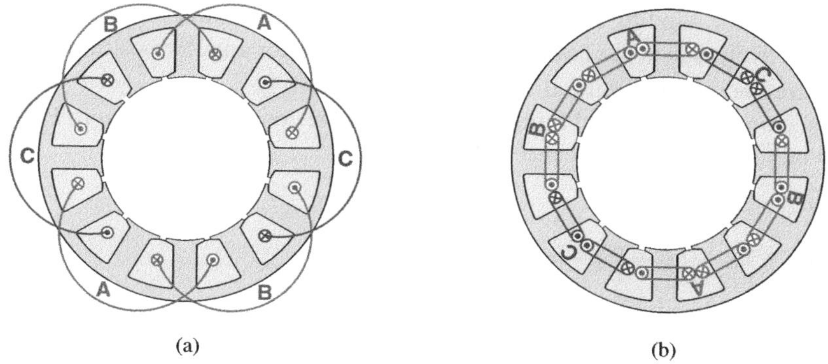

Figure 1.4 Stator winding topologies: (a) overlapping windings (12-slot stator for 4-pole rotor) and (b) non-overlapping windings (12-slot stator for 10-pole rotor).

negligible. Fractional slot PM machines (Figure 1.4b) are often used in domestic appliances and electric-assisted power steering.

1.3 Basic Equations and Machine Parameters

The mathematical modelling of PM machines is essential for control and parameter estimation. This section introduces the fundamental mathematical models of three-phase SPMSM and IPMSM. Meanwhile, the influences of magnetic saturation and temperature on the variation of parameters are also discussed.

1.3.1 Fundamental Mathematical Model for PMSMs

To simplify the mathematical modelling, several assumptions are made as follows:

1. The magnetic saturation, thermal effect, and stator iron loss are ignored.
2. The permeabilities of the PM and the air are the same.
3. The three-phase windings are distributed symmetrically in space. Both the magnetomotive forces (MMFs) generated by the stator windings and the back EMFs in the stator windings induced by the rotor PMs are sinusoidal.

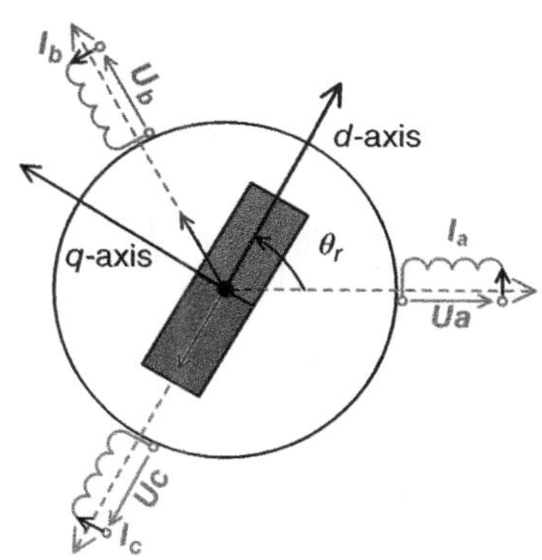

Figure 1.5 Representation of the three-phase PM machine model.

A three-phase PMSM model is presented in Figure 1.5, where $u_{a,b,c}$ can be calculated as follows:

$$\begin{bmatrix} u_a \\ u_b \\ u_c \end{bmatrix} = R_s \begin{bmatrix} i_a \\ i_b \\ i_c \end{bmatrix} + \frac{d}{dt} \begin{bmatrix} \psi_a \\ \psi_b \\ \psi_c \end{bmatrix} \tag{1.1a}$$

$$\begin{bmatrix} \psi_a \\ \psi_b \\ \psi_c \end{bmatrix} = \begin{bmatrix} L_{aa} & L_{ab} & L_{ac} \\ L_{ba} & L_{bb} & L_{bc} \\ L_{ca} & L_{cb} & L_{cc} \end{bmatrix} \begin{bmatrix} i_a \\ i_b \\ i_c \end{bmatrix} + \psi_{pm} \begin{bmatrix} \cos(\theta_r) \\ \cos(\theta_r - 2\pi/3) \\ \cos(\theta_r + 2\pi/3) \end{bmatrix} \tag{1.1b}$$

where R_s is the stator winding resistance. $i_{a,b,c}$ and $\psi_{a,b,c}$ are the three-phase currents and flux linkages, respectively. ψ_{pm} is the PM flux linkage. θ_r is the rotor angular position. L_{ii} and L_{ij} with $\{i, j = a, b, c\}$ are the self- and mutual-inductances, respectively.

For SPM machines, the self- and mutual-inductances are constant as shown in Eq. (1.2) due to the assumed uniform air gap permeance [24]:

$$L_{aa} = L_{bb} = L_{cc} = L_m + L_{s\sigma} \tag{1.2a}$$

$$L_{ab} = L_{ac} = L_{ba} = L_{bc} = L_{ca} = L_{cb} = -\frac{1}{2}L_m \tag{1.2b}$$

where L_m and $L_{s\sigma}$ are the phase self-inductance and the leakage inductance, respectively.

For IPM machines, the self- and mutual-inductances depend on the rotor position due to the saliency of the rotor:

$$\begin{cases} L_{aa} = L_m + L_{s\sigma} + L_{s2}\cos(2\theta_r) \\ L_{bb} = L_m + L_{s\sigma} + L_{s2}\cos(2\theta_r + 2\pi/3) \\ L_{cc} = L_m + L_{s\sigma} + L_{s2}\cos(2\theta_r - 2\pi/3) \end{cases} \tag{1.3a}$$

$$\begin{cases} L_{ab} = L_{ba} = M_{s0} + M_{s2}\cos(2\theta_r - 2\pi/3) \\ L_{bc} = L_{cb} = M_{s0} + M_{s2}\cos(2\theta_r) \\ L_{ac} = L_{ca} = M_{s0} + M_{s2}\cos(2\theta_r + 2\pi/3) \end{cases} \tag{1.3b}$$

where L_{s2} is the amplitude of the second-order harmonic of self-inductances. M_{s0} and M_{s2} are the average value and amplitude of the second-order harmonic of mutual-inductances, respectively.

By using the Clarke transformation $\mathbf{T}_{abc-\alpha\beta 0}$ in Eq. (1.4), the three-phase voltage equations (Eq. 1.1) can be transferred to the stationary $\alpha\beta$-axis, i.e.,

$$\mathbf{T}_{abc-\alpha\beta 0} = \frac{2}{3}\begin{bmatrix} 1 & -1/2 & -1/2 \\ 0 & \sqrt{3}/2 & -\sqrt{3}/2 \\ 1/2 & 1/2 & 1/2 \end{bmatrix} \tag{1.4}$$

$$\begin{bmatrix} u_\alpha \\ u_\beta \end{bmatrix} = R_s\begin{bmatrix} i_\alpha \\ i_\beta \end{bmatrix} + \frac{d}{dt}\begin{bmatrix} \psi_\alpha \\ \psi_\beta \end{bmatrix} \tag{1.5a}$$

$$\begin{bmatrix} \psi_\alpha \\ \psi_\beta \end{bmatrix} = \begin{bmatrix} \Sigma L + \Delta L\cos(2\theta_r) & \Delta L\cos(2\theta_r) \\ \Delta L\sin(2\theta_r) & \Sigma L - \Delta L\cos(2\theta_r) \end{bmatrix}\begin{bmatrix} i_\alpha \\ i_\beta \end{bmatrix} + \psi_{pm}\begin{bmatrix} \cos(\theta_r) \\ \sin(\theta_r) \end{bmatrix} \tag{1.5b}$$

where $\Sigma L = (L_d + L_q)/2$ and $\Delta L = (L_d - L_q)/2$. L_d and L_q are the inductances in the dq-axes.

Afterward, the Park transformation $\mathbf{T}_{\alpha\beta-dq}$ (Eq. 1.6) is used to transform the stationary $\alpha\beta$-axis system to the rotating dq-axis system, where the rotor position–related terms could be cancelled:

$$\mathbf{T}_{\alpha\beta-dq} = \begin{bmatrix} \cos(\theta_r) & \sin(\theta_r) \\ -\sin(\theta_r) & \cos(\theta_r) \end{bmatrix} \tag{1.6}$$

$$\begin{bmatrix} u_d \\ u_q \end{bmatrix} = R_s\begin{bmatrix} i_d \\ i_q \end{bmatrix} + \frac{d}{dt}\begin{bmatrix} \psi_d \\ \psi_q \end{bmatrix} \tag{1.7a}$$

$$\begin{bmatrix} \psi_d \\ \psi_q \end{bmatrix} = \begin{bmatrix} L_d & 0 \\ 0 & L_q \end{bmatrix}\begin{bmatrix} i_d \\ i_q \end{bmatrix} + \begin{bmatrix} \psi_{pm} \\ 0 \end{bmatrix} \tag{1.7b}$$

where u_d, u_q, i_d, i_q, L_d, and L_q are the dq-axis voltages, currents, and inductances, respectively. The phasor diagram of a PM machine can be derived from Eq. (1.7), as presented in Figure 1.6, where γ and φ are the current advancing angle and the power factor angle, respectively.

Finally, by introducing Eq. (1.7b) into Eq. (1.7a), the voltage equations $u_{d,q}$ can be written as follows:

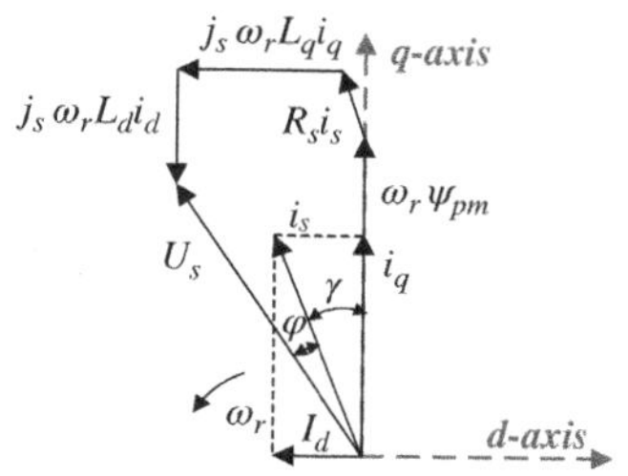

Figure 1.6 Phasor diagram of three-phase PM machine.

$$\begin{cases} u_d = R_s i_d + L_d^{\text{inc}} \dfrac{di_d}{dt} - \omega_r L_q^{\text{ap}} i_q \\ u_q = R_s i_q + L_q^{\text{inc}} \dfrac{di_q}{dt} + \omega_r \left(L_d^{\text{ap}} i_d + \psi_{pm} \right) \end{cases} \tag{1.8}$$

$$L_{d,q}^{\text{inc}} = \frac{d\psi_{d,q}}{di_{d,q}} \tag{1.9a}$$

$$L_{d,q}^{\text{ap}} = \frac{\psi_{d,q}}{i_{d,q}} \tag{1.9b}$$

where $L_{d,q}^{\text{inc}}$ and $L_{d,q}^{\text{ap}}$ are the dq-axis incremental and apparent inductances, respectively. ω_r is the rotor angular velocity. When the magnetic material is not saturated, i.e., linear, the incremental inductance is the same as the apparent inductance. Otherwise, the incremental inductances become smaller once the magnetic material is saturated due to the non-linear change of the flux linkage with the current, which will be discussed in the next section in detail.

Besides, the electromagnetic torque and the motion equation are expressed as follows:

$$T_e = \frac{3}{2} p \left[\psi_{pm} i_q + (L_d - L_q) i_d i_q \right] \tag{1.10}$$

$$J \frac{d\omega_r}{dt} = p \left[T_e - T_L - \frac{F}{p} \omega_r \right] \tag{1.11}$$

where p, J, T_L, and F are the number of pole pairs, the rotor moment of inertia, the load torque, and the friction coefficient, respectively.

1.3.2 Mathematical Model Considering Magnetic Saturation, Thermal Effect, and Iron Loss

1.3.2.1 Influence of Magnetic Saturation

As mentioned above, the ideal mathematical model does not consider the magnetic saturation effect. However, in reality, the magnetic materials are less than ideal, i.e., they will be saturated with the increase in current or magnetic flux. Hence, the magnetic permeabilities and the effectiveness in the contribution of overall flux density will be decreased [25]. Consequently, in the parameter estimation, the magnetic saturation significantly affects the flux linkages, and thus, the dq-axis inductances. Accounting for the magnetic saturation, the correlation between the incremental and apparent inductances is illustrated in Figure 1.7.

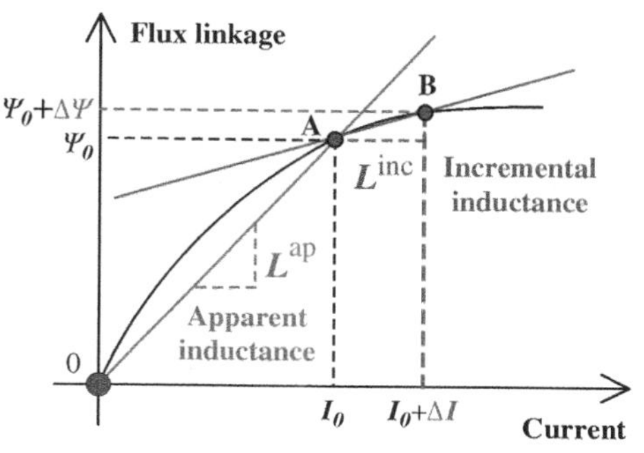

Figure 1.7 Incremental and apparent inductances considering magnetic saturation.

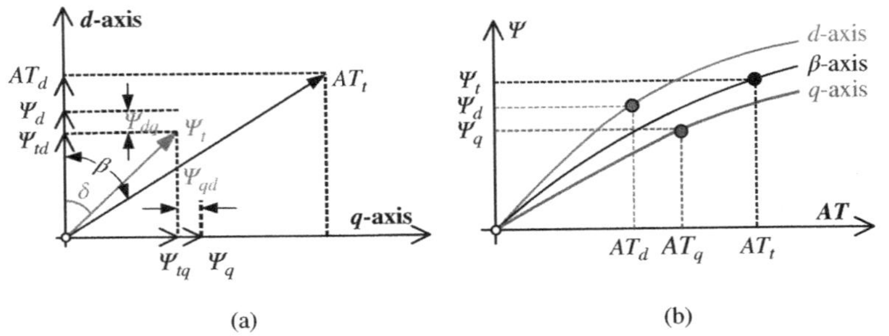

Figure 1.8 Illustration of cross-coupling effect based on a salient-pole synchronous machine: (a) flux vector diagram and (b) saturation curves.

As can be seen, when the magnetic material becomes saturated, both the apparent and incremental inductances will gradually decrease, and the value of incremental inductance will be smaller than the corresponding apparent inductance. Besides the self-saturation, the cross-coupling effect also plays an important role in the electromagnetic performance when the machine is saturated [26, 27], as illustrated in Figure 1.8.

In Figure 1.8a, when the machine is excited by the ampere-turns AT_d and AT_q in the d- and q-axis directions separately, the generated fundamental components of the air gap magnetic fluxes are Y_d and Y_q, respectively. In contrast, when the machine is excited by AT_d and AT_q along the d- and q-axis directions simultaneously, the resultant magnetic flux Y_t corresponding to the total ampere-turns AT_t can be found from the saturation curve (β-axis) in Figure 1.8b. Consequently, the resolved Y_{td} and Y_{tq} are unequal to Y_d and Y_q, respectively. δ is the angle between the total airgap flux and the d-axis when the machine is excited from the d- and q-axis directions simultaneously. The variations of fluxes in the d- and q-axis, Y_{dq} and Y_{qd}, are attributed to the magnetic cross-coupling effect.

Meanwhile, PM flux linkage is also affected by magnetic saturation [28]. Therefore, derived from the ideal equations (Eqs. 1.7 and 1.8), the dq-axis voltage equations and flux linkages accounting for the magnetic saturation effect could be extended and rewritten as follows:

$$\begin{cases} u_d = R_s i_d + L_d^{\text{inc}} \dfrac{di_d}{dt} + L_{dq}^{\text{inc}} \dfrac{di_q}{dt} - \omega_r \psi_q \\ u_q = R_s i_q + L_q^{\text{inc}} \dfrac{di_q}{dt} + L_{qd}^{\text{inc}} \dfrac{di_d}{dt} + \omega_r \psi_q \end{cases} \tag{1.12a}$$

$$\begin{bmatrix} \psi_d \\ \psi_q \end{bmatrix} = \begin{bmatrix} L_d(i_d, i_q) & L_{dq}(i_d, i_q) \\ L_{qd}(i_d, i_q) & L_q(i_d, i_q) \end{bmatrix} \begin{bmatrix} i_d \\ i_q \end{bmatrix} + \begin{bmatrix} \psi_{pm}(i_d, i_q) \\ 0 \end{bmatrix} \tag{1.12b}$$

where L_{dq}, L_{qd}, L_{dq}^{inc}, and L_{qd}^{inc} are the dq-axis apparent and incremental mutual inductances, respectively. The relationship between the incremental and apparent inductance can be found in Ref. [29], where the dq-axis incremental inductance matrix $\boldsymbol{L}_{\boldsymbol{dq}}^{\textbf{inc}}$ is adjusted to the diagonal matrix $\boldsymbol{L}_{\boldsymbol{dq}\textbf{,adj}}^{\textbf{inc}}$:

$$\boldsymbol{L}_{\boldsymbol{dq}}^{\textbf{inc}} = \begin{bmatrix} L_d^{\text{inc}} & L_{dq}^{\text{inc}} \\ L_{qd}^{\text{inc}} & L_q^{\text{inc}} \end{bmatrix} \tag{1.13}$$

$$\boldsymbol{L}_{\boldsymbol{dq}\textbf{,adj}}^{\textbf{inc}} = \begin{bmatrix} L_{d\text{,adj}}^{\text{inc}} & 0 \\ 0 & L_{q\text{,adj}}^{\text{inc}} \end{bmatrix} \tag{1.14}$$

where

$$\begin{cases} L_{d,\text{adj}}^{\text{inc}} = \dfrac{L_d^{\text{inc}}\,\Delta i_d + L_{dq}^{\text{inc}}\,\Delta i_q}{\Delta i_d} \\ L_{q,\text{adj}}^{\text{inc}} = \dfrac{L_q^{\text{inc}}\,\Delta i_q + L_{qd}^{\text{inc}}\,\Delta i_d}{\Delta i_q} \end{cases} \tag{1.15}$$

Thus, the equivalent apparent inductances, L_d^{ap} and L_q^{ap}, can be calculated as follows:

$$\begin{cases} L_d^{\text{ap}} = \dfrac{\int L_{d,\text{adj}}^{\text{inc}}(i_d)\,di_d}{i_d} \\ L_q^{\text{ap}} = \dfrac{\int L_{q,\text{adj}}^{\text{inc}}(i_q)\,di_q}{i_q} \end{cases} \tag{1.16}$$

It is worth mentioning that there is a third type of inductance definition [30–32], which is often called energy inductance.

The energy inductances are derived based on the energy perturbation method and also involve incremental and apparent inductances. In general, the self- and mutual inductances, L_{ii} and L_{ij}, are calculated as follows:

$$L_{ii} = \frac{\partial^2 W}{\partial(\Delta i_i)^2} \tag{1.17a}$$

$$L_{ij} = \frac{\partial^2 W}{\partial(\Delta i_i)\,\partial(\Delta i_j)} \tag{1.17b}$$

where W is the total stored energy and Δi_i and Δi_j are the current perturbations in the ith and jth coils, respectively. For the calculation of apparent inductance, the apparent reluctivity is used to determine the perturbations in energies (excess or deficit), whereas the incremental reluctivity is used to calculate incremental inductance, as presented in Figure 1.9 [30–32].

1.3.2.2 Influence of Temperature

The stator winding resistance and PM flux linkage are also dependent on the temperature. With the rise in the operating temperature, the stator winding resistances will be increased, while the PM flux linkage be decreased:

$$R_{s,s} = R_{s,0}[1 + \alpha_{cu}(\vartheta_s - \vartheta_r)] \tag{1.18}$$

$$\psi_{pm,s} = \psi_{pm,0}\left[1 + \alpha_{pm}(\vartheta_{pm} - \vartheta_r)\right] \tag{1.19}$$

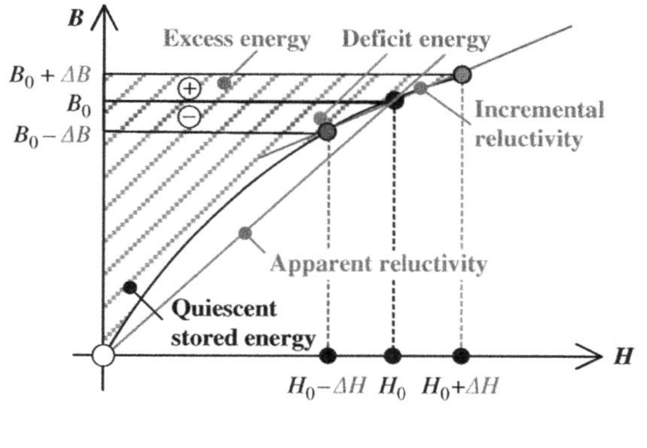

Figure 1.9 Representation of apparent and incremental reluctivities in non-linear magnetic material including energy perturbation.

where $R_{s,s}$, $R_{s,0}$, $\psi_{pm,s}$, and $\psi_{pm,0}$ are the stator winding resistance and the PM flux linkage at the actual average temperatures ϑ_s and ϑ_{pm}, as well as at the reference temperature ϑ_r. α_{cu} and α_{pm} are the temperature coefficients for copper and PM, respectively.

Table 1.1 tabulates the characteristics of typical PM materials, including ferrite, AlNiCo, SmCo, and NdFeB magnet materials, together with their temperature coefficients.

A commercial NdFeB PM material, i.e., N4514 [34], is taken as an example; its intrinsic and normal *BH* curves accounting for different temperatures are presented in Figure 1.10.

As can be seen, the main indicators for the performance assessment of PMs include the remanence B_r, the coercivity H_c, the intrinsic coercivity H_{cl}, and the maximum energy product BH_{max}. Once the external demagnetization field exceeds the knee point, the magnetic flux density

Table 1.1 Physical properties of PM materials.

Parameters	Ferrite	AlNiCo	$SmCo_5$	NdFeB
Remanence B_r (T)	0.38–0.42	0.61–1.35	0.85–1.0	1.0–1.23
Coercivity H_C (kA/m)	280–390	50–59	1000–1200	960–1600
Maximum energy product BH_{max} (kJ/m^3)	10–40	35–60	150–240	200–480
Temperature coefficient of B_r (%/°C)	−0.2	−0.03	−0.06 to −0.02	−0.15 to −0.1
Temperature coefficient of H_c (%/°C)	0.2 to 0.5	<0.1	−0.4 to −0.2	−0.6 to −0.4
Relative permeability μ_r	1.05	<5	1.05	<1.2
Electrical resistivity ρ_{ele} (Ω·m)	$>10^4$	5×10^{-3}	0.8×10^{-6}	1.5×10^{-6}

Source: Adapted from [33].

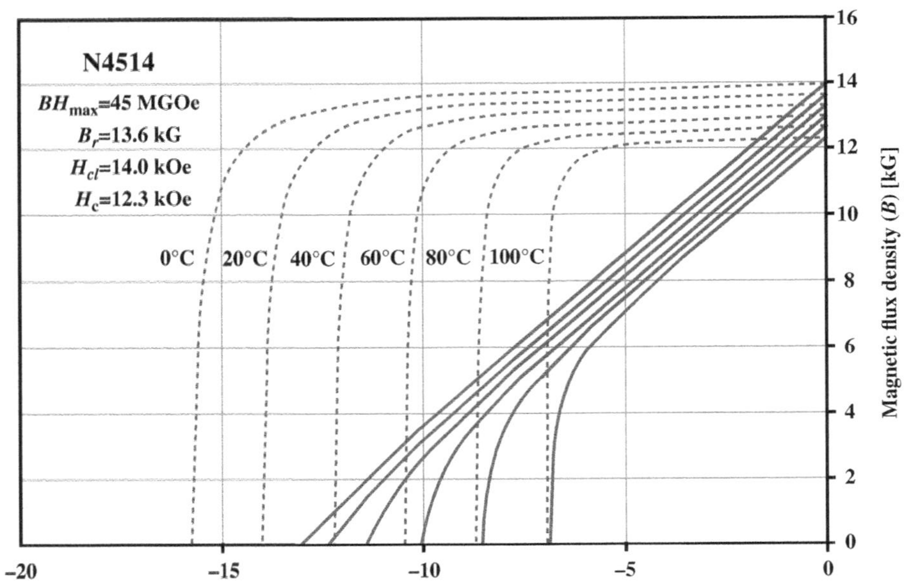

Figure 1.10 Intrinsic and normal *BH* curves of NdFeB magnet at different temperatures.

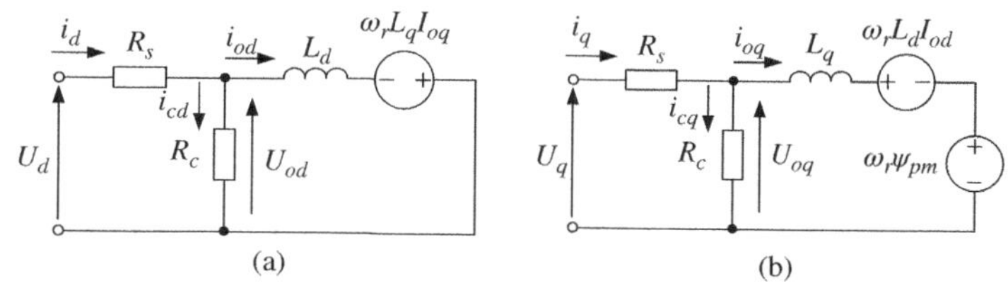

Figure 1.11 *dq*-axis equivalent circuits of PMSM: (a) *d*-axis and (b) *q*-axis.

decreases rapidly, which will cause irreversible demagnetization. Meanwhile, with the increasing temperature in the NeFeB magnets, both the remanence and the coercivity decrease with the rise in temperature, which will deteriorate the output torque and the demagnetization withstand capability, respectively.

Meanwhile, the temperature dependencies of the stator resistance and the rotor flux linkage are also widely used as thermal indicators to monitor thermal conditions indirectly. The detailed applications of condition monitoring will be introduced in Chapter 11.

1.3.2.3 Influence of Iron Loss

As previously mentioned, iron loss is not considered in an ideal PMSM model, but it does influence the flux linkage and output torque capability. Furthermore, the modelling of iron loss becomes critical when the machine is operated at high speed, low flux mode, or at variable speed [35]. The widely used PMSM equivalent circuit accounting for iron losses [36, 37] is presented in Figure 1.11, where it is assumed that the iron loss resistance R_c is greater than the product of the dq-axis reactances, i.e., $\omega_r^2 L_d L_q$, and the winding resistance R_s.

$$\frac{\omega_r^2 L_d L_q}{R_c^2} \ll 1 \text{ and } \frac{R_s}{R_c} \ll 1 \tag{1.20}$$

Therefore, the steady-state equations are as follows:

$$\begin{bmatrix} u_d \\ u_q \end{bmatrix} = \begin{bmatrix} R_s + \dfrac{\omega_r^2 L_d L_q}{R_c} & -\omega_r L_q \\ \omega_r L_d & R_s + \dfrac{\omega_r^2 L_d L_q}{R_c} \end{bmatrix} \begin{bmatrix} i_d \\ i_q \end{bmatrix} + \begin{bmatrix} \dfrac{\omega_r^2 L_q \psi_{pm}}{R_c} \\ \omega_r \psi_{pm} \end{bmatrix} \tag{1.21}$$

As can be seen, in the equivalent circuits, the iron loss is represented by a parallel connected iron loss resistance $R_c = \omega^2 L_d L_q / R_c$ in the stator circuit, which is a function of the magnetic flux and the electrical frequency. Thus, the output torque is strictly proportional to the magnetizing current. The magnetizing current is indirectly calculated based on the line current and the iron loss resistance.

1.4 Drives and Control Strategies

1.4.1 Drive System of PMSM

The schematic diagram of an electrical drive system for PMSM, including a voltage source inverter (VSI), a PMSM, and a control unit, is shown in Figure 1.12a and b. The control strategies play a critical role for PMSM drive systems and should achieve the following performance, e.g., high efficiency, lower torque ripple, wide speed controllability, fast dynamic response, high robustness and reliability, as well as ensuring system stability. For PMSMs, the phase currents need to be precisely controlled,

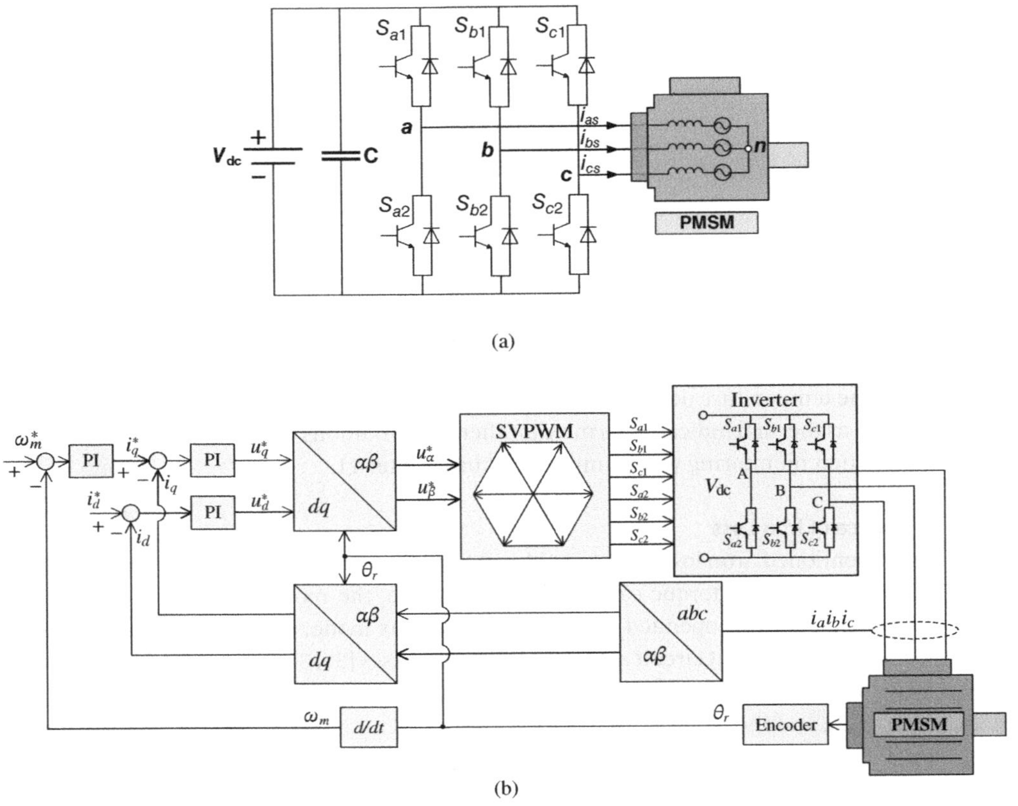

Figure 1.12 Schematic diagram of an electrical drive system for PMSM: (a) VSI-based three-phase PMSM drive system and (b) FOC system.

which can be achieved by employing space vector pulse width modulation (SVPWM), and high performance can be achieved by employing control methods in the controller, e.g., FOC, DTC, and MPC. The most popular control method is FOC, as shown in Figure 1.12.

FOC was first proposed in the 1970s and utilizes a similar control concept as for a dc motor. For PMSMs, the three-phase alternating current (ac) currents are decomposed into two orthogonal components in the rotating reference frame, i.e., the dc i_d and the quadrature current i_q, which are used to control the flux and torque separately. The relevant Park and Clarke transformations have been introduced in Section 1.3.1. In the FOC system, the accurate rotor position must be known and is usually obtained by using either a hardware rotor position sensor such as a resolver or encoder or software-based rotor position sensorless techniques [9]. In addition, two proportional–integral (PI) regulators are used to control both components of the current vector separately, where the accurate machine parameters are essential to determine the current vector and the PI constants for the regulators of both current and speed loops.

1.4.2 Space Vector Pulse Width Modulation

A three-phase VSI is used to provide variable voltage and frequency supply. PWM techniques have been studied extensively in the last few decades, and the SVPWM is most popular due to its high utilization of dc-link voltage and low harmonic distortion [38].

The three numbers for each voltage vector in Figure 1.13 represent the switching states of the three legs of the VSI (Figure 1.12a). To avoid the short circuit, the upper and lower switches in one leg work in the complementary mode with a dead time. Therefore, if the upper switch is on and the lower switch is off, the value of the switching state is "1," and if the upper switch is off and the lower switch is on, the value of the switching state is "0." In the SVPWM, there are six non-zero vectors (100, 110, 010, 011, 001, and 101) and two zero vectors (000 and 111). The whole voltage plane is evenly divided into six sectors by the non-zero voltage vectors, as presented in Figure 1.13, and the three-phase voltages $u_{a,b,c}$ and their corresponding space vectors are summarized in Table 1.2. The sinusoidal reference space vector follows a circular trajectory within the hexagon, and the maximum output phase voltage $u^*_{s,\max}$ achieved by applying SVPWM is $\sqrt{3}/3V_{dc}$, where V_{dc} is the dc bus voltage.

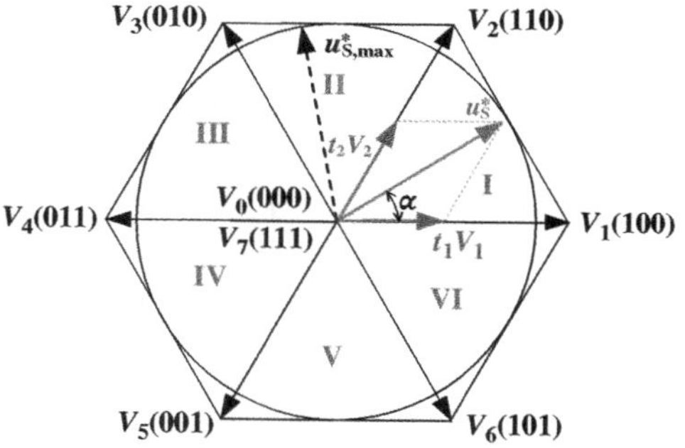

Figure 1.13 Phasor representation of eight basic space vectors.

Therefore, any voltage can be represented by two adjacent fundamental vectors together with zero vectors, and its amplitude is adjusted by the active durations of those fundamental vectors in a sampling period T_s. Taking the voltage vector reference u_s^* in Sector I as an example, Figure 1.13, the active durations (t_1, t_2, and t_0) of space vectors $\boldsymbol{V}_1$, $\boldsymbol{V}_2$, and $\boldsymbol{V}_0$ can be calculated as follows:

$$t_1 = \frac{u_s^*}{V_1}\frac{\sin(\pi/3 - \alpha)}{\sin(2\pi/3)} \tag{1.22a}$$

$$t_2 = \frac{u_s^*}{V_2}\frac{\sin(\alpha)}{\sin(2\pi/3)} \tag{1.22b}$$

$$t_0 = T_s - t_1 - t_2 \tag{1.22c}$$

where V_1 and V_2 are the amplitudes of voltage vectors, i.e.,$V_1 = V_2 = 2/3\ V_{dc}$. α is the angle of voltage vector reference.

Table 1.2 Correlation between voltages and corresponding space vectors.

Vector	Switching pattern	Phase voltage/V_{dc}			Line voltage/V_{dc}		
	$S_aS_bS_c$	u_a	u_b	u_c	u_{ab}	u_{bc}	u_{ca}
$\boldsymbol{V}_0$	000	0	0	0	0	0	0
$\boldsymbol{V}_1$	100	2/3	−1/3	−1/3	1	0	−1
$\boldsymbol{V}_2$	110	1/3	1/3	−2/3	0	1	−1
$\boldsymbol{V}_3$	010	−1/3	2/3	−1/3	−1	1	0
$\boldsymbol{V}_4$	011	−2/3	1/3	1/3	−1	0	1
$\boldsymbol{V}_5$	001	−1/3	−1/3	2/3	0	−1	1
$\boldsymbol{V}_6$	101	1/3	−2/3	1/3	1	−1	0
$\boldsymbol{V}_7$	111	0	0	0	0	0	0

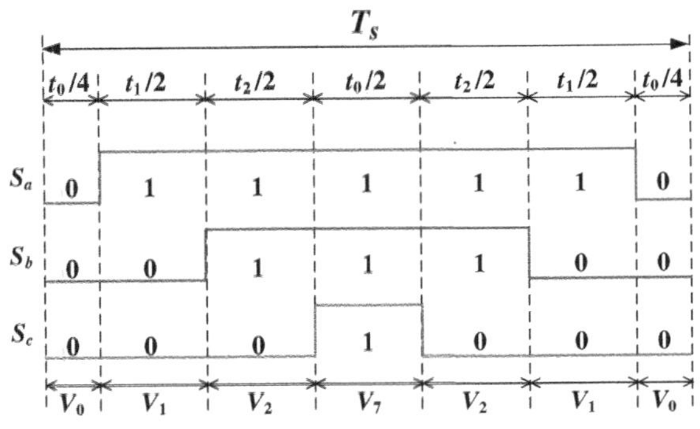

Figure 1.14 Switching states and gate signals within one sampling period in Sector I.

It is well-known that each inverter leg should change only once between 0 and 1 in one switching period to minimize switching loss [39]. Thus, this switching pattern can be achieved by applying the zero vectors followed by two active state vectors in a half-switching period, and then, the switching pattern is repeated in the next half-switching period. Figure 1.14 depicts the switching states within one sampling period in Sector I and the gate signals for each phase leg of the inverter.

In practice, to avoid damage to the power switching devices caused by the short circuit, a proper length of time delay, i.e., dead time, must be performed between the turn-on signal for one power switching device and the corresponding turn-off signal for the other power switching device in the same leg of the inverter. However, the introduced dead-time will cause a difference between the commanded and actual voltages, which will result in the distortions of output voltages and currents. To date, the compensation of dead-time voltage distortion [40, 41], together with the voltage distortions caused by the non-ideal characteristic of the switching device [42, 43], have been extensively developed. The detailed introduction of the influence of VSI non-linearity on the electrical parameter identification and the identification of VSI non-linearity will be described in Chapters 2 and 3, respectively.

1.5 Outline of Parameter Estimation Techniques

The state-of-the-art offline and online parameter estimation methods are outlined in Figure 1.15, and a comprehensive literature review (up to 2022) can be found in Ref. [19]. In this section, both offline and online parameter estimation methods are only briefly highlighted, as they will be comprehensively described later in this book.

1.5.1 Offline Parameter Estimation

The FEA is the most common offline method to estimate PMSM parameters but requires detailed knowledge of geometric and material information. It is widely reported that FEA can be used to calculate a rotor flux linkage map and incremental/apparent *dq*-axis inductances accounting for magnetic saturation and cross-coupling. An example of using FEA to analyse the electromagnetic performance of PMSM is presented in Appendix A. However, sometimes, FEA methods [30, 31] are limited since some material properties are unknown or not completely acquirable. Hence, offline experimental estimation methods have been extensively developed in the last few decades. Some methods are listed in the following.

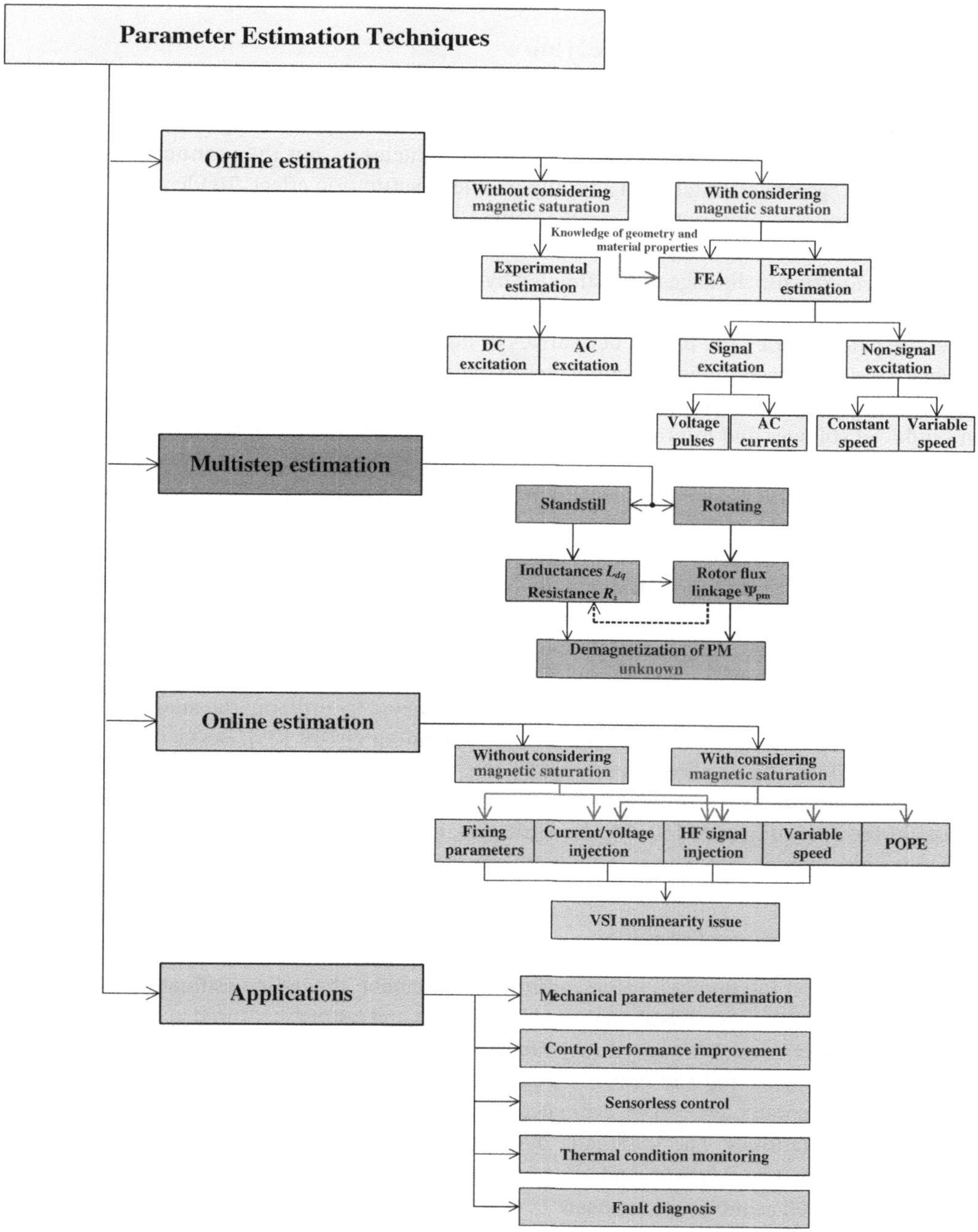

Figure 1.15 Overview of parameter estimation techniques.

1. Voltage excitation [44].
2. Magnetic model estimation:
   a. *dq*-axis voltage pulse injection [45].
   b. Fast Fourier transform (FFT) analysis [46].

c. AC signal excitation [3, 47].
d. Estimation under constant speed [10].
e. Estimation under variable speed [48, 49].

According to the time constant at the transient state, the step dc voltage excitation tests at a standstill in Ref. [44] are used to predict apparent inductance, but they cannot determine the rotor flux linkage or accurately consider the magnetic saturation effect. To identify *dq*-axis flux linkages and incremental inductances, the voltage pulses are applied to the *d*-axis and *q*-axis in turn while the other axis current remains constant [45]. Accounting for self- and cross-saturation effects, the *dq*-axis flux linkages are estimated by using the FFT method in Ref. [46] based on the measured *dq*-axis currents at constant rotor speed, and then, the *dq*-axis incremental inductances are calculated by the partial derivatives of *dq*-axis flux linkages with respect to currents. Besides, ac signal excitations are employed to estimate apparent [3] and incremental inductances [47], respectively, where the dc bias point sets the magnetic operating point in Ref. [47], and induced voltages and internal load angles are measured in Ref. [3]. Moreover, the magnetic models are estimated based on the FOC schemes at constant [10] and at variable speed [48, 49]. Specifically, an additional voltage measurement circuit is required in Ref. [10]. The variations of stator resistance and VSI-caused distorted voltage is determined at a standstill [48]. The *dq*-axis flux-linkage map is modelled as second-order polynomials under variable speed control [49]. It should be mentioned that the estimated *dq*-axis flux linkages can derive both incremental and apparent inductances based on Eqs. (1.15) and (1.16). However, the values of two types of inductances may be different due to magnetic saturation, and the above papers select either incremental or apparent inductance. In addition to the direct calculation by means of inductance definitions, the apparent inductances can also be estimated by utilizing the analysis of a phasor diagram [2] or the time constant of the armature winding circuit [25]. Besides, the multi-step estimation utilizes the acquirable parameters, e.g., stator resistance and distorted voltage drop, at standstill, and then, the rotor flux linkage and the stator apparent inductances can be estimated in the operating state [50–52]. Nevertheless, it cannot accurately track online parameter variations due to neglecting the magnet temperature rise and the core loss.

1.5.2 Online Parameter Estimation

Without relying on the nominal values/offline measurement, the online estimation methods are conducted during operation based on the available input and output measured quantities, i.e., current (i), voltage (v), and rotor speed (ω_r). Generally, the online parameter estimation process is carried out under sensored and sensorless control of the PMSM drive system. Both the electrical and mechanical parameters can be estimated by utilizing the various estimation algorithms and the measured quantities. The schematic diagram of online parameter estimation is shown in Figure 1.16, where $\Delta\omega_r$, Δi_d, Δu_d, and $\pm\Delta\theta_p$ are the variables of rotor speed, injected *d*-axis current and voltage, as well as the position offsets.

The general issues in online parameter estimation, including the rank-deficient problem, the VSI non-linearity, and the influence of signal injection, must be addressed first. To ensure the full-rank estimation schemes, the PMSMs usually need to be excited by one or more additional sets of signals, and the state-of-the-art methods include the injections of currents, voltages, HF signals, and rotor position offsets, along with intentionally changing the rotor speed. Although the online parameter estimation techniques incorporate high levels of computational efficiency, with the improvement of the computing capability of the processor, some AI or modern control theory-based algorithms have

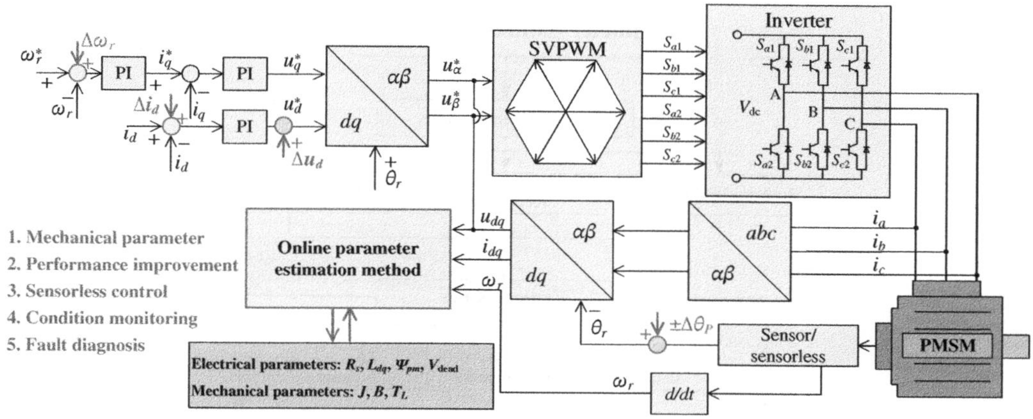

Figure 1.16 Schematic diagram of online parameter estimation.

been applied. Most online methods are concerned with the apparent inductance estimation since the terms of time derivation of dq-axis flux linkages are cancelled in the steady-state equation. Besides, the incremental inductances are also important for HF signal injection-based sensorless rotor position estimation, particularly at zero and low speed [9].

The main offline and online modelling techniques, including HF signal injection, will be illustrated later in this book in greater detail.

1.6 Scope of This Book

This book provides a comprehensive resource on the basic and state-of-the-art online and offline parameter estimation techniques for PMSMs, as summarized in Figure 1.17.

It starts with a brief description of PMSMs and drives and then introduces the basic parameters and estimation techniques in Chapter 1 and the issues in parameter estimation in Chapter 2, such as rank deficiency and non-linearity of inverter. It is followed by various online parameter estimation techniques based on rotor position sensored operation, including those (a) with the aid of thermocouples in the stator windings in Chapter 3, (b) based on current/voltage injections in Chapter 4, and (c) based on the position offset injection in Chapter 5, under constant and variable speeds and/or torque in Chapter 6. The effect of magnetic saturation and the cross-coupling effect in PMSMs, along with the corresponding parameter estimation methods, are reported in Chapter 7. Offline parameter estimation by signal injection at standstill and by multi-step methods is also described in Chapter 8. In addition, three different techniques for estimating mechanical parameters are described in Chapter 9. The book also introduces many modern control theory-based parameter estimation techniques, including recursive least squares, Kalman filter, model reference adaptive system, Adaline neural network, gradient-based methods, particle swarm optimization, and genetic algorithm in Chapter 10. Finally, applications of parameter estimation techniques for (a) improvement of control performance, (b) sensorless control, (c) thermal condition monitoring, and (d) fault diagnosis are presented in Chapter 11. The finite element method for calculating winding inductances is given in the appendix and the classic methods for estimating and testing the parameters of PMSMs are embodied in the main body for comparison.

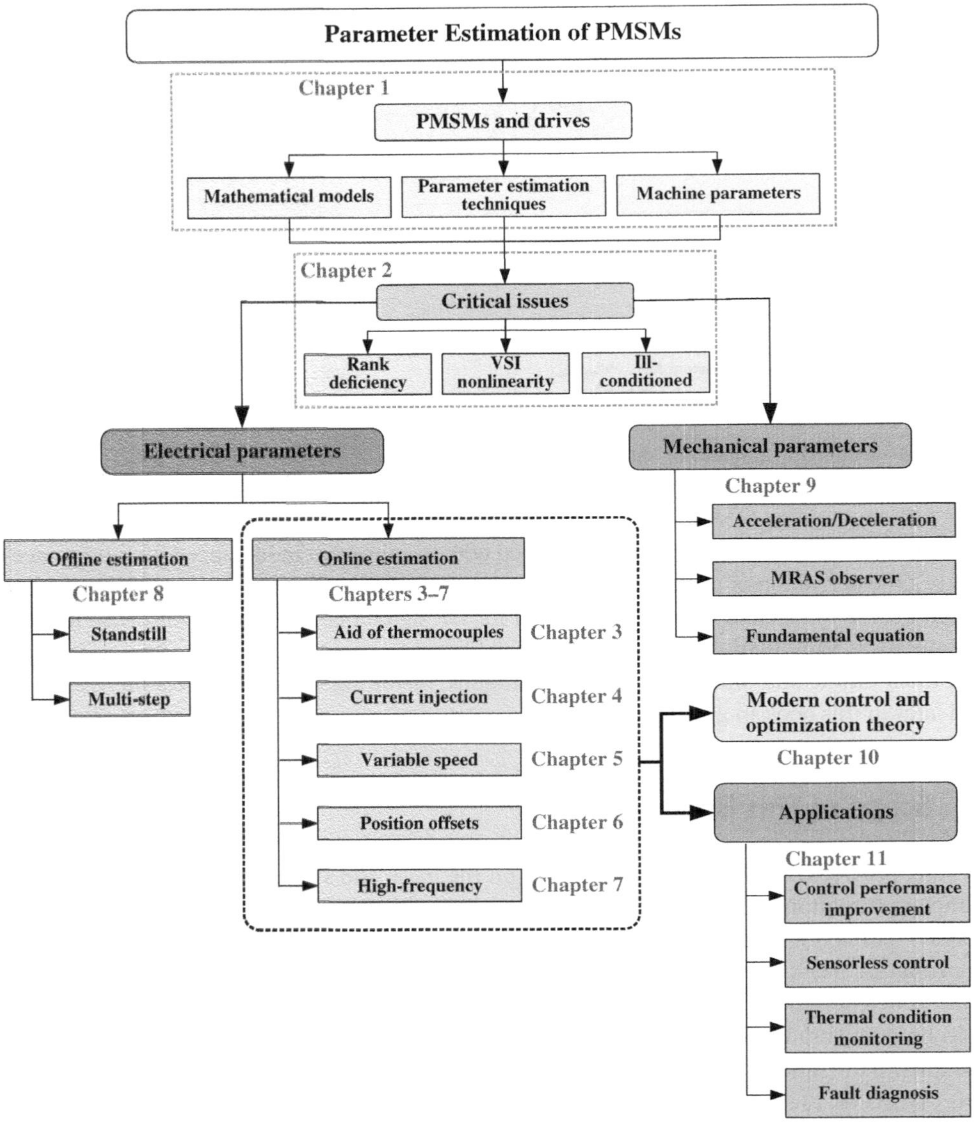

Figure 1.17 Outline of the book.

References

1 US Energy Information Administration, "Manufacturing Energy Consumption Survey." Energy Efficiency Indicators, 2020.

2 B. Chalmers, S. Hamed, and G. Baines, "Parameters and performance of a high-field permanent-magnet synchronous motor for variable-frequency operation," *IEE Proc. B*, vol. 132, no. 3, p. 117, May 1985.

3 B. Stumberger, B. Kreca, and B. Hribernik, "Determination of parameters of synchronous motor with permanent magnets from measurement of load conditions," *IEEE Trans. Energy Convers*, vol. 14, no. 4, pp. 1413–1416, Dec. 1999.

4 F. Fernandez-Bernal, A. Garcia-Cerrada, and R. Faure, "Determination of parameters in interior permanent-magnet synchronous motors with iron losses without torque measurement," *IEEE Trans. Ind. Appl.*, vol. 37, no. 5, pp. 1265–1272, Sept./Oct. 2001.

5 S.K. Sul, Control of Electric Machine Drive Systems. New Jersey: Wiley, 2011.

6 S. Li and Z. Liu, "Adaptive speed control for permanent-magnet synchronous motor system with variations of load inertia," *IEEE Trans. Ind. Electron*, vol. 56, no. 8, pp. 3050–3059, Aug. 2009.

7 K. Liu, "Research on the online multi-parameter identification of permanent magnet synchronous machines," PhD dissertation, Hunan Univ., Changsha, China, 2010.

8 J. Ziegler and N. Nichols, "Optimum settings for automatic controllers," *J. Fluids Eng.*, vol. 64, no. 8, pp. 759–765, 1942.

9 Z. Q. Zhu and X. M. Wu, *Sensorless Control of Permanent Magnet Synchronous Machine Drives*. New York: Wiley and IEEE Press, 2023.

10 E. Armando, R. Bojoi, P. Guglielmi, G. Pellegrino, and M. Pastorelli, "Experimental identification of the magnetic model of synchronous machines," *IEEE Trans. Ind. Appl.*, vol. 49, no. 5, pp. 2116–2125, Sept./Oct. 2013.

11 Y. Inoue, Y. Kawaguchi, S. Morimoto, and M. Sanada, "Performance improvement of sensorless IPMSM drives in a low-speed region using online parameter identification," *IEEE Trans. Ind. Appl.*, vol. 47, no. 2, pp. 798–804, Mar./Apr. 2011.

12 Z. Q. Zhu, P. Wang, N. M. A. Freire, Z. Azar, and X. Wu, "A novel rotor position-offset injection-based online parameter estimation of sensorless controlled surface-mounted PMSMs," *IEEE Trans. Energy Convers.*, vol. 39, no. 3, pp. 1930–1946, Sept. 2024.

13 Y. Li, Z. Q. Zhu, D. Howe, C. M. Bingham, and D. A. Stone, "Improved rotor-position estimation by signal injection in brushless AC motors, accounting for cross-coupling magnetic saturation," *IEEE Trans. Ind. Appl.*, vol. 45, no. 5, pp. 1843–1850, Sept./Oct. 2009.

14 D. D. Reigosa, F. Briz, P. García, J. M. Guerrero, and M. W. Degner, "Magnet temperature estimation in surface PM machines using high-frequency signal injection," *IEEE Trans. Ind. Appl.*, vol. 46, no. 4, pp. 1468–1475, July/Aug. 2010.

15 S. Wilson, P. Stewart, and B. Taylor, "Methods of resistance estimation in permanent magnet synchronous motors for real-time thermal management," *IEEE Trans. Energy Convers.*, vol. 25, no. 3, pp. 698–707, Sept. 2010.

16 K. Liu, "Novel techniques for parameter estimation of permanent magnet synchronous machines," PhD dissertation, University of Sheffield, Sheffield, U.K., Dec. 2013.

17 D. Liang, "Critical temperature estimation for permanent magnet synchronous machines," PhD dissertation, University of Sheffield, Sheffield, U.K., Feb. 2022.

18 Z. Q. Zhu and D. Liang, "Perspective of thermal analysis and management for permanent magnet machines, with particular reference to hotspot temperatures," *Energies*, vol. 15, no. 21, pp. 1–53, Nov. 2022.

19 Z. Q. Zhu, D. Liang, and K. Liu, "Online parameter estimation for permanent magnet synchronous machines: an overview," *IEEE Access*, vol. 9, pp. 59059–59084, 2021.

20 A. Binder, T. Schneider, and M. Klohr, "Fixation of buried and surface-mounted magnets in high-speed permanent-magnet synchronous machines," *IEEE Trans. Ind. Appl.*, vol. 42, no. 4, pp. 1031–1037, Jul./Aug. 2006.

21 S. Morimoto, M. Sanada, and Y. Takeda, "Wide-speed operation of interior permanent magnet synchronous motors with high-performance current regulator," *IEEE Trans. Ind. Appl.*, vol. 30, no. 4, pp. 920–926, Jul./Aug. 1994.

22 Y. Sato, S. Ishikawa, T. Okubo, M. Abe, and K. Tamai, "Development of high response motor and inverter system for the Nissan LEAF electric vehicle," SAE Technical Paper, 2011-01-0350, 2011.

23 M. Kamiya, "Development of traction drive motors for the Toyota hybrid system," *IEEJ Trans. Ind. Appl.*, vol. 126, no. 4, pp. 473–479, 2006.

24 A. E. Fitzgerald, C. Kingsley, and S. D. Umans, Electric Machinery. New York: McGraw-Hill, 1989.

25 Y. S. Chen, "Motor topologies and control strategies for permanent magnet brushless AC drives," Ph.D. dissertation, Univ. Sheffield, U.K., 1999.

26 A. M. El-Serafi, A. S. Abdallah, M. K. El-Sherbiny, and E. H. Badawy, "Experimental study of the saturation and the cross-magnetizing phenomenon in saturated synchronous machines," *IEEE Trans. Energy Convers.*, vol. 3, no. 4, pp. 815–823, Dec. 1988.

27 A. M. El-Serafi and J. Wu, "Determination of the parameters representing the cross-magnetizing effect in saturated synchronous machines," *IEEE Trans. Energy Convers.*, vol. 8, no. 3, pp. 333–342, Sept. 1993.

28 Z. Azar, Z.Q. Zhu, and G. Ombach, "Influence of electric loading and magnetic saturation on cogging torque, back-EMF and torque ripple of PM machines," *IEEE Trans. Magn.*, vol. 48, no. 10, pp. 2650–2658, Oct. 2012.

29 W. Xu and R. D. Lorenz, "High-frequency injection-based stator flux linkage and torque estimation for DB-DTFC implementation on IPMSMs considering cross-saturation effects," *IEEE Trans. Ind. Appl.*, vol. 50, no. 6, pp. 3805–3815, Nov./Dec. 2014.

30 N. A. Demerdash, T. M. Hijazi, and A. A. Arkadan, "Computation of winding inductances of permanent magnet brushless DC motors with damper windings by energy perturbation," *IEEE Trans. Energy Convers.*, vol. 3, no. 3, pp. 705–713, Sept. 1988.

31 T. W. Nehl, F. A. Fouad, and N. A. Demerdash, "Determination of saturated values of rotating machinery incremental and apparent inductances by an energy perturbation method," *IEEE Power Eng. Rev.*, vol. PER-2, no. 12, pp. 28–29, Dec. 1982.

32 N. A. Demerdash and T. W. Nehl, "Electric machinery parameters and torques by current and energy perturbations from field computations. I. Theory and formulation," *IEEE Trans. Energy Convers.*, vol. 14, no. 4, pp. 1507–1513, Dec. 1999.

33 S. Liu and G. Kuhl, "Temperature coefficients of rare earth permanent magnets," *IEEE Trans. Magn.*, vol. 35, no. 5, pp. 3271–3273, Sept. 1999.

34 *ND Magnets-N4514*, Dexter Magnet Technologies, [Online]. Available https://www.dextermag.com/products/permanent-magnets/nd-magnets-n4514/.

35 M. Popescu, D. G. Dorrell, and D. M. Ionel, "A study of the engineering calculations for iron losses in 3-phase AC motor models," *IECON 2007 – 33rd Ann. Conf. IEEE Ind. Electron. Soc.*, 2007, pp. 169–174.

36 N. Urasaki, T. Senjyu, and K. Uezato, "A novel calculation method for iron loss resistance suitable in modeling permanent-magnet synchronous motors," *IEEE Trans. Energy Convers.*, vol. 18, no. 1, pp. 41–47, March 2003.

37 S. Morimoto, Y. Tong, Y. Takeda, and T. Hirasa, "Loss minimization control of permanent magnet synchronous motor drives," *IEEE Trans. Ind. Electron.*, vol. 41, no. 5, pp. 511–517, Oct. 1994.

38 D Holmes and T. Lipo, *Pulse width modulation for power converters*. Hoboken, NJ: John Wiley, *2003*.

39 S. Bowes and Y. Lai, "The relationship between space-vector modulation and regular-sampled PWM," *IEEE Trans. Ind. Electron.*, vol. 44, no. 5, pp. 670–679, Oct. 1997.

40 Y. Murai, T. Watanabe, and H. Iwasaki, "Waveform distortion and correction circuit for PWM inverters with switching lag-times," *IEEE Trans. Ind. Appl.*, vol. IA-23, no. 5, pp. 881–886, Sept. 1987.

41 S. Jeong and M. Park, "The analysis and compensation of dead-time effects in PWM inverters," *IEEE Trans. Ind. Electron.*, vol. 38, no. 2, pp. 108–114, April 1991.

42 J. Choi and S. K. Sul, "Inverter output voltage synthesis using novel dead time compensation," *IEEE Trans. Power Electron.*, vol. 11, no. 2, pp. 221–227, March 1996.

43 N. Urasaki, T. Senjyu, K. Uezato, and T. Funabashi, "An adaptive dead-time compensation strategy for voltage source inverter fed motor drives," *IEEE Trans. Power Electron.*, vol. 20, no. 5, pp. 1150–1160, Sept. 2005.

44 S. Weisgerber, A. Proca, and A. Keyhani, "Estimation of permanent magnet motor parameters," In *Proc. IEEE Ind. Appl. Soc. Annu. Meeting*, 1997, vol. 1, pp. 29–34.

45 B. Stumberger, G. Stumberger, D. Dolinar, A. Hamler, and M. Trlep, "Evaluation of saturation and cross-magnetization effects in interior permanent-magnet synchronous motor," *IEEE Trans. Ind. Appl.*, vol. 39, no. 5, pp. 1264–1271, Sept./Oct. 2003.

46 K. M. Rahman and S. Hiti, "Identification of machine parameters of a synchronous motor," *IEEE Trans. Ind. Appl.*, vol. 41, no. 2, pp. 557–565, Mar./Apr. 2005.

47 S. A. Odhano, R. Bojoi, S. G. Rosu, and A. Tenconi, "Identification of the magnetic model of permanent-magnet synchronous machines using DC-biased low-frequency AC signal injection," *IEEE Trans. Ind. Appl.*, vol. 51, no. 4, pp. 3208–3215, Jul./Aug. 2015.

48 G. Pellegrino, B. Boazzo, and T. M. Jahns, "Magnetic model self-identification for PM synchronous machine drives," *IEEE Trans. Ind. Appl.*, vol. 51, no. 3, pp. 2246–2254, May/Jun. 2015.

49 K. Liu, J. Feng, S. Guo, L. Xiao, and Z.Q. Zhu, "Identification of flux linkage map of permanent magnet synchronous machines under uncertain circuit resistance and inverter nonlinearity," *IEEE Trans. Ind. Informat.*, vol. 14, no. 2, pp. 556–568, Feb. 2018.

50 M. Rashed, P. Macconnell, A. Stronach, and P. Acarnley, "Sensorless indirect-rotor-field-orientation speed control of a permanent-magnet synchronous motor with stator-resistance estimation," *IEEE Trans. Ind. Electron.*, vol. 54, no. 3, pp. 1664–1675, Jun. 2007.

51 S. Morimoto, M. Sanada, and Y. Yakeda, "Mechanical sensorless drives of IPMSM with online parameter identification," *IEEE Trans. Ind. Appl.*, vol. 42, no. 5, pp. 1241–1248, Sep./Oct. 2006.

52 S. Lee, "Closed-loop estimation of permanent magnet synchronous motor parameters by PI controller gain tuning," *IEEE Trans. Energy Convers.*, vol. 21, no. 4, pp. 863–870, Dec. 2006.

2

Critical Issues with Online Parameter Estimation

2.1 Rank-Deficient Problem

In this section, the rank-deficient problem in the multi-parameter estimation of permanent magnet synchronous machines (PMSMs) is highlighted and investigated at first. Although the electrical parameters can be estimated based on different reference frames, i.e., the *abc* three-phase reference frame, the stationary $\alpha\beta$-axis reference frame, and the rotating dq-axis frame, the same rank-deficient problem always exists. By way of example, the following analysis is based on the dq-axis reference frame.

2.1.1 Rank-Deficient Issue

The PMSM model in the dq-axis reference frame can be expressed as follows:

$$\begin{cases} \dfrac{di_d}{dt} = -\dfrac{R_s}{L_d} i_d + \dfrac{L_q}{L_d} \omega_r i_q + \dfrac{u_d}{L_d} \\ \dfrac{di_q}{dt} = -\dfrac{R_s}{L_q} i_q - \dfrac{L_d}{L_q} \omega_r i_d + \dfrac{u_q}{L_d} - \dfrac{\psi_{pm}}{L_q} \omega_r \end{cases} \tag{2.1}$$

where i_d, i_q, u_d, and u_q are the dq-axis currents and voltages. ω_r is the electrical angular speed. R_s, L_d, L_q, and ψ_{pm} are the stator resistance, the dq-axis inductances, and the rotor PM flux linkage, respectively.

At steady state, Eq. (2.1) can be discretized by using Euler's method:

$$\begin{cases} u_d(k) = R_s i_d(k) - L_q \omega_r(k) i_q(k) \\ u_q(k) = R_s i_q(k) + L_d \omega_r(k) i_d(k) + \psi_{pm} \omega_r(k) \end{cases} \tag{2.2}$$

where "k" is the index of the discrete sampling instant.

Under the widely used $i_d = 0$ control, Eq. (2.2) can be further simplified as follows:

$$\begin{cases} u_d(k) = -L_q \omega_r(k) i_q(k) \\ u_q(k) = R_s i_q(k) + \left[L_d i_q(k) + \psi_{pm}\right] \omega_r(k) \end{cases} \tag{2.3}$$

It can be easily found that there are four electrical parameters in Eq. (2.3), including winding resistance R_s, dq-axis inductances L_d and L_q, and rotor flux linkage ψ_{pm}. Thus, it is impossible to simultaneously estimate all four parameters from the above two linear equations, which is called the rank-deficient problem. Consequently, no matter which estimation method is employed for

Parameter Estimation of Permanent Magnet Synchronous Machines, First Edition. Zi Qiang Zhu, Kan Liu, and Dawei Liang.

multi-parameter estimation, e.g., extended Kalman filter (EKF), model reference adaptive system (MRAS), Adaline neural network (ANN), or recursive least squares (RLS), it cannot ensure that the estimation converges to the correct point under a rank-deficient system.

It can be found in the literature that some early research [1–6] tried to solve this problem and have made good contributions. For instance, an additional torque equation is introduced in Ref. [1] by a load test and combined with online estimation methods so that three parameters, i.e., the dq-axis inductances and the rotor flux linkage, could be estimated simultaneously. In Ref. [2], the inductance is fixed to a nominal value, and an error model was employed for the resistance and rotor flux linkage estimation, while the influence of inductance variation cannot be considered. In Ref. [3], a fast estimation for the rotor flux linkage based on PMSM reactive power by using $i_d < 0$ control is proposed, whereas the winding resistance value and the variations of inductances are not considered. An RLS-based method is proposed in Ref. [4] to estimate the winding resistance, the rotor flux linkage, and the inductance, which requires periodic changes in torque and could not be effective when the machine is in a steady state. Moreover, by switching between the low and high speeds, the stator winding resistance and the rotor flux linkage can be estimated in Ref. [5], but they are estimated subsequently instead of simultaneously. In addition, the stator winding resistance, the dq-axis inductances, and the rotor flux linkage can be estimated simultaneously in Ref. [6] by injecting a three-level perturbation in the d-axis current, whereas the estimation is still based on a rank-deficient model, and thus, the correct convergence of identified parameters may be not guaranteed.

Besides, it is worth mentioning that the winding resistance, the inductances, and the rotor flux linkage are affected by magnetic saturation and temperature rise, and consequently, it will cause inaccurate estimation if they are not estimated at the same time. However, due to rank deficiency, if the convergence speed of the algorithm is not appropriately designed and/or the initial values are chosen not close to the actual values, a severe ill-convergence may occur since there are infinite sub-optimums. To overcome the rank-deficient issue and realize simultaneous estimation of multiple parameters under real-time conditions, the key point is the modelling of a full-rank reference system, which can be performed by several recently developed approaches, including current/voltage injection [7, 8], with the aid of additional thermal sensor [9], rotor position offset [10], variable rotor speed control [11], etc., which will be introduced in the following chapters in detail.

2.1.2 Experimental Analysis and Results

In this sub-section, four estimation schemes for an SPMSM ($L_d = L_q$) are designed to further illustrate its influence on the multi-parameter estimation [7], where ANN estimators are employed, and their introduction can be found in Chapter 10.

Scheme I:
Estimation of the inductances and the rotor flux linkage, while the winding resistance is fixed to its measured value.

Scheme II:
Estimation of the inductances and the winding resistance, while the rotor flux linkage is fixed to its nominal value.

Scheme III:
Simultaneous estimation of the inductance, the winding resistance, and the rotor flux linkage. The estimated values of the winding resistance and the rotor flux linkage value will be transferred to each other at every step of estimation.

Scheme IV:

Step 1: The inductance value ($L_d = L_q$) is estimated under $i_d = 0$.

Step 2: The inductance is set to the value estimated from the first step, and a negative i_d is injected. Since the inductance is set to be constant, the dq-axis equation when $i_d < 0$ is of full rank for simultaneously estimating the winding resistance and rotor flux linkage. When $i_d < 0$, the winding resistance can be estimated from the d-axis equation.

Step 3: With the estimated winding resistance, the rotor flux linkage can be further estimated.

As mentioned above, Schemes I and II are similar to the method proposed in Ref. [12]. Scheme III is similar to the rank-deficient estimation method proposed in Refs. [13, 14], while Scheme IV is similar to the schemes proposed in Refs. [2, 3]. In addition, scheme II is also similar to the employed method in Ref. [15] when it is used for non-salient-pole PMSM ($L_d = L_q$) parameter estimation. Otherwise, the setup in Ref. [15] will be similar to scheme III if it is used for the parameter estimation of salient-pole PMSM ($L_d \neq L_q$).

The aforementioned four schemes are verified on a vector-controlled SPMSM drive system [7]. The digital signal processor (DSP) hardware drive platform (TMS TI 2812) is presented in Figure 2.1, and the design parameters of Prototype Machine 1 are listed in Table 2.1. The experimental results under all schemes are presented in Figures 2.2–2.5 and summarized in Table 2.2. Additionally, the experimental results in the following sections are all based on the same machine and drive system.

Figure 2.1 DSP platform and experimental test rig.

Table 2.1 Design parameters and specification of Prototype Machine 1.

Parameter	Value	Parameter	Value
Rated current (A)	4	Rated speed (r/min)	400
dc-link voltage (V)	36	Number of slot/pole number	12/10
Nominal phase resistance* (Ω)	0.33	Nominal terminal wire resistance (Ω)	0.043
Nominal self-inductance (mH)	2.91	Nominal mutual inductance (mH)	−0.33
Nominal d-axis inductance (mH)	3.24	Nominal q-axis inductance (mH)	3.24
Nominal rotor flux linkage (mWb)	77.6	Rotor moment of inertia (kg m^2)	0.8 10^{-5}

Note: Nominal values are measured. The nominal phase resistance* is measured at 25°C.

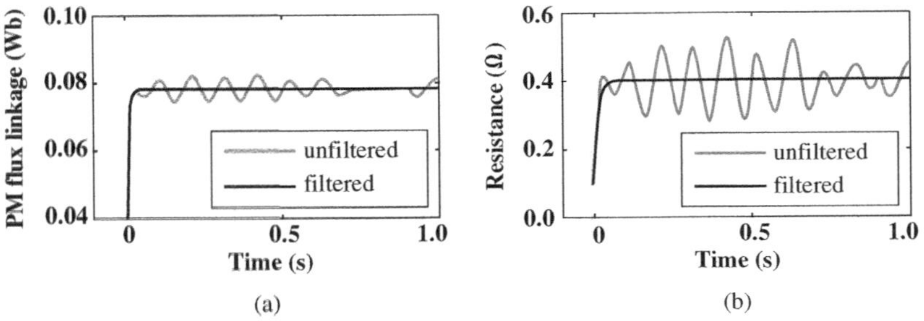

Figure 2.2 Experimental results with and without filters [7], IEEE: (a) Scheme I and (b) Scheme II.

In the experiment, the sampling period of the drive control system and the update interval of all the estimation methods are set to 83.3 μs. Since the measured signals, e.g., rotor speed, *dq*-axis currents, and voltages, fluctuate around the values specified by the proportional–integral (PI) regulators, these signals are filtered by the first-order Butterworth low-pass filters before estimation. Figure 2.2 shows the results of Schemes I and II with and without first-order Butterworth low-pass filters. As can be seen, the estimated result with low-pass filtering has no oscillation and converges very fast to the final value.

First, the estimated parameters in Schemes I and II are presented in Figure 2.3, which are similar to the two schemes in Ref. [12], and it is evident that the two schemes are accurate compared with the nominal values given in Table 2.1 when the PMSM is at nominal temperature. However, after heating for 20 min, contrary to the actual variation trend that would be expected, the estimated rotor flux linkage becomes larger as a result of the mismatching between the actual and nominal winding resistance values, as shown in Figure 2.3b. The same problem also occurs in Scheme II, with the estimated winding resistance value (0.421 Ω) being less than the actual one (0.456 Ω) due to the mismatching between the actual and nominal rotor flux linkages, as shown in Figure 2.3d.

Second, the experimental results of Scheme III are presented in Figure 2.4. It can be found from Figure 2.4a that when the initial values of the three parameters to be estimated are far from the nominal values, the estimated winding resistance and rotor flux linkage are ill-converged into 0.895 Ω and 67.0 mWb, respectively, which are significantly different from the nominal values in Table 2.1.

In comparison, when the initial values are set to be the nominal values, as shown in Figure 2.4b and c, it is still hard to ensure that the estimated solutions are always correct since there inevitably exist fluctuations in the measured signals, which may cause the estimation solution to slowly diverge into the sub-optimum if the measured *dq*-axis data are not filtered. Meanwhile, the estimated final results of R_s and ψ_{pm} are still very sensitive to the convergence speed (η) and the low-pass filters and may be converged to the wrong values or even diverged. Even if the *dq*-axis data are filtered by low-pass filters, the convergence speed of estimators will also affect the estimation results.

In addition, in terms of the estimated *q*-axis inductance L_q in Figure 2.4a–c, they are all converged to the same results, which can be explained that L_q can be estimated independently from the *d*-axis equation in Eq. (2.3), irrespective of the estimated values of winding resistance and rotor flux linkage, i.e., it is of full rank. Therefore, it can be uniquely determined by the measured *d*-axis voltage, *q*-axis current, and rotor speed.

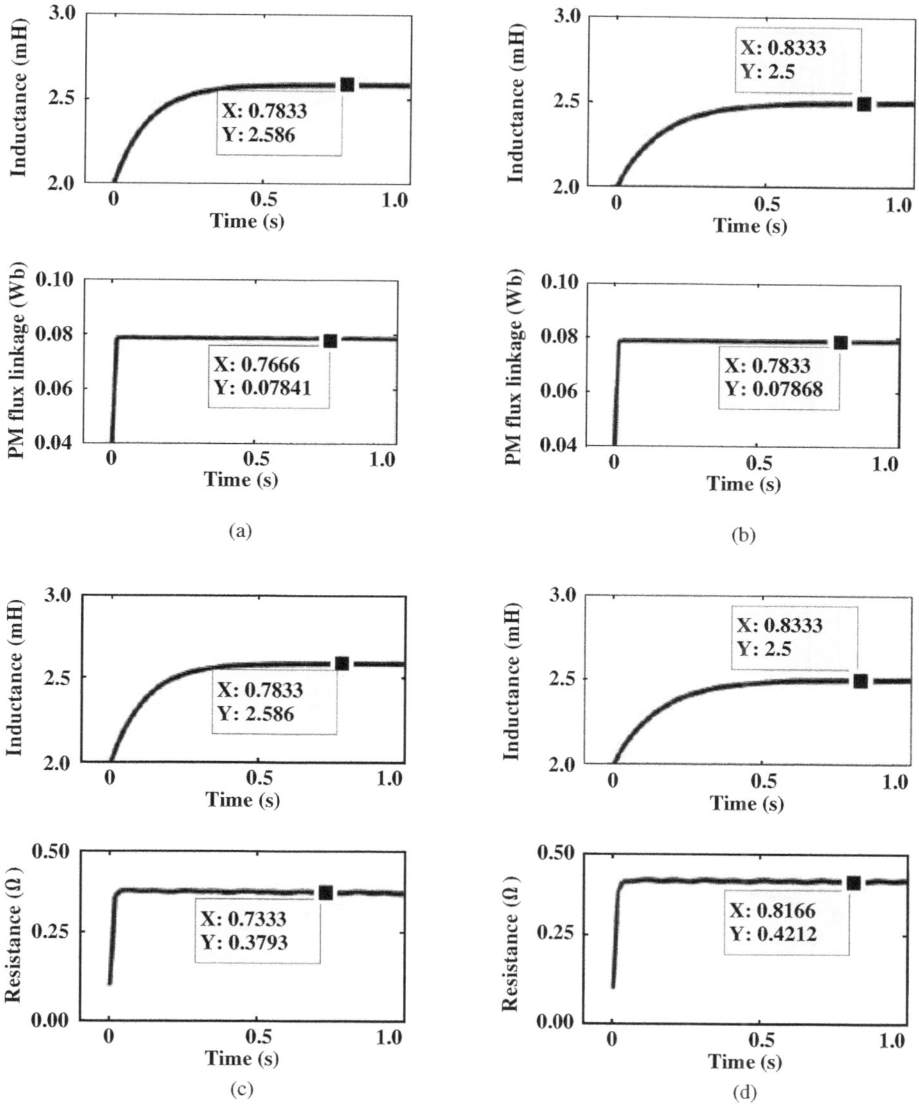

Figure 2.3 Estimated parameters of schemes I and II [7], IEEE: (a) scheme I (normal temperature), (b) scheme I (after heating for 20 min), (c) scheme II (normal temperature), and (d) scheme II (after heating for 20 min).

Finally, the experimental results of Scheme IV, Figure 2.5, show that if the dq-axis inductances are both set to L_q ($i_d = 0$), the estimated winding resistance and rotor flux linkage will be different from the nominal value and vary due to different i_d. This is due to the inductance variation and the difference between L_d and L_q caused by i_d ($\neq 0$). In addition, Refs. [2] and [3] are based on the same principle of Scheme IV, both assuming that $L_d = L_q$ and inductance will not change with i_d. Consequently, they will have the same problem as Scheme IV. In addition, when the PMSM machine is modelled in the stator abc or $\alpha\beta$-axis reference frame, the same rank-deficient problem also exists.

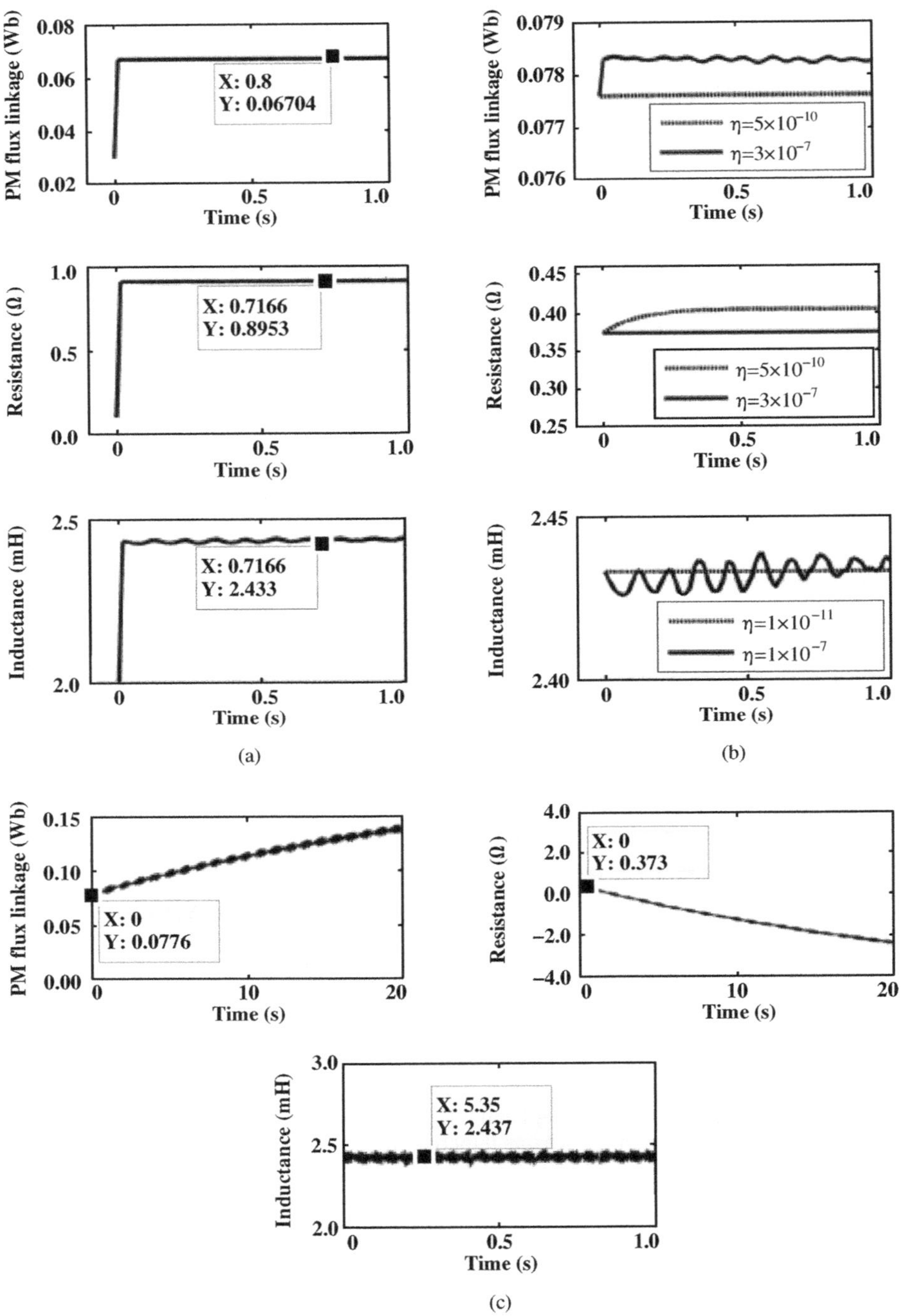

Figure 2.4 Estimated parameters of Scheme III [7], IEEE: (a) initial values far from actual values, (b) initial values close to actual values with low-pass filtering, and (c) similar to (b) but without low-pass filtering for sampled signals.

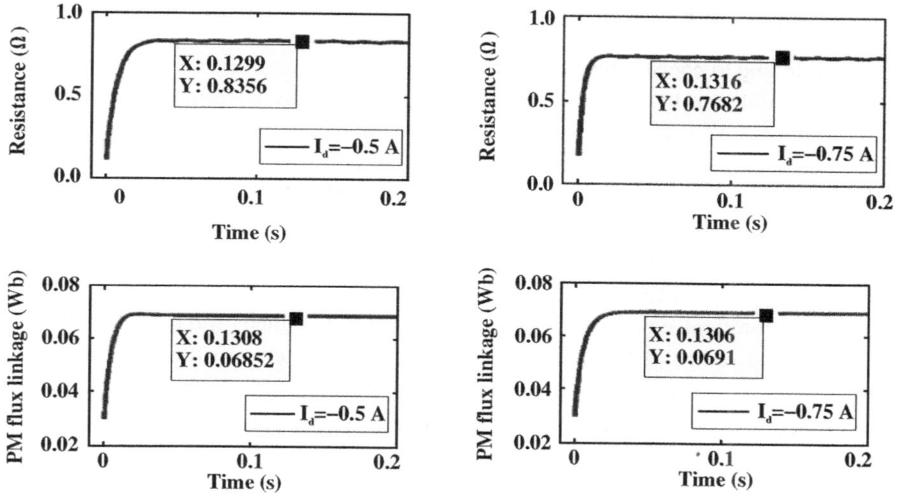

(a) (b)

Figure 2.5 Estimated parameters of Scheme IV [7], IEEE: (a) $I_d = -0.5$ A and $I_d = -0.75$ A.

Table 2.2 Summary of experimental results in four designed schemes.

	Resistance (Ω)	Inductance (mH)	Rotor flux linkage (Wb)	Condition
Scheme I	Fixed to nominal	2.586	0.0784	Normal temperature
		2.500	0.0787	After heating
Scheme II	0.379	2.586	Fixed to nominal	Normal temperature
	0.421	2.500		After heating
Scheme III	0.895	2.433	0.0670	Normal temperature
Scheme IV	0.768	Set to L_q ($I_d = 0$)	0.0685	$I_d = -0.5$ A
	0.836		0.0691	$I_d = -0.75$ A
Nominal	0.33	3.24	0.0776	Normal temperature

2.2 Non-linearity of VSI

For most drive systems, stator voltages are usually indirectly calculated from the reference voltage of the PI regulator, whereas voltage source inverter (VSI) non-linearities such as switch on/off delay, dead time, and voltage drops of diode/insulated-gate bipolar transistor (IGBT) will be introduced into the stator voltages and appear like disturbance voltages [16–25]. Thus, there will be a difference between the reference and real stator voltages. It is usually demanded in commercial drives that these non-linearities should be properly compensated. Otherwise, the control performance may be seriously influenced. For parameter estimation of PMSMs, this issue also needs to be taken into account since the estimation algorithm usually needs stator voltages for iterative calculation. Therefore, in this section, the issue of VSI non-linearity will be introduced from different perspectives, including the mathematical modelling, the online estimation and compensation, as well as its influences on the parameter estimation.

2.2.1 Modelling of VSI Non-linearity

2.2.1.1 VSI Modelling in *abc* Stationary Reference Frame

The modelling of VSI non-linearity has been widely reported in existing literature. Usually, for high-voltage applications, the influence of dead time is more significant than the voltage drops in IGBT and diode, whereas for low-voltage applications, it will be vice versa [26].

Figure 2.6 shows the circuit of a VSI-based three-phase PMSM drive system, where V_{dc} is the direct current (dc)-link voltage and i_a, i_b, and i_c are *abc* three-phase currents. Figure 2.7 shows the typical switch on/off signals of space vector pulse width modulation (SVPWM), where T_s is the sampling time and S_a, S_b, and S_c are the gate signals of three-phase IGBTs.

Moreover, the theoretical, equivalent, and real IGBT output voltages and drive signals are shown in Figure 2.8, where T_d and T_{on}/T_{off} are the dead time and the switch on/off time delay of the power device, respectively. T_{com} is the theoretical compensation time. Taking the phase "*a*" as an example, the difference between the durations of effective time T_a and commanded time T_a^* can be expressed as follows:

$$T_a = T_a^* + \text{sgn}(i_{as})M \tag{2.4}$$

where $\text{sgn}(i_{as}) = \begin{cases} +1 & \geq 0 \\ -1 & < 0 \end{cases}$. $M = T_{com} + T_{off} - T_{on} - T_d$ is the overall time error [27].

The line to neutral voltage of the phase "*a*" can be expressed as follows:

$$\begin{cases} V_{as} = V_{an} - V_{sn} \\ V_{as} = V_{as}^* + V_{as}' - \frac{1}{2}(r_{ce} + r_d)i_{as} \end{cases} \tag{2.5}$$

where V_{as}^* is the reference voltage of the phase "*a*". V_{as}', r_{ce}, and r_d are the distorted voltage caused by the VSI non-linearity, the on-state resistances of IGBT, and the freewheeling diode, respectively.

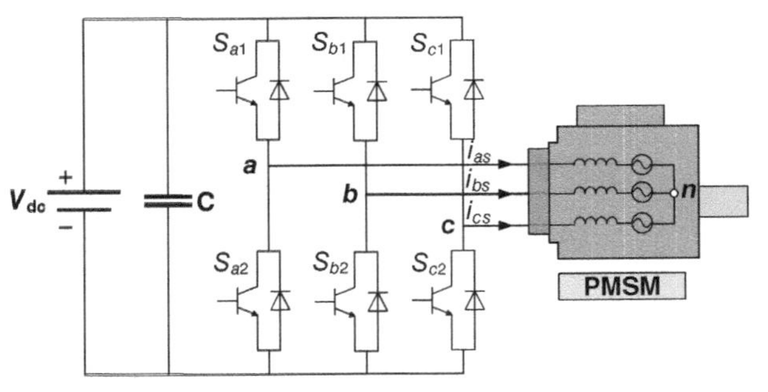

Figure 2.6 VSI-based three-phase PMSM drive system.

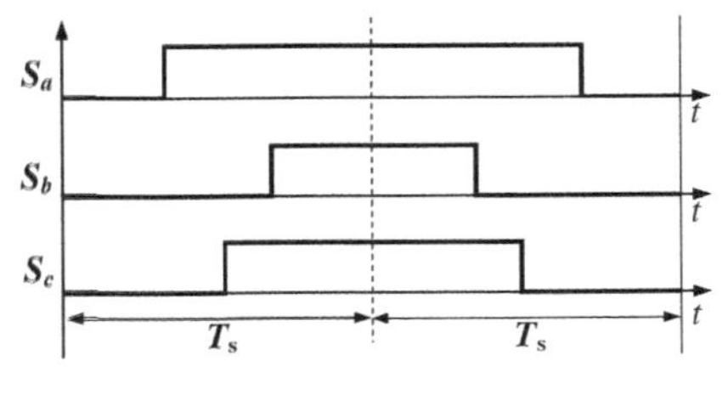

Figure 2.7 Switch on/off signals of space vector pulse width modulation.

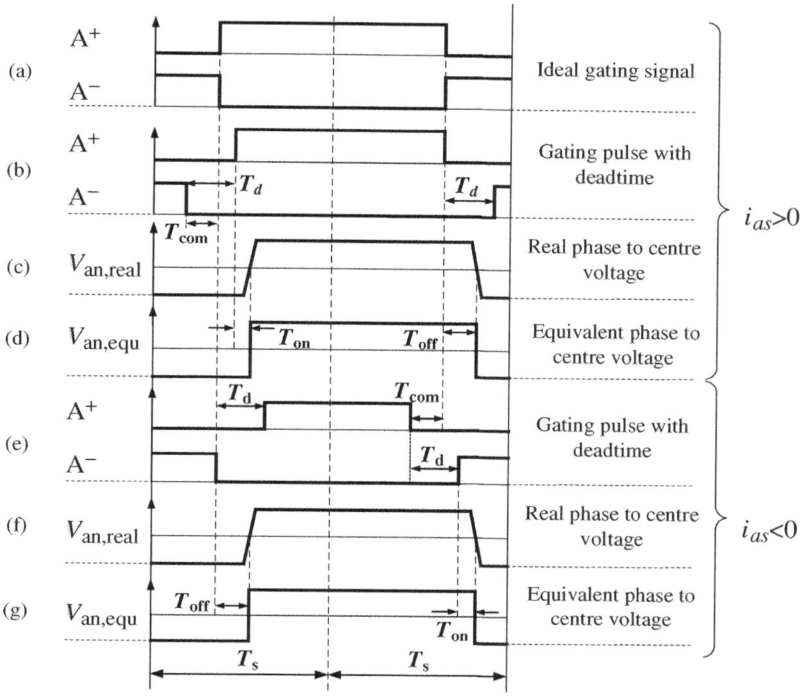

Figure 2.8 Theoretical, equivalent, and real IGBT output voltages and drive signals. (a)–(d) correspond to $i_{as} > 0$, while (e)–(g) correspond to $i_{as} < 0$.

In addition, the average voltage of phase "a" to point "n" in two sampling periods can be expressed as follows [27]:

$$V_{an} = (V_{dc} - V_{ce} + V_d)\left(\frac{T_a}{2T_s} - \frac{1}{2}\right) - \frac{1}{2}\text{sgn}(i_{as})(V_{ce0} + V_{d0}) - \frac{1}{2}(r_{ce} + r_d)\,i_{as} \tag{2.6}$$

where $V_{ce} = V_{ce0} + r_{ce}|i_{as}|$ and $V_d = V_{d0} + r_d|i_{as}|$ are the on-state voltage drop of IGBT and freewheeling diode, respectively. V_{ce0} and V_{d0} are the threshold voltage of the IGBT and the freewheeling diode. Thus, the average voltage of the phase "a" in two sampling periods can be calculated as follows:

$$V_{as}^* = \frac{1}{3}(V_{dc} - V_{ce} + V_d)\frac{2T_a^* - T_b^* - T_c^*}{2T_s} \tag{2.7}$$

Similarly, V_{bn}, V_{cn}, V_{bs}^*, and V_{cs}^* can be expressed as follows:

$$\begin{cases} V_{bn} = (V_{dc} - V_{ce} + V_d)\left(\frac{T_b}{2T_s} - \frac{1}{2}\right) - \frac{1}{2}\text{sgn}(i_{bs})(V_{ce0} + V_{d0}) - \frac{1}{2}(r_{ce} + r_d)\,i_{bs} \\ V_{cn} = (V_{dc} - V_{ce} + V_d)\left(\frac{T_c}{2T_s} - \frac{1}{2}\right) - \frac{1}{2}\text{sgn}(i_{cs})(V_{ce0} + V_{d0}) - \frac{1}{2}(r_{ce} + r_d)\,i_{cs} \end{cases} \tag{2.8}$$

$$\begin{cases} V_{bs}^* = \frac{1}{3}(V_{dc} - V_{ce} + V_d)\frac{2T_b^* - T_a^* - T_c^*}{2T_s} \\ V_{cs}^* = \frac{1}{3}(V_{dc} - V_{ce} + V_d)\frac{2T_c^* - T_b^* - T_a^*}{2T_s} \end{cases} \tag{2.9}$$

In addition, since $V_{as} + V_{bs} + V_{cs} = 0$, V_{sn} can be derived as follows:

$$\begin{aligned} V_{sn} &= \frac{(V_{an} + V_{bn} + V_{cn})}{3} \\ &= \frac{(V_{dc} - V_{ce} + V_{d})}{3}\left(\frac{T_a + T_b + T_c}{2T_s} - \frac{3}{2}\right) \\ &\quad - \frac{(V_{ce0} + V_{d0})}{6}[\text{sgn}(i_{bs}) + \text{sgn}(i_{bs}) + \text{sgn}(i_{bs})] \end{aligned} \tag{2.10}$$

From Eqs. (2.5) to (2.10), the distorted voltage V'_{as} in the phase "a" can be derived as follows:

$$\begin{aligned} V'_{as} &= V_{an} - V_{sn} - V'_{as} + \frac{1}{2}(r_{cd} + r_d)i_{as} \\ &= \frac{1}{6}\left(\frac{M}{T_s}V_{dc} - V_{ce0} - V_{d0}\right)[2\text{sgn}(i_{as}) - \text{sgn}(i_{bs}) - \text{sgn}(i_{cs})] \end{aligned} \tag{2.11}$$

Similarly, V'_{bs} and V'_{cs} can be derived as follows:

$$\begin{cases} V'_{bs} = \frac{1}{6}\left(\frac{M}{T_s}V_{dc} - V_{ce0} - V_{d0}\right)[2\text{sgn}(i_{bs}) - \text{sgn}(i_{as}) - \text{sgn}(i_{cs})] \\ V'_{cs} = \frac{1}{6}\left(\frac{M}{T_s}V_{dc} - V_{ce0} - V_{d0}\right)[2\text{sgn}(i_{cs}) - \text{sgn}(i_{as}) - \text{sgn}(i_{bs})] \end{cases} \tag{2.12}$$

It is obvious that the distorted voltage V' can be minimized to be 0 if the value of M is set to a proper value in Eq. (2.13). Meanwhile, the theatrical compensation time T_{com} for VSI non-linearity is calculated in Eq. (2.14):

$$V' = \frac{1}{6}\left(\frac{M}{T_s}V_{dc} - V_{ce0} - V_{d0}\right) = 0 \Rightarrow M = \frac{T_s}{V_{dc}}(V_{d0} + V_{ce0}) \tag{2.13}$$

$$T_{com} = T_d - T_{off} + T_{on} + T_s\frac{V_{ce0} + V_{d0}}{V_{dc}} \tag{2.14}$$

where T_s and T_d are set up manually in software, and T_{on}, T_{off}, V_{ce0}, and V_{d0} vary with the working condition.

2.2.1.2 VSI Modelling in *dq* Rotating Reference Frame

For the SVPWM, the relationship among reference voltages, real output voltages, and distorted voltages in the *abc* stationary reference frame can be expressed as follows [27]:

$$\begin{cases} V_{as} = V^*_{as} + V'_{as} - \frac{1}{2}(r_{ce} + r_d)i_{as} \\ V_{bs} = V^*_{bs} + V'_{bs} - \frac{1}{2}(r_{ce} + r_d)i_{bs} \\ V_{cs} = V^*_{cs} + V'_{cs} - \frac{1}{2}(r_{ce} + r_d)i_{cs} \end{cases} \tag{2.15}$$

where V^*_{as}, V^*_{bs}, V^*_{cs}, V'_{as}, V'_{bs}, and V'_{cs} are the reference and distorted *abc* phase voltages, respectively. r_{ce} and r_d are the on-state resistances of IGBT and freewheeling diode, respectively.

The transformation from the *abc* stationary reference frame to the *dq*-axis reference frame is expressed as follows:

$$\begin{bmatrix} u_d \\ u_q \end{bmatrix} = T_{abc-dq}\begin{bmatrix} u_{as} \\ u_{bs} \\ u_{cs} \end{bmatrix} \tag{2.16a}$$

with

$$T_{abc-dq} = \frac{2}{3}\begin{bmatrix} \cos(\theta_r) & \cos(\theta_r - 2\pi/3) & \cos(\theta_r + 2\pi/3) \\ -\sin(\theta_r) & -\sin(\theta_r - 2\pi/3) & -\sin(\theta_r + 2\pi/3) \end{bmatrix} \tag{2.16b}$$

where θ_r is the rotor position.

Thus, the dq-axis voltages can be expressed as follows:

$$\begin{bmatrix} u_d \\ u_q \end{bmatrix} = T_{abc-dq} \begin{bmatrix} V_{as}^* + V'_{as} - \frac{1}{2}(r_{ce} + r_d) i_{as} \\ V_{bs}^* + V'_{bs} - \frac{1}{2}(r_{ce} + r_d) i_{bs} \\ V_{cs}^* + V'_{cs} - \frac{1}{2}(r_{ce} + r_d) i_{cs} \end{bmatrix} = T_{abc-dq} \begin{bmatrix} V_{as}^* + V'_{as} \\ V_{bs}^* + V'_{bs} \\ V_{cs}^* + V'_{cs} \end{bmatrix} - \frac{1}{2}(r_{ce} + r_d) \begin{bmatrix} i_d \\ i_q \end{bmatrix} \tag{2.17}$$

The distorted voltages in the abc three phases can be expressed as follows:

$$\begin{bmatrix} V'_{as} \\ V'_{bs} \\ V'_{cs} \end{bmatrix} = \frac{1}{6}\left(\frac{M}{T_s} V_{dc} - V_{ce0} - V_{d0}\right) \begin{bmatrix} 2 & -1 & -1 \\ -1 & 2 & -1 \\ -1 & -1 & 2 \end{bmatrix} \begin{bmatrix} \text{sgn}(i_{as}) \\ \text{sgn}(i_{bs}) \\ \text{sgn}(i_{cs}) \end{bmatrix} = V_{dead} \boldsymbol{D} \tag{2.18}$$

where $V_{dead} = 1/6(M/T_s(V_{dc} - V_{ce0} - V_{d0}))$ is the voltage term due to the VSI non-linearity. The overall time error M and the distorted voltage term V_{dead} are calculated as follows:

$$M = T_{\text{com}} + T_{\text{off}} - T_{\text{on}} - T_d \tag{2.19a}$$

$$V_{\text{dead}} = \frac{M}{T_s} V_{\text{dc}} - V_{\text{ce0}} - V_{\text{d0}} \tag{2.19b}$$

where T_s is the sampling time. T_d and $T_{\text{on}}/T_{\text{off}}$ are the dead time and the switch on/off time delay of the power device, respectively. V_{ce0} and V_{d0} are the threshold voltages of IGBT and freewheeling diode [27].

The matrix $\boldsymbol{D}$ is the gain matrix of distorted voltage due to VSI non-linearity in the abc three-phase reference frame and will vary with the directions of three-phase currents:

$$\begin{bmatrix} Dd \\ Dq \end{bmatrix} = T_{abc-dq} \boldsymbol{D} = T_{abc-dq} \begin{bmatrix} 2 & -1 & -1 \\ -1 & 2 & -1 \\ -1 & -1 & 2 \end{bmatrix} \begin{bmatrix} \text{sgn}(i_{as}) \\ \text{sgn}(i_{bs}) \\ \text{sgn}(i_{cs}) \end{bmatrix} = 2 \begin{bmatrix} \sin(\theta_e) & \sin(\theta_e - 2\pi/3) & \sin(\theta_e + 2\pi/3) \\ \cos(\theta_e) & \cos(\theta_e - 2\pi/3) & \cos(\theta_e + 2\pi/3) \end{bmatrix} \begin{bmatrix} \text{sgn}(i_{as}) \\ \text{sgn}(i_{bs}) \\ \text{sgn}(i_{cs}) \end{bmatrix} \tag{2.20}$$

where $\theta_e = \theta_r + \pi/2$. Furthermore, accounting for the non-linearity of VSI, the PMSM model in the dq-axis is represented as follows:

$$\frac{d}{dt} \begin{bmatrix} L_d i_d \\ L_q i_q \end{bmatrix} = -\left[R_s + \frac{(r_{ce} + r_d)}{2}\right] \begin{bmatrix} i_d \\ i_q \end{bmatrix} + \begin{bmatrix} \omega_r L_q i_q \\ -\omega_r (L_q i_q + \psi_{pm}) \end{bmatrix} + \begin{bmatrix} u_d^* \\ u_q^* \end{bmatrix} + V_{\text{dead}} \begin{bmatrix} Dd \\ Dq \end{bmatrix} \tag{2.21}$$

It is obvious that the on-state resistances of IGBT and freewheeling diode are presented as part of the winding resistance. Furthermore, the voltage term V_{dead} due to VSI non-linearity is presented as a constant under a certain working point. Thus, in real applications, the distorted voltage due to VSI non-linearity can be well compensated if V_{dead} can be accurately minimized. Table 2.3 shows the relationship between current directions and dq-axis distorted voltages (V'_d and V'_q) due to VSI non-linearity, where δ is the angle between the current vector and q-axis.

Table 2.3 Relationship between current directions and *dq*-axis voltages.

Mode	θ_e	sgn(i)			$\theta_e = \theta_r + \pi/2$	
		i_a	i_b	i_c	V_d'	V_q'
1	$-\pi/6 - \delta < \theta_e < \pi/6 - \delta$	1	−1	−1	$4V_{\text{dead}}\sin(\theta_e)$	$4V_{\text{dead}}\cos(\theta_e)$
2	$\pi/6 - \delta < \theta_e < \pi/2 - \delta$	1	1	−1	$4V_{\text{dead}}\sin(\theta_e - \pi/3)$	$4V_{\text{dead}}\cos(\theta_e - \pi/3)$
3	$\pi/2 - \delta < \theta_e < 5\pi/6 - \delta$	−1	1	−1	$4V_{\text{dead}}\sin(\theta_e - 2\pi/3)$	$4V_{\text{dead}}\cos(\theta_e - 2\pi/3)$
4	$5\pi/6 - \delta < \theta_e < 7\pi/6 - \delta$	−1	1	1	$4V_{\text{dead}}\sin(\theta_e - \pi)$	$4V_{\text{dead}}\cos(\theta_e - \pi)$
5	$7\pi/6 - \delta < \theta_e < 3\pi/2 - \delta$	−1	−1	1	$4V_{\text{dead}}\sin(\theta_e - 4\pi/3)$	$4V_{\text{dead}}\cos(\theta_e - 4\pi/3)$
6	$3\pi/2 - \delta < \theta_e < 11\pi/6 - \delta$	1	−1	1	$4V_{\text{dead}}\sin(\theta_e - 5\pi/3)$	$4V_{\text{dead}}\cos(\theta_e - 5\pi/3)$

Under $i_d = 0$ control, the distorted voltages V_d' and V_q' can be expressed as follows:

$$\begin{cases} V_d' = V_{\text{dead}}Dd = 4V_{\text{dead}}\sin\left(\theta_e - \frac{\pi}{3}\text{Mode}\right) \\ V_q' = V_{\text{dead}}Dq = 4V_{\text{dead}}\cos\left(\theta_e - \frac{\pi}{3}\text{Mode}\right) \end{cases} \tag{2.22}$$

It can be found from Eq. (2.22) that V_d' is a zero-mean sixth harmonic component, while V_q' is a sixth harmonic component with dc offset. The simulated and real waveforms of Dd and Dq under $i_d = 0$ control of Prototype Machine 1 (see Table 2.1) are depicted in Figure 2.9. In addition, it is obvious that the *dq*-axis distorted voltages due to VSI non-linearity will introduce sixth harmonic components into *dq*-axis currents, as shown in Figure 2.10.

2.2.2 VSI Non-Linearity Estimation and Compensation

2.2.2.1 Estimation of VSI Non-linearity

Accounting for the distorted voltage caused by the VSI non-linearity, the steady-state PMSM model (2.2) can be rewritten as follows:

$$\begin{cases} u_d^*(k) + Dd(k)V_{\text{dead}} = R_s i_d(k) - L_q\omega_r(k)i_q(k) \\ u_q^*(k) + Dq(k)V_{\text{dead}} = R_s i_q(k) + L_d\omega_r(k)i_d(k) + \psi_{pm}\omega_r(k) \end{cases} \tag{2.23}$$

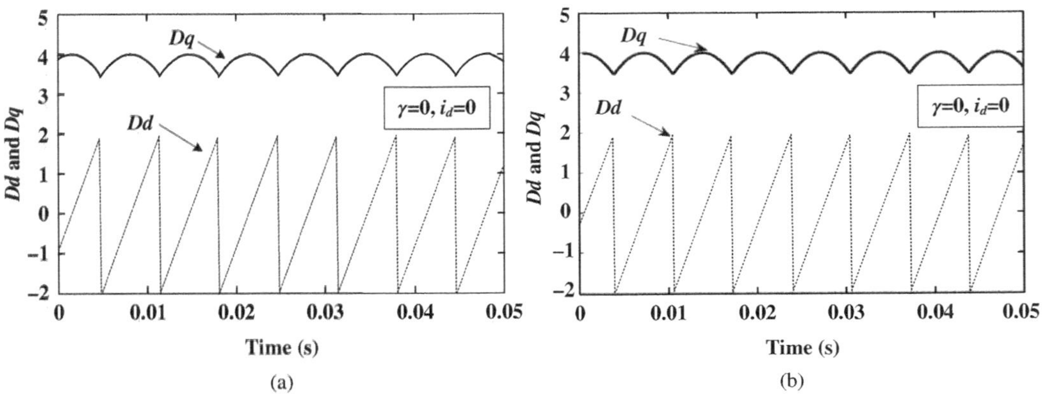

Figure 2.9 Waveforms of Dd and Dq under $i_d = 0$ control [9], IEEE: (a) simulated and (b) measured.

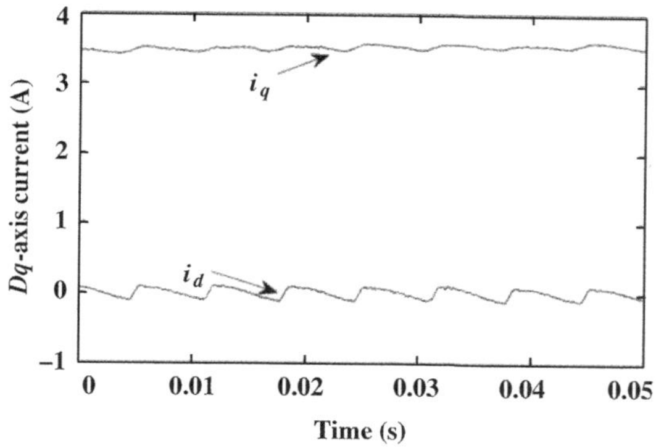

Figure 2.10 Measured i_d and i_q under $i_d = 0$ control [9], IEEE.

where u_d^*, u_q^*, i_d, i_q, L_d, and L_q are the reference voltages, the currents, and the inductances in the d- and q-axis. V_{dead}, R_s, ψ_{pm}, and ω_r are the distorted voltage, the stator resistance, the rotor PM flux linkage, and the electrical angular speed, respectively.

The distorted voltage V_{dead} in Eq. (2.23) not only deteriorates the control performance in the PMSM drive system [16–24] but also introduces non-negligible errors in the machine parameter estimation [9, 25]. However, most VSI estimation/compensation methods need accurate parameters of PMSM and VSI, and consequently, it is impossible to estimate the distorted voltage prior to estimating PMSM parameters. Therefore, this sub-section will introduce an online estimation method to estimate the distorted voltage V_{dead} under $i_d = 0$ control, where the machine parameters and the perturbance signal injection are not required.

First, it has been proved in Figure 2.9 that the d-axis distorted voltage DdV_{dead} is a sixth zero-mean harmonic and reflects as a voltage disturbance around 0 V. Thus, taking into account the other high-frequency component under $i_d = 0$ control, Eq. (2.23) can be rewritten as follows:

$$R_s i_d(k) = R_s \overline{i_d}(k) + R_s i_{dh}(k) = R_s i_{dh}(k) \tag{2.24a}$$

$$\overline{Dd}(k) V_{\text{dead}} = 0 \tag{2.24b}$$

$$u_d^*(k) = -Dd(k) V_{\text{dead}} - L_q \omega_r(k) i_q(k) \tag{2.24c}$$

$$u_{dh}^*(k) = -Dd(k) V_{\text{dead}} - L_q \omega_h(k) i_{qh}(k) \tag{2.24d}$$

$$L_q \omega_r(k) i_q(k) = L_q \overline{\omega_r}(k) \overline{i_q}(k) + L_q \overline{\omega_r}(k) i_{qh}(k) + L_q \omega_h(k) \overline{i_q}(k) + L_q \omega_h(k) i_{qh}(k) \tag{2.24e}$$

where $\overline{i_d}$, $\overline{Dd}$, $\overline{\omega_r}$, and $\overline{i_q}$ are the dc components of respective quantities, while u_{dh}^*, i_{dh}, ω_h, i_{qh} are their high-frequency components. At the steady state of $i_d = 0$ control, i_{dh} and i_{qh} are mainly generated by the sixth harmonic components caused by the VSI non-linearity. ω_h consists of the harmonic components due to the VSI non-linearity and the mechanical issues accounting for the rotating friction, load oscillation, rotor eccentricities, etc., whereas the harmonics of the rotor speed caused by the mechanical issues are at relatively low-frequency range with small amplitudes. In addition, the sixth harmonic component in $L_q \omega_h(k) \overline{i_q}(k)$ is also caused by the VSI non-linearity and can be regarded as part of $Dd(k) V_{\text{dead}}$.

By way of example, Figure 2.11 shows the waveforms and fast Fourier transform (FFT) results of rotor speed at 300 r/min for Prototype Machine 1 accounting for the VSI non-linearity compensation.

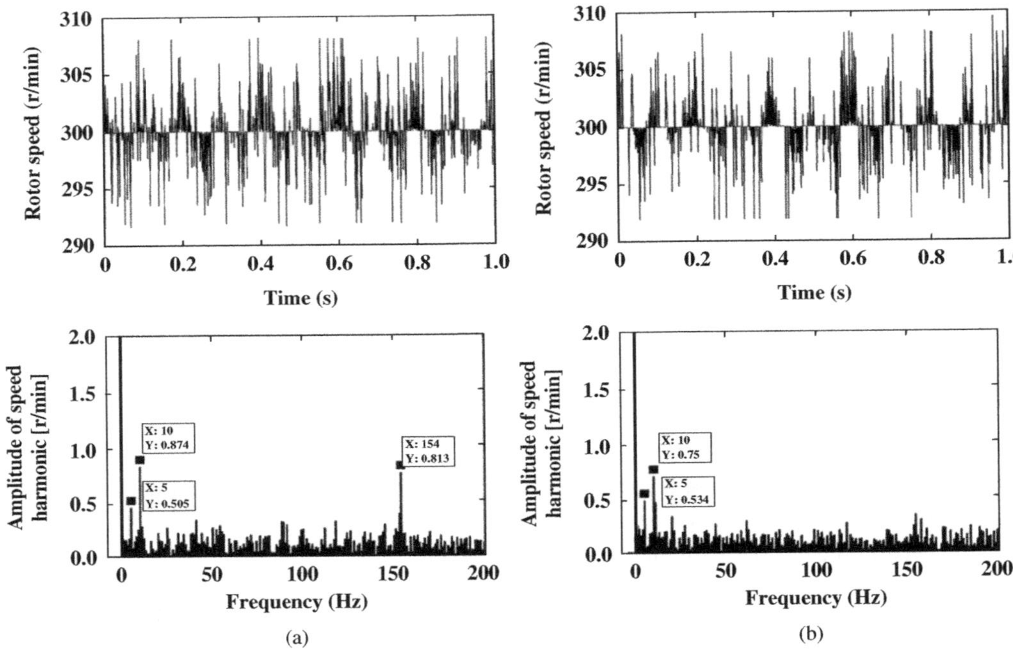

Figure 2.11 Waveforms and FFT results of measured rotor speed with and without VSI non-linearity compensation ($i_q = 4$ A and $i_d = 0$ A at 300 r/min) [9], IEEE: (a) rotor speed waveform and FFT result without compensation and (b) rotor speed waveform and FFT result with compensation.

As can be seen, when the VSI non-linearity is not compensated in Figure 2.11a, the main harmonic components of the rotor speed include 5, 10, and 154 Hz. On the one hand, the 5 and 10 Hz harmonics are related to mechanical issues, since the rotor speed is 300 r/min, and the pole pair number is five. On the other hand, the 154 Hz harmonic component is mainly correlated to the VSI non-linearity, since the VSI non-linearity is a sixth harmonic component, and its frequency is 150 Hz at 300 r/min. In contrast, when the VSI non-linearity is well compensated in Figure 2.11b, the 154 Hz harmonic has been reduced significantly.

Second, the acquisition of the high-frequency component in u_d^* is shown in Figure 2.12, where a first-order low-pass filter is employed. Therefore, there are only two high-frequency terms left in Eq. (2.24d), i.e., $Dd(k)V_{\text{dead}}$ and $L_q\omega_h(k)i_{qh}(k)$, while $L_q\omega_h(k)i_{qh}(k)$ is negligible compared to $Dd(k)$ V_{dead}. Thus, $u_{dh}^*(k)$ can be approximately equal to $Dd(k)V_{\text{dead}}$.

$$u_{dh}^*(k) \approx -Dd(k)V_{\text{dead}} \tag{2.25}$$

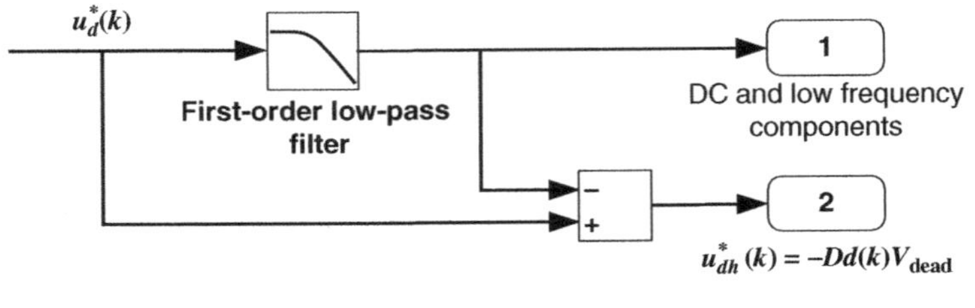

Figure 2.12 Acquisition of high-frequency component from u_d^* by using a first-order low-pass filter [9], IEEE.

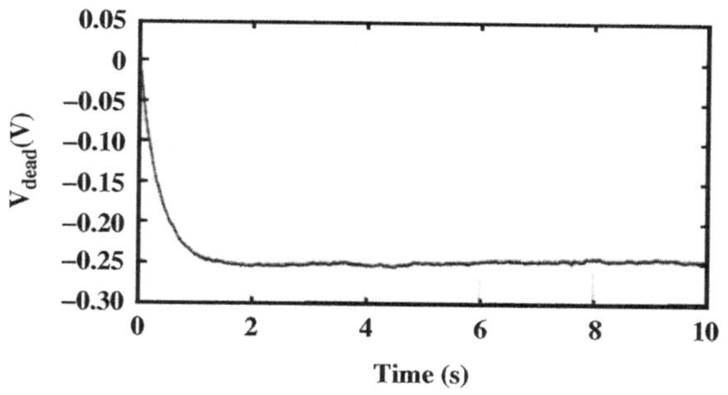

Figure 2.13 Estimated V_{dead} caused by VSI non-linearity under $i_d = 0$ control [9], IEEE.

Furthermore, since $L_q \omega_h(k)\, i_{qh}(k)$ is much smaller than $Dd(k)V_{dead}$, the influence of the variation of L_q on the estimation of V_{dead} can also be negligible. The Prototype Machine 1 (see Table 2.1) is taken as an example for further illustration, whose nominal q-axis inductance is 3.24 mH. When there is a 50% increase in L_q under the conditions of $\omega_h(k) = 5$ rad/s and $i_{qh}(k) = 0.3$ A, the amplitudes of $L_q\omega_h(k)\, i_{qh}(k)$ and $Dd(k)\, V_{\text{dead}}$ are 0.0073 V ($5 \times 0.3 \times 1.5 \times 3.24\text{e}^{-3}$) and 0.5 V ($2 \times 0.25$), respectively.

Finally, V_{dead} can be directly estimated from Eq. (2.25) and shown in Figure 2.13. In this case, an ANN estimator is designed and expressed in Eq. (2.26):

$$\widehat{V}_{\text{dead}}(k + 1) = \widehat{V}_{\text{dead}}(k) + 2\eta Dd(k)[d(k) - O(k)] \tag{2.26}$$

where η is the convergence factor. $O(k)$ and $\widehat{V}_{\text{dead}}$ are the output of the variable model and estimated V_{dead}, respectively.

2.2.2.2 Online Compensation of VSI Non-linearity

In real applications, the compensation of VSI non-linearity V_{dead} relies on the well-estimated values. However, in fact, the response of the drive system to the compensation of VSI non-linearity is non-linear, and thus, an online tuning process minimizing V_{dead} is usually required. By way of example, an online tuning method is proposed in Ref. [23] to minimize the sixth harmonics in the dq-axis currents for the compensation of VSI non-linearity. Similarly, the method proposed in Ref. [24] utilizes a trapezoidal voltage to compensate for the VSI non-linearity, where the trapezoidal angle is tuned by an integral controller under real-time conditions.

Since V_{dead} can be estimated directly from Eq. (2.25), the VSI non-linearity can be effectively compensated by minimizing the estimated $\widehat{V}_{\text{dead}}$. The schematic diagram of VSI non-linearity compensation method is shown in Figure 2.14a based on the field-oriented control (FOC) system in Figure 2.14b, where V_{thr}, ζ, and γ are the threshold voltage of minimized V_{dead} and the tuning and the gain factors of V_{dead}, respectively.

The adjustments of the tuning factor ζ and the gain factor γ are performed as follows.

1. *Initialize V_{thr}, ζ, and γ.* For instance, ζ and γ are initialized to be 1.0e^{-4} and 0.1 in this study, and V_{thr} is set to 1.0e^{-4} V, which is close to 0.
2. *If* $|\widehat{V}_{dead}| < V_{\text{thr}}$, the VSI non-linearity has been well compensated and the value of γ will be kept constant during the compensation.
3. *If* $|\widehat{V}_{\text{dead}}| > V_{\text{thr}}$, the VSI non-linearity is not well compensated. The value of γ will be increased or decreased in each step of computation till $[\widehat{V}_{\text{dead}}] < V_{\text{thr}}$. The rule for decreasing or increasing γ is described in Figure 2.14a.

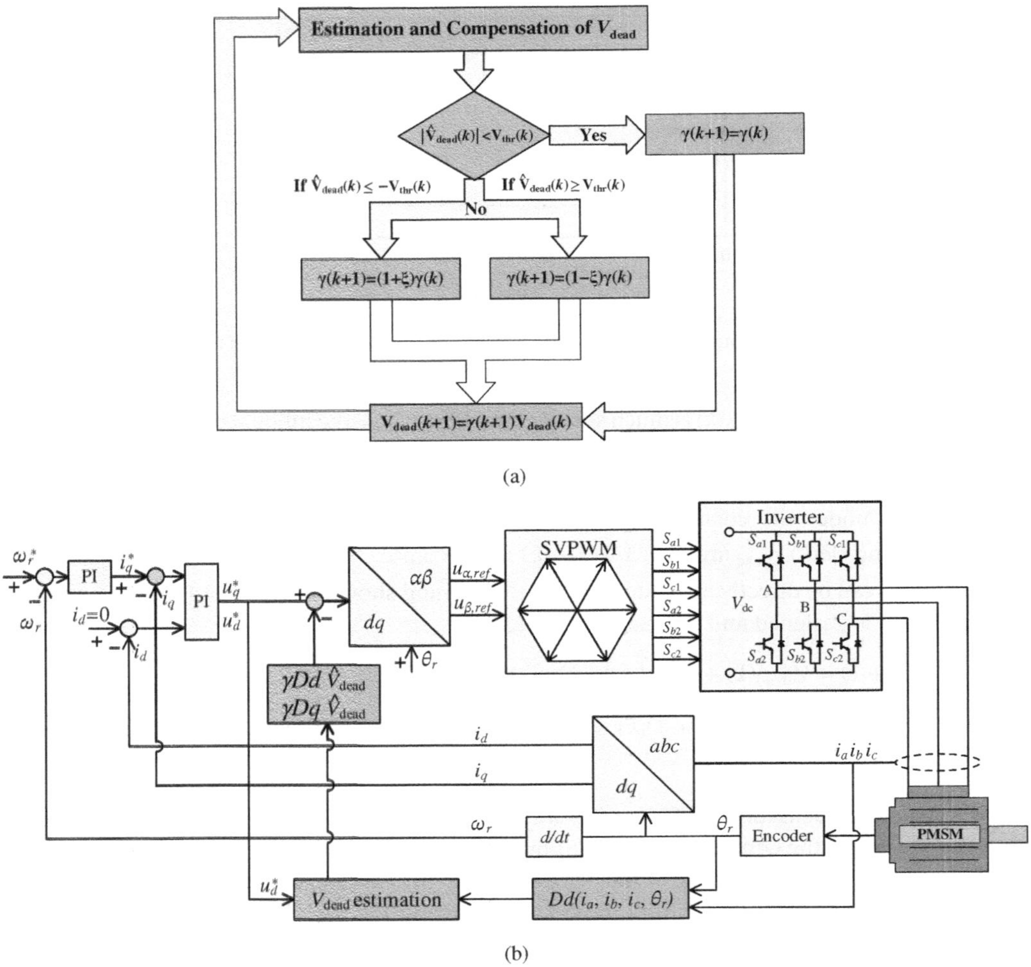

Figure 2.14 Process of complete VSI non-linearity estimation/compensation method: (a) flowchart of real-time compensation of V_{dead} and (b) VSI non-linearity estimation/compensation method based on FOC system.

4. The $Dd\gamma\hat{V}_{dead}$ and $Dq\gamma\hat{V}_{dead}$ will be computed and then added to the output of dq-axis PI regulators for VSI non-linearity compensation under real-time conditions.
5. Go back to step 2.

First, the setup of threshold voltage V_{thr} depends on the voltage drop of the winding resistance. For instance, the nominal value of the winding resistance of Prototype Machine 1 (see Table 2.1) is 0.320 Ω. Under the conditions of $i_d = 0$ and $i_q = 1$ A, the voltage drop due to the winding resistance is ~320 mV. In order to minimize the estimation error of winding resistance to be less than 1% (3.20 mV), the minimized V_{dead} should be smaller than 3.20 mV and the minimized threshold value of V_{dead} will be therefore set to 3.20 mV.

Second, the tuning factor ζ (>0) controls the minimization speed of the $\hat{V}_{dead}$. An abrupt change may be introduced into the control system if ζ is too large. Therefore, it is usually set to a small value, i.e., $1.0e^{-4}$ in this study, to smooth the minimizing process.

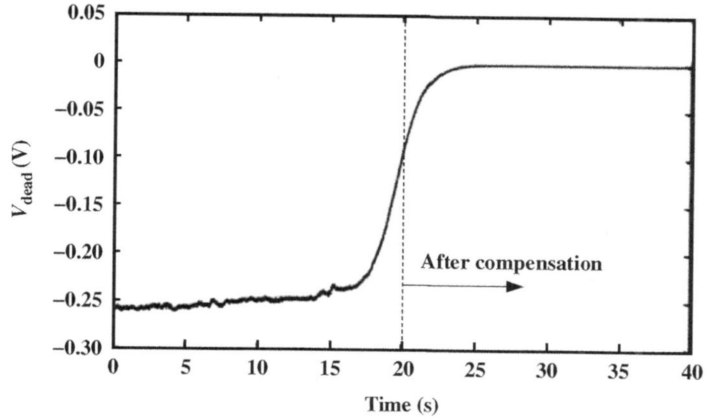

Figure 2.15 Estimated V_{dead} before and after VSI non-linearity compensation [9], IEEE.

Finally, the VSI non-linearity estimation/compensation method is applied to Prototype Machine 1 for validation, as shown in Figure 2.15, where V_{dead} can be minimized to 0 after compensation. Moreover, Figures 2.16–2.18 show the measured currents at 300 r/min in different reference frames with and without the compensation of VSI non-linearity. Obviously, after appropriate compensation, the *dq*-axis currents contain fewer sixth harmonics, and the phase current will be more sinusoidal. In addition, it is evident that the circle of the $\alpha\beta$-axis currents shown in Figure 2.17a and b will be more orbicular after compensation. It is also manifested from the FFT results depicted in Figure 2.17c and d that the fifth and seventh harmonics in the α-axis current are minimized to be negligible, of which the sum is the sixth harmonic.

2.2.3 Influences of VSI Non-linearity on Parameter Estimation

This sub-section will further investigate the influences of VSI non-linearity on the parameter estimation based on the Prototype Machine 1 (Table 2.1 and Figure 2.1). The employed VSI is Mitsubishi PS21867, whose typical electrical characteristics are referred from the Mitsubishi datasheet [28], as tabulated in Table 2.4. The diagram of the employed FOC system is presented in Figure 2.19. In the experiments, the dc-link voltage is fixed to 36 V, and the rotor speed and load are constant. In addition, the waveforms of the three-phase currents are shown in Figure 2.20.

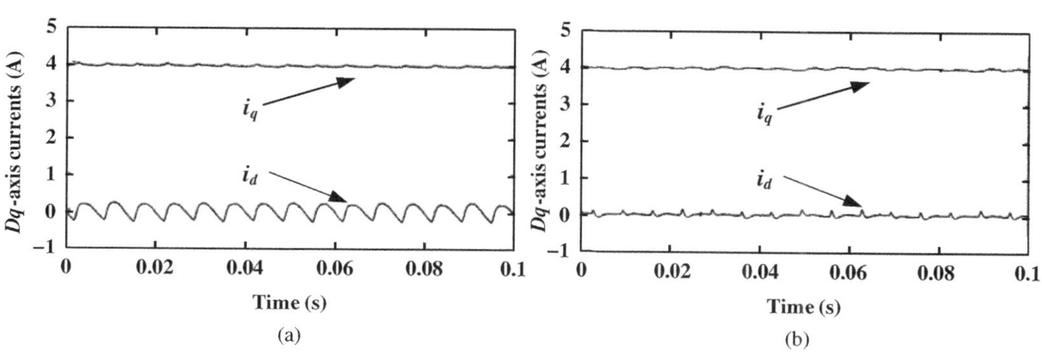

Figure 2.16 *Dq*-axis currents (a) without and (b) with compensation of VSI non-linearity [9], IEEE.

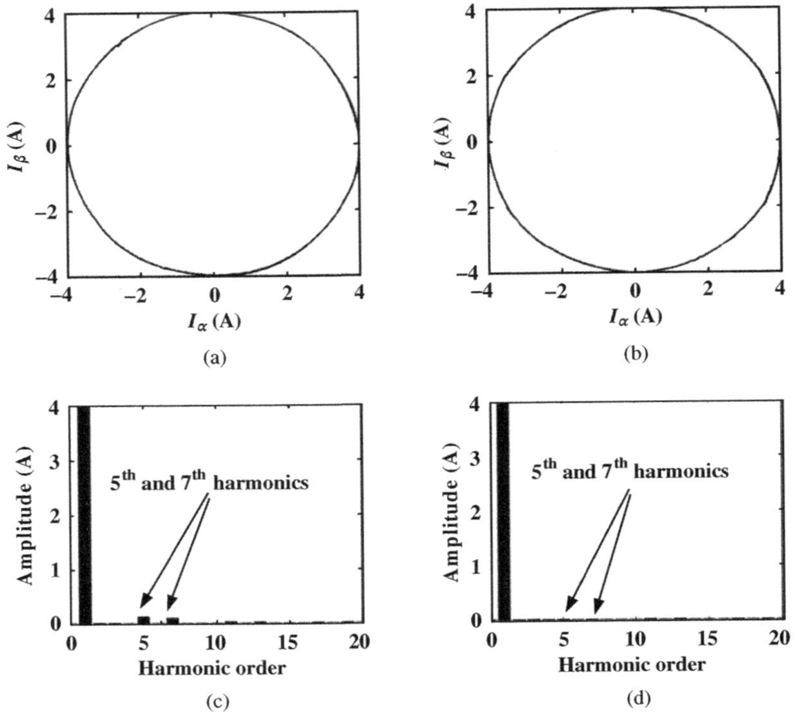

Figure 2.17 $\alpha\beta$-axis currents (a) without and (b) with compensation of VSI non-linearity and FFT of α-axis current (c) without compensation and (d) with compensation.

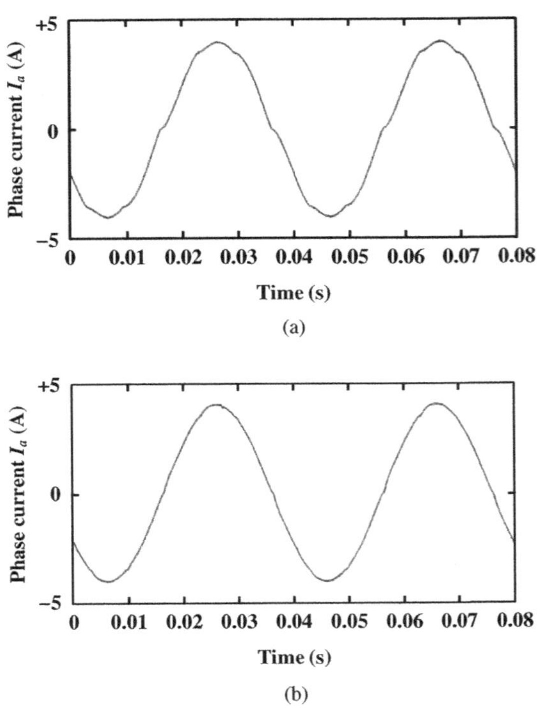

Figure 2.18 Currents of phase-a (I_a) (a) without and (b) with compensation of VSI non-linearity [9], IEEE.

Table 2.4 Typical electrical parameters of VSI from Mitsubishi PS21867 datasheet.

Parameter	Value	Parameter	Value
Turn off delay T_{off} (μs)	1.7	Turn on delay T_{on} (μs)	1.3
Voltage drop of active switch V_{ce0} (V) at 25°C	1.5	Voltage drop of freewheeling diode V_{d0} (V) at 25°C	1.6
Switch control dead time T_d (μs)	2.0	Supply voltage range (V)	0–400

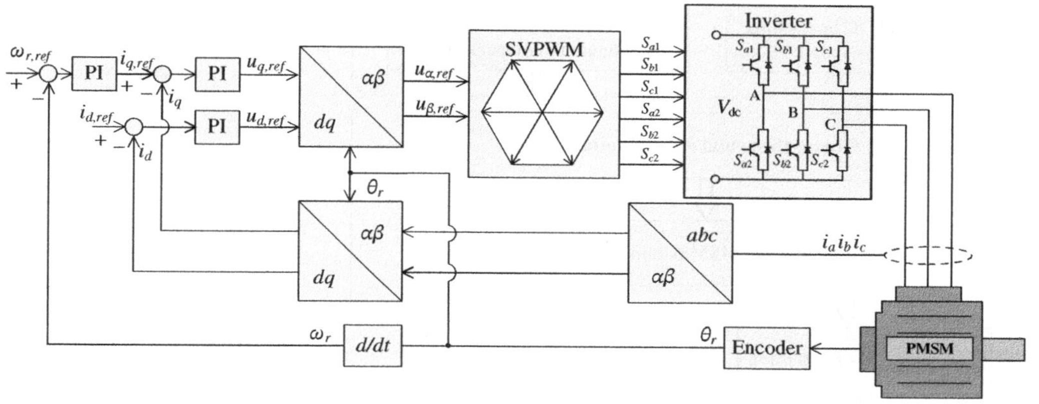

Figure 2.19 System control diagram for PMSM.

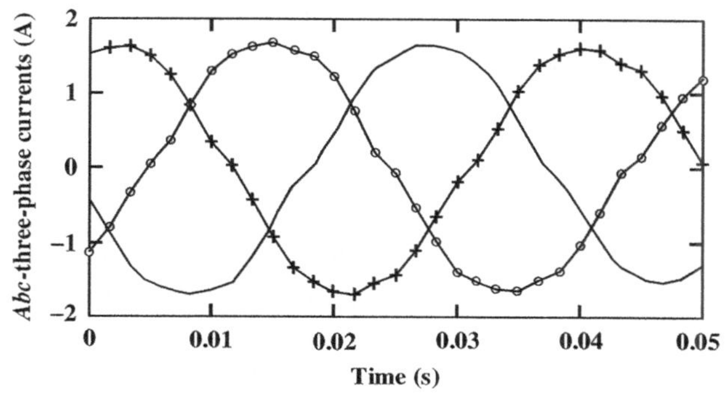

Figure 2.20 Measured three-phase current waveforms [25], IEEE.

As illustrated in Section 2.1, under $i_d = 0$ control, it is impossible to simultaneously estimate the stator winding resistance and the rotor flux linkage due to the rank-deficient issue, while the q-axis inductance can be individually estimated from the d-axis equation. Afterwards, a negative d-axis current impulse of -2 A is transiently injected to activate the winding resistance term in the d-axis equation. Under the assumption of $L_d = L_q$, the rotor PM flux linkage can be estimated subsequently. A detailed introduction of online parameter estimation based on current impulse will be introduced in Chapter 4.

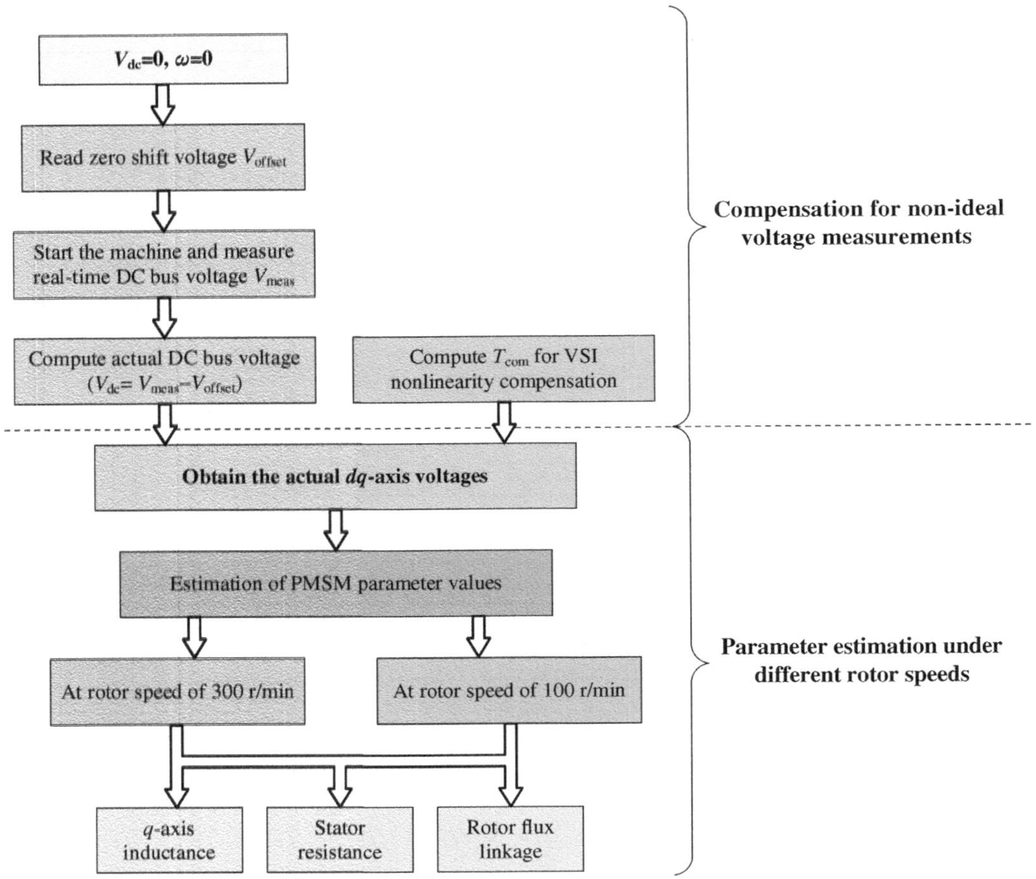

Figure 2.21 Schematic diagram of designed experimental investigation.

The schematic diagram of the designed experiment procedure is shown in Figure 2.21 [25]. In the experiments, the dc-link voltage will be measured in real time via a dc voltage sensor to avoid the misestimation caused by the on-load voltage drop. Meanwhile, the non-zero dc offset V_{offset} of the dc voltage sensor caused by the zero shift in the amplifier will be measured with the PMSM at a standstill and then compensated after the machine is started.

Under $i_d = 0$ control, the estimated L_q at the rotor of 300 r/min is presented in Figure 2.22a and b without/with VSI non-linearity compensation. It can be found that after taking into account the non-ideal voltage measurements, the estimated results of 3.1 mH in Figure 2.22b are much closer to the nominal value of 3.24 mH. Meanwhile, the influence of VSI non-linearity compensation on the q-axis inductance estimation is negligible. It can be explained that the average value of distorted voltage in the d-axis reference frame is zero (see Figure 2.9).

Moreover, the estimated stator resistance and rotor flux linkage at the rotor speed of 300 r/min without/with considering the influences of the non-ideal voltage measurement and VSI non-linearity are also compared [25], as shown in Figure 2.23, where $R_0 = 0.45\ \Omega$, $R_1 = 0.39\ \Omega$, $\psi_{pm0} = 78.8$ mWb, and $\psi_{pm1} = 78.9$ mWb.

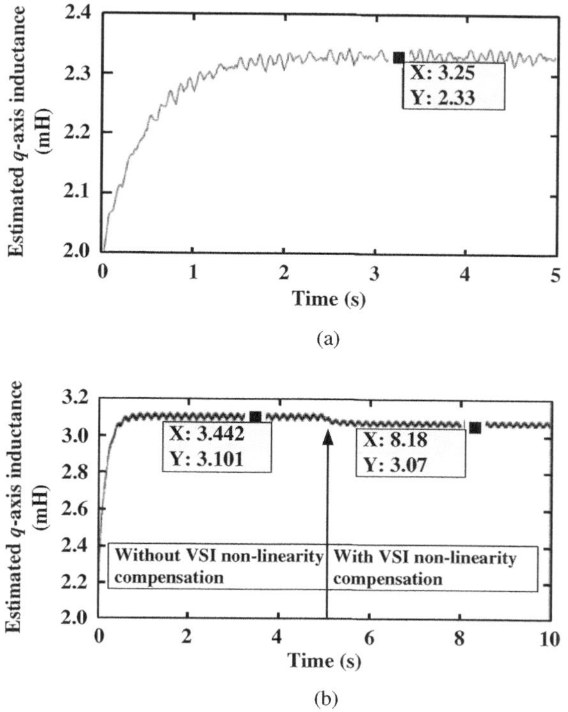

Figure 2.22 Estimated q-axis inductance under $i_d = 0$ control at 300 r/min [25], IEEE: (a) without considering non-ideal voltage measurements and (b) with considering VSI non-linearity, measurement of bus voltage variation, and dc offset due to zero shift.

As can be seen from Figure 2.23a, the estimated parameter values are far from their nominal values ($R_s = 0.33\ \Omega$ and $\psi_{pm} = 0.0776$ Wb), since the non-ideal voltage measurement is not considered. In comparison, the estimated rotor flux linkage in Figure 2.23b has better estimation results, while the estimated stator winding resistance in Figure 2.23b is still far from its nominal value since the distorted voltage due to VSI non-linearity is not well compensated. However, when the non-ideal voltage measurements and the VSI non-linearity are fully considered, the estimated parameters in Figure 2.23c are much closer to the nominal values. It can be concluded that the VSI non-linearity compensation has a significant influence on the winding resistance estimation but a negligible influence on the rotor flux linkage estimation at high speeds.

The corresponding dq-axis voltages and currents with/without VSI non-linearity compensation are shown in Figure 2.24, where the compensation is added to the drive system after 1.5 s. It is evident that the dq-axis voltages slightly change following the addition of compensation. However, the dq-axis currents are maintained almost constant since the load and specified d-axis current do not change.

At the low rotor speed of 100 r/min, the dq-axis voltages and current, as well as the estimated stator resistance and rotor flux linkages, are shown in Figures 2.25 and 2.26, respectively, where $R_2 = 0.42\ \Omega$, $R_3 = 0.39\ \Omega$, $\psi_{pm2} = 78.6$ mWb, and $\psi_{pm3} = 80.0$ mWb.

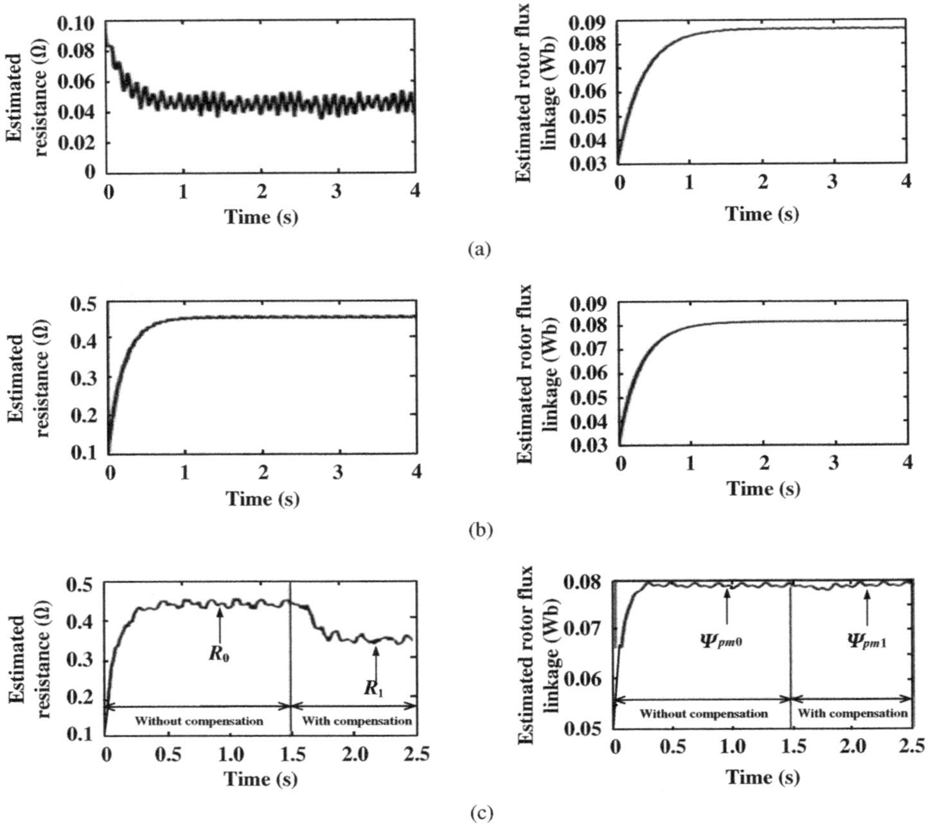

Figure 2.23 Estimated stator winding resistance and rotor flux linkage at 300 r/min ($i_d = -2$ A): (a) without considering the non-ideal voltage measurement, (b) with considering the measurement of dc-link voltage variation and dc offset, and (c) with considering the measurement of dc-link voltage variation, dc offset, and VSI non-linearity compensation.

As can be seen, the variation of estimated rotor flux linkage with/without the compensation of VSI non-linearity is significant. Therefore, the estimated rotor flux linkage and stator winding resistance at 100 r/min are different from those estimated at 300 r/min, and both vary significantly.

It can be further concluded that for most PMSMs, the voltage drop of the stator winding resistance ($R_s i_q$) in the q-axis voltage equation is usually quite small compared with the back-electromotive-force (EMF) voltage ($\omega_r \psi_{pm}$) when operating at the rated speed. This is attributed to that a large stator winding resistance will increase the heat losses due to the stator current and will cause decreased efficiency. Although some special PMSMs, for high-speed applications, have small rotor flux linkage, its back-EMF voltage at the rated speed is still much larger than $R_s i_q$. Thus, the above phenomena are suitable for most PMSMs in real applications. Furthermore, for those special PMSMs, the PMSM with a large stator winding resistance and a small back-EMF voltage, the VSI non-linearity compensation will have significant influences on both the rotor flux linkage and winding resistance estimations irrespective of low- or high-speed operation.

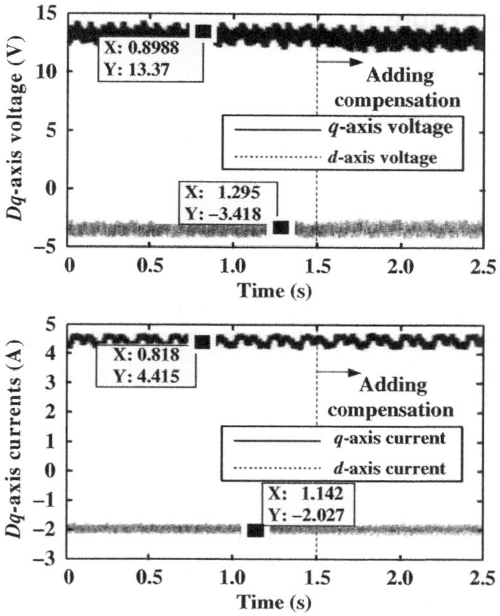

Figure 2.24 *Dq*-axis currents and voltage with/without VSI compensation at 300 r/min [25], IEEE.

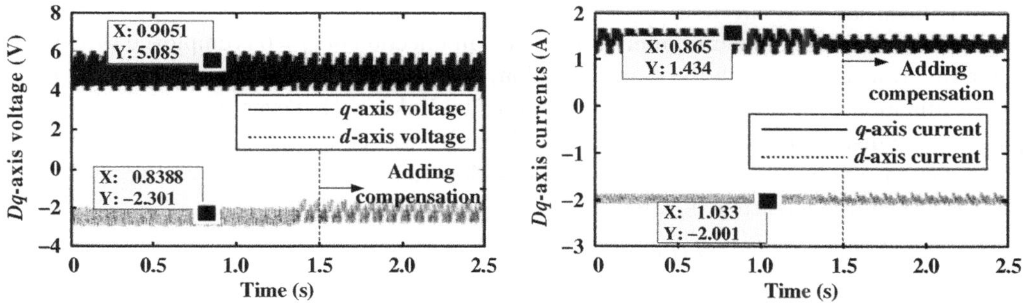

Figure 2.25 *Dq*-axis currents and voltages with/without VSI compensation at 100 r/min [25], IEEE.

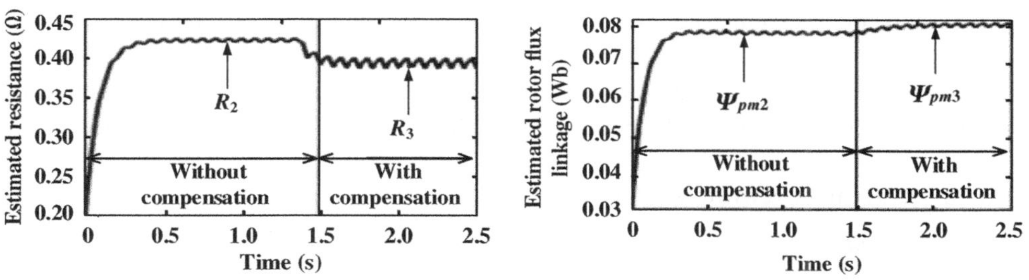

Figure 2.26 Estimated resistance and rotor flux linkage considering the dc-link voltage variation, the dc offset due to zero shift, and VSI nonlinearity compensation at 100 r/min [25], IEEE.

2.3 Ill-Conditioned Problem

The ill-conditioned problem is one of the fundamental problems in system identification technologies [29], where a small error in the measured data or in subsequent calculation results in much larger errors in the solutions. By way of way, considering the following systems:

$$\begin{cases} x + y = 2 \\ x + 1.001y = 2 \end{cases} \text{ and } \begin{matrix} x + y = 2 \\ x + 1.001y = 2.001 \end{matrix} \tag{2.27}$$

The system on the left has solution $x = 2, y = 0$, while the one on the right has solution $x = 1, y = 1$. The coefficient matrix is called ill-conditioned because a small change in the constant coefficients or the variables (x, y) will result in a large change in the solution. For a coefficient matrix $\boldsymbol{A}$, the condition number (cond($\boldsymbol{A}$)) is utilized to judge whether it is an ill-conditioned matrix, which is defined as follows:

$$\text{cond}(\boldsymbol{A})_v = \|\boldsymbol{A}^{-1}\|_v \|\boldsymbol{A}\|_v \tag{2.28}$$

where "v" is the operator norm of matrix $\boldsymbol{A}$, which can be 1, 2, or ∞. The system equations are called ill-conditioned if cond$(\boldsymbol{A})_v$ is much larger than 1, or else, they are called well-conditioned.

In system identification, the ill-conditioned problem widely exists, especially in applications such as power system state estimation [30], identification of electric circuits [31], and GPS positioning [32]. Thus, for multi-parameter estimation, this problem needs to be taken into account before the design of estimators. Otherwise, the estimation result will converge to a wrong point even if the estimation is based on a full-rank reference system.

In fact, the ill-conditioned problem also exists in the parameter estimation of PMSM, especially for the methods based on injection of $i_d \neq 0$. For example, it is proposed in Refs. [2] and [14] to fix the dq-axis inductances to their nominal values, and an injection of $i_d \neq 0$ is employed to activate the winding resistance term in the d-axis equation, which makes the dq-axis equation full rank for the estimation of winding resistance and rotor flux linkage. However, as can be seen from Eq. (2.29), the distorted voltage due to VSI non-linearity will be introduced into the measured dq-axis reference voltages, of which the compensation cannot be perfect in real applications. Thus, the ill-conditioned problem will occur.

$$\begin{cases} u_d^*(k) + Dd(k)V_{\text{dead}} = Ri_d(k) - L_q\omega_r(k)i_q(k) \\ u_d^*(k) + Dd(k)V_{\text{dead}} = Ri_q(k) + L_d\omega_r(k)i_d(k) + \psi_{pm}\omega_r(k) \end{cases} \tag{2.29}$$

This issue is analysed experimentally based on Prototype Machine 1 (see Table 2.1), and the estimated results are shown in Figure 2.27, where $L_d = L_q$ are estimated at $i_d = 0$ control. i_q is fixed to 4 A. The winding resistance and rotor flux linkage are estimated under two sets of negative d-axis current injections, i.e., −0.5 A and −2 A. The nominal values $R_s = 0.33\ \Omega$ and $\psi_{pm} = 0.0707$ Wb.

It can be easily found that the estimation results will be closer to their nominal values if a larger amplitude of $i_d \neq 0$ is injected, which can be explained based on the ill-conditioned issue as follows.

First, Eq. (2.29) can be rewritten as follows:

$$\begin{bmatrix} u_d^*(k) + Dd(k)V_{\text{dead}} + L_q\omega_r(k)\,i_q(k) \\ u_q^*(k) + Dd(k)V_{\text{dead}} - L_d\omega_r(k)\,i_d(k) \end{bmatrix} = \underbrace{\begin{bmatrix} i_d(k) & 0 \\ i_q(k) & \omega_r(k) \end{bmatrix}}_{\boldsymbol{A}} \begin{bmatrix} R \\ \psi_{pm} \end{bmatrix} \tag{2.30}$$

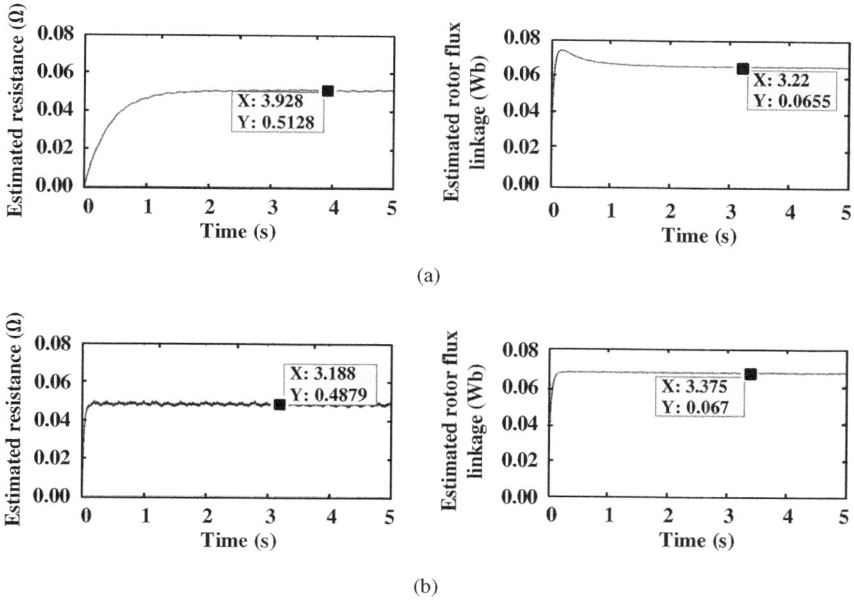

Figure 2.27 Estimated winding resistance and rotor flux linkage at 300 r/min [26] at (a) $i_d = -0.5$ A and (b) $i_d = -2.0$ A.

For the coefficient matrix $\boldsymbol{A}$, its condition number ($v = 1$) can be obtained by Eq. (2.28) and expressed as follows:

$$\begin{aligned}\operatorname{cond}(\boldsymbol{A})_1 &= \|\boldsymbol{A}^{-1}\|_1 \|\boldsymbol{A}\|_1 \\ &= \frac{|\omega_r(k)| + |i_q(k)|}{|\omega_r(k) i_d(k)|}\left(|i_d(k)| + |\omega(k)|\right) \\ &\approx \frac{|\omega_r(k)|}{|i_d(k)|}\left[\text{when}\,|\omega_r(k)| \gg |i_q(k)| > |i_d(k)|\right]\end{aligned} \tag{2.31a}$$

where

$$\begin{aligned}\|\boldsymbol{A}^{-1}\|_1 &= \frac{\max\left(|\omega_r(k)|, 0\right) + \max\left(|-i_q(k)|, |i_d(k)|\right)}{|\omega_r(k)\, i_d(k)|} \\ &= \frac{|\omega_r(k)| + |i_q(k)|}{|\omega_r(k)\, i_d(k)|}\end{aligned} \tag{2.31b}$$

$$\begin{aligned}\|\boldsymbol{A}\|_1 &= \max\left(|i_d(k)|, 0\right) + \max\left(|i_q(k)|, |\omega_r(k)|\right) \\ &= |i_d(k)| + |\omega_r(k)|\end{aligned} \tag{2.31c}$$

Above all, at 300 r/min, $\operatorname{cond}(\boldsymbol{A})_1$ with $i_d = -0.5$ A and $i_d = -2.0$ A are 323.0 and 81.5, respectively. Thus, the case of $i_d = -0.5$ A may suffer from the ill-conditioned problem, which also underpins the misestimated results shown in Figure 2.27a. In comparison, a larger amplitude of $|i_d|$ can make the estimation results closer to the real parameter values, as presented in Figure 2.27b. Therefore, in real applications, it is necessary to conduct an error analysis to overcome this issue, by which the amplitude of injected i_d can be minimized and the accuracy of estimation can also be ensured, which will be described in Chapter 4 in detail.

2.4 Summary

In this chapter, three critical issues in online parameter estimation have been introduced together with the experimental results based on a PMSM drive system.

First, the principle of the rank-deficient problem and its influence on the multi-parameter estimation have been discussed. Both the theoretical analysis and experimental results have demonstrated that the estimated parameters in a rank-deficient system could be mis- and even non-converged, especially when the initial values are set far from the actual values.

Second, the VSI non-linearity issue has been introduced from the perspectives of modelling, online estimation/compensation, as well as its influences on parameter estimation for PMSMs. An online VSI estimation/compensation method via the d-axis voltage equation under $i_d = 0$ control has been described, where both the machine parameters and the perturbation signal injections are not required. The experimental results have underpinned that the current harmonics caused by the distorted voltage can be eminently restrained after compensation. Besides, it has also been revealed that the VSI non-linearity has a significant impact on both the estimation of rotor flux linkage and winding resistance at low speeds. However, at high speed, its influence on the rotor flux linkage is negligible due to the dominant voltage drop of the back EMF. In addition, although the estimation of the q-axis inductance under $i_d = 0$ control is less affected by the VSI non-linearity, it is strongly affected by the non-ideal measurement of the variation of the dc-link voltage and zero shift in the amplifier.

Third, the ill-conditioned problem also deteriorates the estimation accuracy even for a full-rank reference system of PMSM, particularly when the current impulse needs to be injected to activate the resistance term in the d-axis equation. The experiments also underpin that a small amplitude of the d-axis current injection easily leads to the ill-conditioned problem, which will be further discussed in Chapter 4.

References

1 M. Jabbar, J. Dong, and Z. Liu, "Determination of machine parameters for internal permanent magnet synchronous motors," In *Proc. Power Electron., Mach. Drives*, Nov. 2004, vol. 2, pp. 805–810.

2 K. Lee, D. Jung, and I. Ha, "An online identification method for both stator resistance and back-EMF coefficient of PMSMs without rotational transducers," *IEEE Trans. Ind. Electron.*, vol. 51, no. 2, pp. 507–510, Apr. 2004.

3 R. Krishnan and P. Vijayraghavan, "Fast estimation and compensation of rotor flux linkage in permanent magnet synchronous machines," In *Proc. IEEE Int. Symp. Ind. Electron.*, July 1999, vol. 2, pp. 661–666.

4 R. Ramakrishnan, R. Islam, M. Islam, and T. Sebastian, "Real time estimation of parameters for controlling and monitoring permanent magnet synchronous motors," In *Proc. IEEE Int. Elect. Mach. Drives Conf.*, Miami, FL, June 2009, pp. 1194–1199.

5 A. Piippo, M. Hinkkanen, and J. Luomi, "Adaptation of motor parameters in sensorless PMSM drives," *IEEE Trans. Ind. Appl.*, vol. 45, no. 1, pp. 203–212, Jan./Feb. 2009.

6 S. Underwood and I. Husain, "Online parameter estimation and adaptive control of permanent-magnet synchronous machines," *IEEE Trans. Ind. Electron.*, vol. 57, no. 7, pp. 2435–2443, Nov. 2010.

7 K. Liu, Q. Zhang, J. Chen, Z. Q. Zhu, and J. Zhang, "Online multiparameter estimation of nonsalient-pole PM synchronous machines with temperature variation tracking," *IEEE Trans. Ind. Electron.*, vol. 58, no. 5, pp. 1776–1788, May 2011.

8 K. Liu, Z. Q. Zhu, and D. A. Stone, "Parameter estimation for condition monitoring of PMSM stator winding and rotor permanent magnets," *IEEE Trans. Ind. Electron.*, vol. 60, no. 12, pp. 5902–5913, Dec. 2013.

9 K. Liu and Z. Q. Zhu, "Online estimation of the rotor flux linkage and voltage-source inverter nonlinearity in permanent magnet synchronous machine drives," *IEEE Trans. Power Electron.*, vol. 29, no. 1, pp. 418–427, Jan. 2014.

10 K. Liu and Z. Q. Zhu, "Position-offset-based parameter estimation using the Adaline NN for condition monitoring of permanent-magnet synchronous machines," *IEEE Trans. Ind. Electron.*, vol. 62, no. 4, pp. 2372–2383, Apr. 2015.

11 K. Liu and Z. Q. Zhu, "Quantum genetic algorithm-based parameter estimation of PMSM under variable speed control accounting for system identifiability and VSI nonlinearity," *IEEE Trans. Ind. Electron.*, vol. 62, no. 4, pp. 2363–2371, Apr. 2015.

12 M. Rashed, P. F. A. Macconnell, A. F. Stronach, and P. Acarnley, "Sensorless indirect-rotor-field-orientation speed control of a permanent-magnet synchronous motor with stator-resistance estimation," *IEEE Trans. Ind. Electron.*, vol. 54, no. 3, pp. 1664–1675, Jun. 2007.

13 S. Moreau, R. Kahoul, and J.-P. Louis, "Parameters estimation of permanent magnet synchronous machine without adding extra-signal as input excitation," In *Proc. IEEE Int. Symp. Ind. Electron.*, Jan. 2004, vol. 1, pp. 371–376.

14 Z. Q. Zhu, X. Zhu, and P. D. Sun, "Estimation of winding resistance and PM flux-linkage in brushless AC machines by reduced-order extended Kalman filter," In *Proc. IEEE Int. Conf. Net., Sensing Control*, June 2007, pp. 740–745.

15 S. Ichikawa, M. Tomita, S. Doki, and S. Okuma, "Sensorless control of permanent-magnet synchronous motors using online parameter identification based on system identification theory," *IEEE Trans. Ind. Electron.*, vol. 53, no. 2, pp. 363–372, Apr. 2006.

16 H. Kim, K. Kim, and M. Youn, "On-line dead-time compensation method based on time delay control," *IEEE Trans. Contr. Syst. Technol.*, vol. 11, no. 2, pp. 279–285, Mar./Apr. 2003.

17 R. Raute, C. Caruana, and C. Staines, "Analysis and compensation of inverter nonlinearity effect on a sensorless PMSM drive at very low and zero speed operation," *IEEE Trans. Ind. Electron.*, vol. 57, no. 12, pp. 4065–4074, Dec. 2010.

18 R. Munoz and T. Lipo, "On-line dead-time compensation technique for open-loop PWM-VSI drives," *IEEE Trans. Power Electron.*, vol. 14, no. 4, pp. 683–689, Jul. 1999.

19 H. Kim, H. Moon, and M. Youn, "On-line dead-time compensation method using disturbance observer," *IEEE Trans. Power Electron.*, vol. 18, no. 6, pp. 1336–1345, Nov. 2003.

20 H. Kim, M. Youn, K. Cho, and H. Kim, "Nonlinearity estimation and compensation of PWM VSI for PMSM under resistance and flux linkage uncertainty," *IEEE Trans. Contr. Syst. Technol.*, vol. 14, no. 4, pp. 589–601, July. 2006.

21 H. Kim, M. Youn, and K. Cho, "New voltage distortion observer of PWM VSI for PMSM," *IEEE Trans. Ind. Electron.*, vol. 52, no. 4, pp. 1188–1192, Aug. 2005.

22 S. Kim, W. Lee, M. Rho, and S. Park, "Effective dead-time compensation using a simple vectorial disturbance estimator in PMSM drives," *IEEE Trans. Ind. Electron.*, vol. 57, no. 5, pp. 1609–1614, May 2010.

23 H. Zhao, Q. Wu, and A. Kawamura, "An accurate approach of nonlinearity compensation for VSI inverter output voltage," *IEEE Trans. Power Electron.*, vol. 19, no. 4, pp. 1029–1035, July 2004.

24 Y. Park and S. Sul, "A novel method utilizing trapezoidal voltage to compensate for inverter nonlinearity," *IEEE Trans. Power Electron.*, vol. 27, no. 12, pp. 4837–4846, Dec. 2012.

25 K. Liu, Z. Q. Zhu, Q. Zhang, and J. Zhang, "Influence of nonideal voltage measurement on parameter estimation in permanent magnet synchronous machines," *IEEE Trans. Ind. Electron.*, vol. 59, no. 6, pp. 2438–2447, June 2012.

26 K. Liu, "Novel techniques for parameter estimation of permanent magnet synchronous machines," Ph.D. dissertation, Dept. Elect. Eng., Univ. Sheffield, U.K., Dec. 2013.

27 J. Choi and S. Sul, "Inverter output voltage synthesis using novel dead-time compensation," *IEEE Trans. Power Electron.*, vol. 11, no. 2, pp. 221–227, Mar. 1996.

28 *DIPIPM Ver. 3 Application note*, Mitsubishi Electric Corporation, Aug. 2009. [Online]. Available: https://www.mitsubishielectric.com/semiconductors/files/manuals/dipv3-e.pdf.

29 V. Solo, "Limits to estimation in stochastic ill-conditioned inverse problems," *IEEE Trans. Inf. Theory*, vol. 46, no. 5, pp. 1872–1880, Aug. 2002.

30 J. Gu, K. Clements, G. Krumpholz, and P. Davis, "The solution of ill-conditioned power system state estimation problems via the method of Peters and Wilkinson," *IEEE Trans. Power Appar. Syst.*, vol. PAS-102, no. 10, pp. 3473–3480, Oct. 1983.

31 A. Adalev, N. Korovkin, and M. Hayakawa, "Identification of electric circuits described by ill-conditioned mathematical models," *IEEE Trans. Circuits and Syst. I: Regul. Pap.*, vol. 53, no. 1, pp. 78–91, Jan. 2006.

32 T. Chang, L. Wang, and F. Chang, "A solution to the ill-conditioned GPS positioning problem in an urban environment," *IEEE Trans. Intell. Transp. Syst.*, vol. 10, no. 1, pp. 135–145, Mar. 2009.

3

Online Estimation of Rotor Flux Linkage with the Aid of Thermocouples in Stator Windings

3.1 Introduction

As demonstrated in Chapter 2, to mitigate the rank-deficient issue in the electrical parameter estimation for a non-salient-pole permanent magnet synchronous machine (PMSM), i.e., the winding resistance, the rotor PM flux linkage, and the inductances ($L_d = L_q$), the designed estimation schemes I and II fix the winding resistance or the rotor PM flux linkage and then estimate the remaining two parameters. However, the experimental results show that the estimated parameters will exhibit significant misestimation issues when the winding resistance and the rotor PM flux linkage are changed due to the temperature variation. It can be simply explained that the temperature dependencies of the winding resistance and the rotor PM flux linkage cannot be considered since they have been fixed to the nominal values. Consequently, the above estimation schemes lose their effectiveness in the important application of thermal condition monitoring.

Nevertheless, in many applications, some machine parameters can be acquired by the measurement instruments that are additionally equipped to overcome this issue. For instance, the torque transducer is employed in Ref. [1] under the load tests to online estimate the rotor PM flux linkage, and subsequently, the winding resistance and *dq*-axis inductances can be estimated. Moreover, the stator winding temperature is measured in Refs [2–4] by buried thermocouples for obtaining the winding resistance under real-time conditions, with which the rotor PM flux linkage can be estimated by using high-frequency signal injection [2] or the *q*-axis voltage equation [3].

In this chapter, a practical method will be introduced for online estimating the rotor PM flux linkage, where the temperature-dependent stator winding resistance is obtained by buried thermocouples and the distorted voltage due to voltage source inverter (VSI) non-linearity is estimated via the *d*-axis equation (described in Chapter 2). As will be shown, the described method is easy to implement, and the perturbance signal injection is not required as well. The experimental results will demonstrate that when the VSI non-linearity can be estimated and compensated delicately, the estimated rotor PM flux linkage is adequate to track the temperature variations of the rotor PM for thermal condition monitoring.

Parameter Estimation of Permanent Magnet Synchronous Machines, First Edition. Zi Qiang Zhu, Kan Liu, and Dawei Liang.

3.2 Online Estimation of Rotor Flux Linkage with the Aid of Thermocouples in Stator Windings

3.2.1 Online Estimation of Rotor Flux Linkage

First, considering the VSI non-linearity, the steady-state PMSM model can be expressed as follows [3]:

$$\begin{cases} u_d^*(k) + Dd(k)\,V_{\text{dead}} = R_s i_d(k) - L_q\,\omega_r(k) i_q(k) \\ u_q^*(k) + Dq(k)\,V_{\text{dead}} = R_s i_q(k) + L_d \omega_r(k) i_d(k) + \psi_{pm}\,\omega_r(k) \end{cases} \tag{3.1}$$

where u_d^*, u_q^*, i_d, i_q, L_d, and L_q are the reference voltages, the currents, and the d- and q-axis inductances, respectively. V_{dead}, R_s, ψ_{pm}, and ω_r are the distorted voltage, the stator resistance, the rotor PM flux linkage, and the electrical angular speed, respectively.

Under $i_d = 0$ control, the dq-axis voltage equation (Eq. 3.1) can be simplified as follows:

$$\begin{cases} u_d^*(k) + Dd(k)\,V_{\text{dead}} = -L_q\,\omega_r(k) i_q(k) \\ u_q^*(k) + Dq(k)\,V_{\text{dead}} = R_s i_q(k) + L_d \omega_r(k) i_d(k) + \psi_{pm}\,\omega_r(k) \end{cases} \tag{3.2}$$

It can be found that there are three unknown parameters in the q-axis voltage equation, i.e., V_{dead}, R_s, and ψ_{pm}. Thus, it is impossible to separate the rotor PM flux linkage from one q-axis voltage equation (Eq. 3.2). However, it has been introduced in Section 2.2 that the distorted voltage V_{dead} can be estimated from the d-axis high-frequency reference voltages u_{dh}^* in Eq. (3.3) by using a first-order low-pass filter.

$$V_{\text{dead}} \approx -\frac{u_{dh}^*(k)}{Dd(k)} \tag{3.3}$$

It can be observed that the rotor PM flux linkage can be estimated once the actual value of stator winding resistance R_s can be obtained. Therefore, this chapter introduces a practical method for estimating rotor PM flux linkage with the aid of buried thermocouples in the stator winding, where the temperature dependency of the resistance (Eq. 3.4) can be considered:

$$R_{s,s} = R_{s,0} + \alpha_{cu}(\vartheta_s - \vartheta_r) \tag{3.4}$$

where $R_{s,0}$ and $R_{s,s}$ are the winding resistance at the reference temperature of ϑ_r and the actual temperature ϑ_s, respectively. α_{cu} is the temperature coefficient of winding resistance. It is noteworthy that in many applications of PMSMs, e.g., aerospace, electrification transportation, and wind power generation, the thermocouples are widely employed and buried during the winding process.

Afterwards, with the determined V_{dead} and R_s, the rotor PM flux linkage ψ_{pm} can be finally estimated by using an Adaline estimator [5], which is expressed as follows:

$$\hat{\psi}_{pm}(k+1) = \hat{\psi}_{pm}(k+1) + 2\eta\omega_r(k)\left[d_q(k) - O(k)\right] \tag{3.5}$$

where the convergence factor η is $1.0e^{-6}$. The detailed instructions can be found in Chapter 10.

The complete estimation procedure based on the field-oriented control (FOC) system is shown in Figure 3.1, where the complete estimation procedure is divided into three steps:

1. The stator winding resistance value $R_{s,0}$ at normal temperature ϑ_r is measured by milliohm meter at a standstill, and two sets of thermocouples are employed for measuring the variation of the stator winding temperature ϑ_s, which will be used for real-time calculation of the actual value of stator winding resistance R_s.

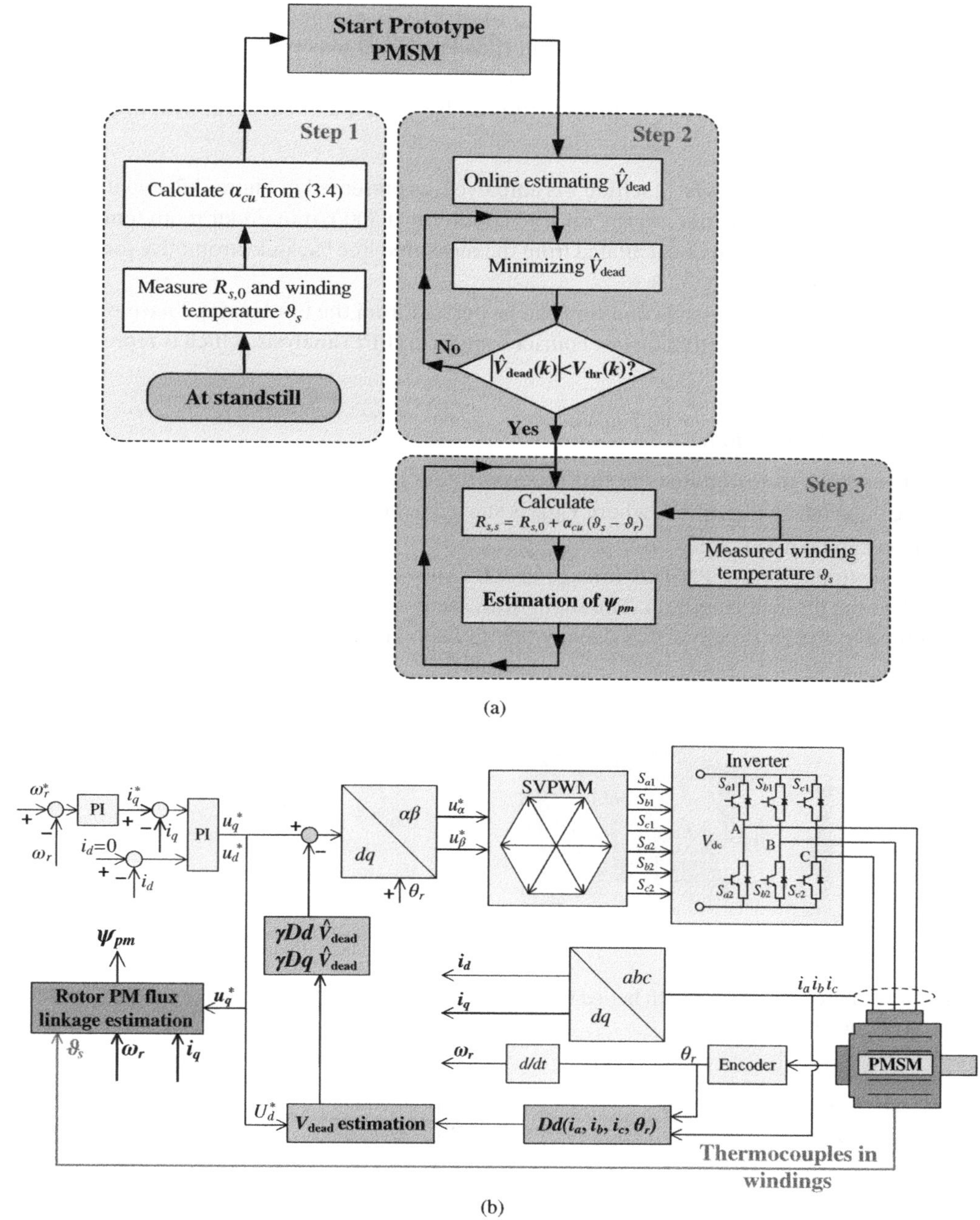

Figure 3.1 Online estimation of rotor PM flux linkage with the aid of thermocouples and estimated V_{dead} caused by VSI non-linearity: (a) complete diagram of estimating VSI non-linearity and PM flux linkage and (b) application of rotor PM flux linkage estimation method in FOC system.

2. The distorted voltage V_{dead} due to VSI non-linearity is online estimated from the d-axis equation (Eq. 3.3) under $i_d = 0$ control and then employed to compensate for the outputs of current proportional–integral (PI) regulators.
3. The rotor PM flux linkage ψ_{pm} is finally estimated from the q-axis equation with the aid of the obtained R_s and V_{dead}.

For verification, the Prototype Machine 1 is employed, as presented in Figure 3.2, and it is loaded by a direct current (dc) machine. At the no-load condition at 300 r/min under room temperature, the rotor PM flux linkage can be calculated from the measured line back-electromotive force (EMF) for reference, as shown in Figure 3.3.

The PM flux linkage at no-load condition can be derived from the fundamental harmonic of the line back-EMF (19.23 V) by utilizing fast Fourier transform (FFT) analysis, which is represented as follows:

$$\frac{19.23\ \text{V} \times 60\ \text{s}}{2\pi\sqrt{3} \times 5 \times 300\ \text{r/min}} = 70.7\ \text{mWb} \tag{3.6}$$

By way of example, the estimated PM flux linkages under $i_d = 0$ control with different q-axis currents i_q are depicted in Figure 3.4, where V_{dead} is started to be compensated after 10 s, and several phenomena can be observed.

First, the estimated rotor PM flux linkage at 300 r/min is 68.3 mWb, which is close to that of 70.7 mWb at the no-load condition. The slight decrease in rotor PM flux linkage by 2.4 mWb may be attributed to the variation of flux density due to load. However, the estimated rotor PM flux

(a) (b)

Figure 3.2 Prototype Machine 1 with buried thermocouples: (a) prototype PMSM and (b) buried thermocouples in PMSM.

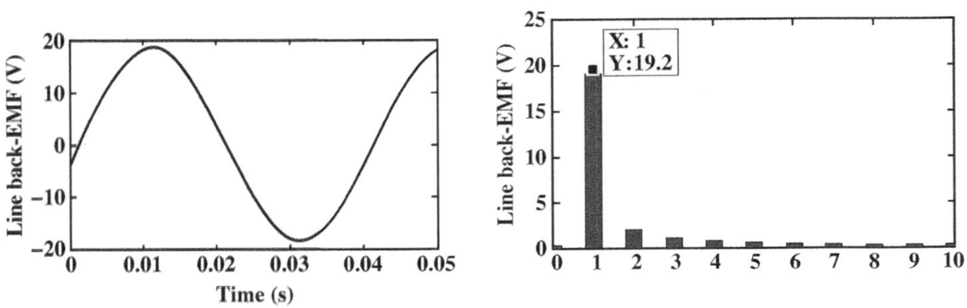

(a) (b)

Figure 3.3 (a) Line back-EMF and its (b) FFT results under no-load condition at 300 r/min under room temperature [3], IEEE.

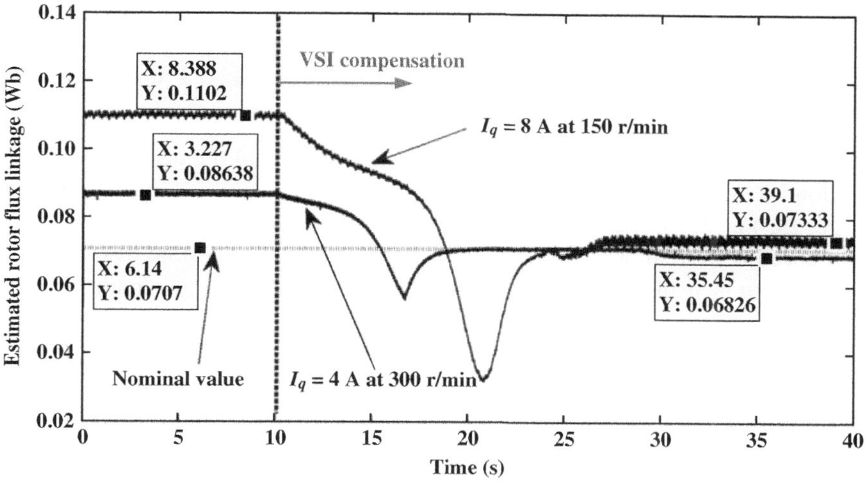

Figure 3.4 Estimated rotor PM flux linkage with and without VSI non-linearity compensation at room temperature [3], IEEE.

linkage at 150 r/min is 73.3 mWb and slightly larger than 70.7 mWb measured at no-load condition, which is contrary to the fact that the rotor PM flux linkage at loaded condition should be smaller than that at no-load condition. This estimation error can be explained that there are still influences from other non-linearities, e.g., non-ideal position measurement, current offsets, and non-ideal measurement of dc-link voltage.

Second, since the back-EMF of the PMSM is relatively small at low speed, i.e., less than 8 V at 150 r/min, the accuracy of estimated rotor PM flux linkage at 150 r/min is 73.3 mWb, which easily suffers from the above-unestimated non-linearities and presents a relatively significant estimation error of 2.6 mWb even if the VSI non-linearity is well compensated. The influences of these unestimated non-linearities can be neglected when the PM flux linkage is estimated at a relatively high speed, i.e., a relatively high ratio of back EMF versus other non-linearities can be achieved. Therefore, the estimated PM flux linkage at 300 r/min exhibits a much higher accuracy.

Finally, the difference in PM flux linkage results with and without the compensation of V_{dead} at 300 r/min is 18.1 mWb, which is almost half of that at 150 r/min of 36.9 mWb. This can be explained that although the variation of V_{dead} at different operating conditions is relatively small, the ratio of $Dq\,V_{\text{dead}}$ to $\psi_{pm}\omega_r$ will be reduced approximately by half when the rotor speed increases from 150 to 300 r/min.

3.2.2 Thermal Condition Monitoring of Rotor PM

In terms of condition monitoring of PM temperature, it is well known that the residual flux density B_r of the NdFeB magnet decreases linearly with temperature [6] and can be written as follows:

$$B_{r,s} = B_{r,0} - \alpha_{pm}(\vartheta_{pm} - \vartheta_r) \tag{3.7}$$

where $B_{r,0}$ and $B_{r,s}$ are the remanent flux density at normal ambient temperature ϑ_r and at actual temperature ϑ_{pm}. α_{pm} is the temperature coefficient. Furthermore, neglecting the saturation effects of the magnetic circuit of the machine and assuming that α_{pm} is constant over the entire temperature range, the back-EMF is proportional to the remanence of PM. Thus, the temperature variation of PM can be reflected by the changes in the rotor PM flux linkage.

To verify the temperature tracking capability by utilizing the estimated rotor PM flux linkage, a comprehensive experiment is designed as follows:

1. The PMSM is heated by an external heat source for 20 min and then cooled down by natural convection.
2. At $t = 20$, 25, 40, 55, 75, and 155 min, the dc machine will be started to draw the PMSM, and its rotor speed is fixed at 300 r/min while the line back-EMF of PMSM at no-load condition will be recorded by using a scope. The rotor PM flux linkage at no-load condition can be computed from these measured data with aiding from FFT.
3. At $t = 20$, 30, 45, 60, 80, and 160 min, the PMSM will be loaded and the rotor PM flux linkage at loaded conditions will be online estimated with the aid of the measured stator winding temperatures.

The stator winding resistances at different temperatures are measured and shown in Figure 3.5, which vary with the measured winding temperature linearly and the coefficient of temperature α_{pm} can be obtained afterwards ($\alpha_{pm} \approx 0.0012\ \Omega/^\circ$C). The experimental results of Steps 2 and 3 are shown in Figure 3.6.

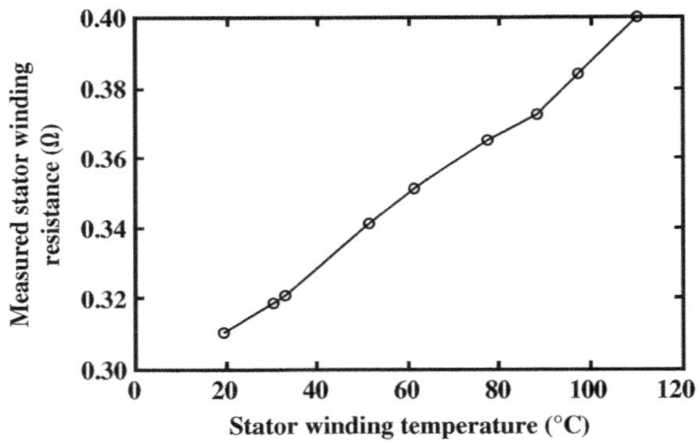

Figure 3.5 Correlation of measured stator winding resistance and temperature [4], IEEE.

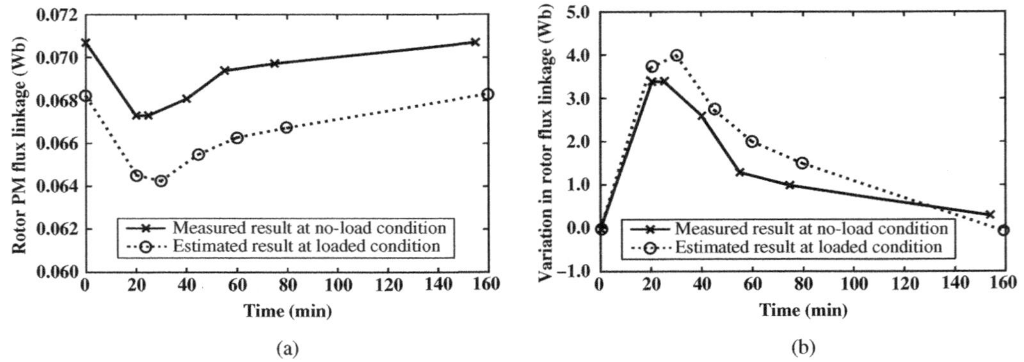

Figure 3.6 Comparison of estimated and measured rotor PM flux linkages at different time points ($i_q = 4$ A and $i_d = 0$ A at 300 r/min) [3]: (a) measured and estimated rotor PM flux linkages and (b) variation in rotor PM flux linkage.

It is evident that the presented method has good performance in tracking the variation of rotor PM flux linkage. As can be seen from Figure 3.6a, the estimated PM flux linkage varies from 68.3 to 64.5 mWb after heating for 20 min. However, since the thermal time constants of the stator and the PM are not the same, the response of the temperature rise in the PM is delayed from that in the stator, which is reflected in the further decrease in PM flux linkage to 64.3 mWb after natural cooling of 10 min. Afterwards, the rotor PM flux linkage will slowly increase with the decrease in PM temperature and will reach its normal value ultimately.

3.3 Summary

This chapter has described a practical method for estimating the rotor PM flux linkage with the aid of the thermocouples buried in the stator winding and the estimated distorted voltage due to VSI non-linearity, which is particularly appropriate for real-time thermal condition monitoring for rotor PM. The introduced method is easy to be implemented since commercial electrical machines are usually equipped with several thermocouples in the winding. Moreover, the experimental results have demonstrated that the real-time variation of the thermal states of PM can be adequately reflected by the estimated rotor PM flux linkage, once the VSI non-linearity has been compensated delicately. Besides, in the following chapters, several other online parameter estimation methods will be introduced, which do not rely on additional measurement instruments and machine nominal parameter values.

References

1 M.N. Uddin and M.M.I. Chy, "Online parameter-estimation-based speed control of PM AC motor drive in flux-weakening region," *IEEE Trans. Ind. Appl.*, vol. 44, no. 5, pp. 1486–1494, Sep./Oct. 2010.

2 D. Reigosa, F. Briz, P. Gacia, J.M. Guerrero, and M.W. Degner, "Magnet temperature estimation in surface PM machines using high-frequency signal injection," *IEEE Trans. Ind. Appl.*, vol. 46, no. 4, pp. 1468–1475, Jul./Aug. 2010.

3 K. Liu and Z. Q. Zhu, "Online estimation of the rotor flux linkage and voltage-source inverter nonlinearity in permanent magnet synchronous machine drives," *IEEE Trans. Power Electron.*, vol. 29, no. 1, pp. 418–427, Jan. 2014.

4 K. Liu, "Novel techniques for parameter estimation of permanent magnet synchronous machines," PhD dissertation, University of Sheffield, UK, 2013.

5 B. Widrow and M. A. Lehr, "30 years of adaptive neural networks: Perceptron, madaline, and backpropagation," *Proc. IEEE*, vol. 78, no. 9, pp. 1415–1442, Sep. 1990.

6 S. Liu and G. E. Kuhl, "Temperature coefficients of rare earth permanent magnets," *IEEE Trans. Magn.*, vol. 35, no. 5, pp. 3271–3273, Sep. 1999.

4

Online Parameter Estimation Based on Current Injections

4.1 Introduction

Accurately acquiring electrical parameters in permanent magnet synchronous machines (PMSMs) is critical for machine control performance and condition monitoring. However, according to the system identification theory, it is impossible to guarantee that the estimated parameters could converge to the actual values when the rank of the reference model is smaller than the number of the parameters to be estimated. More specifically, in the PMSM system, one set of the steady-state dq-axis equations is a rank-deficient reference model for simultaneously estimating stator winding resistance, dq-axis inductances, and rotor PM flux linkage, as described in Chapter 2.

To address this rank-deficient problem, some early studies [1–5] estimate some electrical parameters by fixing other parameters, e.g., the conventional two-step [3] and three-step methods. However, as discussed in Chapter 8, these multi-step methods are sensitive to the mismatch between the nominal and real parameter values.

Besides, the additional physical quantities measured by the measurement instruments [6–8] can help to establish a full-rank reference model, e.g., the rotor PM flux linkage estimation with the aid of buried thermocouples in the stator winding in Chapter 3. Nevertheless, the additionally equipped measurement instruments increase the system complexity and may have a detrimental impact on system reliability [9]. Meanwhile, some measurement requires the machine to operate under the no-load condition due to safety issue. However, it is less possible to guarantee that all required measurement instruments are always available, and some parameter values under on-load conditions are usually different from those under no-load conditions.

Therefore, it is also popular to estimate machine parameters under the on-load conditions by introducing additional disturbance signals [10], e.g., current/voltage injection, rotor position offset, and high-frequency injection, by utilizing online estimation algorithms, e.g., model reference adaptive systems [2], Adaline neural network (ANN) [6], recursive least squares (RLS) [11], and extended Kalman filters [12]. Amongst different methods, current injection methods [13–19] have been widely used to build a full-rank reference model by changing the state of the PMSM, where only the basic electrical quantities, i.e., currents, voltages, and the rotor speeds are required, but the nominal/measured electrical parameters are unnecessary. The schematic diagram of the current injection–based online parameter estimation is shown in Figure 4.1. In theory, the main electrical parameters, i.e., stator winding resistance R_s, dq-axis inductances L_d, L_q, and rotor PM flux linkage ψ_{pm}, can be estimated simultaneously by utilizing the d-axis current injection ΔI_d.

For instance, by injecting an alternating d-axis current i_d as described in the previous study [13], the winding resistance and rotor PM flux linkage can be estimated. However, the estimation

Parameter Estimation of Permanent Magnet Synchronous Machines, First Edition. Zi Qiang Zhu, Kan Liu, and Dawei Liang.

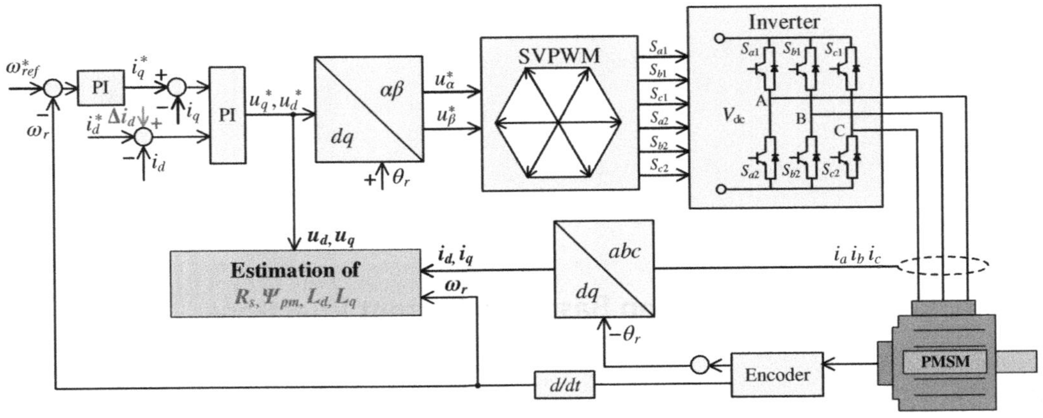

Figure 4.1 Schematic diagram of current injection–based parameter estimation.

performance is sensitive to the amplitude of the injected i_d since the variation of the inductance and the non-ideal voltage measurement are not considered. Furthermore, the *d*-axis current injection–based estimation methods are further employed to estimate the stator winding resistance as described in Ref. [14] and the multi-parameters, i.e., stator winding resistance, the *dq*-axis inductances, and the rotor PM flux linkage, in Refs [15–18]. In addition, the magnetic saturation effect can also be considered [19]. However, it has been highlighted that the current injection method easily suffers from the ill-conditioned problem due to the uncertainties in the machine drive systems, e.g., the non-linearity of the voltage source inverter, the variations in electrical parameters, and the non-ideal measurements of position and direct current (dc)-link voltage. Consequently, it will significantly deteriorate the estimation accuracy.

This chapter describes two typical *d*-axis current injection–based parameter estimation methods for PMSMs. The first method can simultaneously estimate the stator winding resistance, the *dq*-axis inductances, and the rotor flux linkage. The second method focuses on the estimation of the stator winding resistance and the rotor flux linkage under constant torque/speed control, particularly appropriate for thermal condition monitoring. More importantly, the ill-conditioned issue existing in the typical current injection–based methods is investigated theoretically and experimentally in this chapter, where the error analyses are performed in both methods to address this issue.

4.2 Multi-parameter Estimation Based on Current Injection and Error Analysis

4.2.1 Designed Parameter Estimation Scheme

This section describes a typical online multi-parameter estimation method based on an injection of $i_d \neq 0$ for PMSM [15], where any nominal parameter values are unnecessary. The estimation scheme is shown in Figure 4.2 and explained as follows.

1. Two sets of simplified PMSM state equations, i.e., $i_d = 0$ and $i_d \neq 0$, are obtained by injection of a short pulse of non-zero *d*-axis current and solved simultaneously.
2. For the non-salient-pole PMSM, the convergence of solutions will be ensured by assuming $L_q(i_d \neq 0) = L_d(i_d \neq 0)$ and $\psi_{pm}(i_d \neq 0) = \psi_{pm}(i_d = 0)$ since the built PMSM models are of full rank.

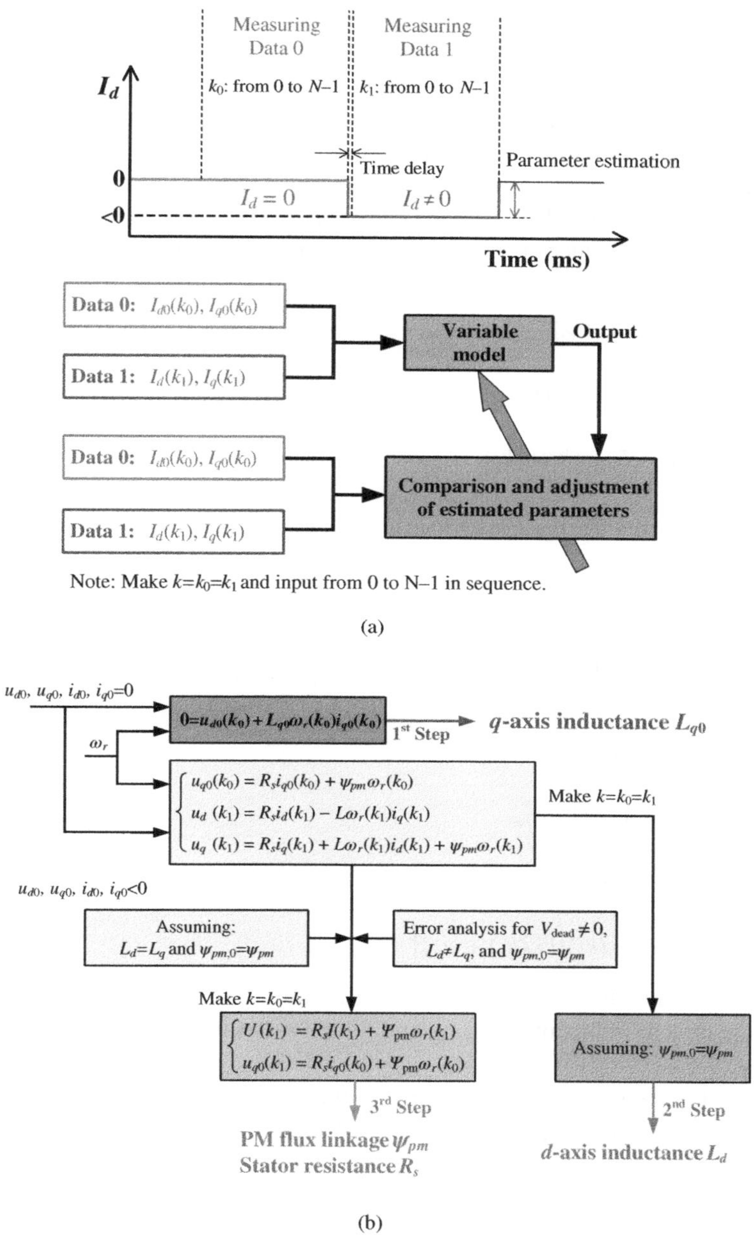

Figure 4.2 Schematic diagram of multi-parameter estimation scheme: (a) process of sampling, recording, and parameter estimation and (b) mathematical models.

3. The error analysis is carried out to determine the minimum $|i_d|$, which ensures the injected current has negligible errors in the estimation of the winding resistance and the rotor PM flux linkage under $i_d = 0$ control.

As observed, the two sets of the steady-state PMSM equations are obtained by injecting a negative d-axis current, corresponding to $i_d = 0$ and $i_d \neq 0$, respectively. The recorded Data 0 ($i_d = 0$) and Data 1 ($i_d \neq 0$) include the steady-state rotor speed and dq-axis currents and voltages, i.e., Data 0 is first measured, and a short pulse of $i_d < 0$ is injected, and Data 1 is measured after a short time delay, e.g., 2 ms in this study. Both data sets are with the same sampling period and sampling number N, as shown in Figure 4.2a, and the introduced time delay before the measuring of Data 1 ensures that the measurement is conducted after settling of step transients correlated to the pulse of i_d. Since the duration of injected pulse current is very short, the rotor speed and winding resistance are assumed to be constant due to the considerably large mechanical and thermal time constants compared to the electrical time constant.

Hence, neglecting the influence of voltage source inverter (VSI) non-linearity, the corresponding two sets of dq-axis equations are as follows:

$$u_{d0}(k_0) = -L_{q0}\,\omega_r(k_0) i_{q0}(k_0) \tag{4.1a}$$

$$u_{q0}(k_0) = R_s i_{q0}(k_0) + \psi_{pm0}\,\omega_r(k_0) \tag{4.1b}$$

$$u_d(k_1) = R_s i_d(k_1) - L_q\,\omega_r(k_1)\, i_q(k_1) \tag{4.1c}$$

$$u_q(k_1) = R_s i_q(k_1) + \omega_r(k_1)\left[L_d i_d(k_1) + \psi_{pm1}\right] \tag{4.1d}$$

where variables and parameters with/without suffix "0" are referred to those associated with Data 0 and Data 1, respectively. k_0 and k_1 are indices of the k_0th and k_1th measured instants for Data 0 and Data 1, respectively.

Under the assumptions of $L = L_d = L_q$ and $\psi_{pm} = \psi_{pm0} = \psi_{pm1}$, Eq. (4.1) can be simplified as follows:

$$u_{d0}(k_0) = -L_{q0}\,\omega_r(k_0) i_{q0}(k_0) \tag{4.2a}$$

$$u_{q0}(k_0) = R_s i_{q0}(k_0) + \omega_r(k_0)\psi_{pm} \tag{4.2b}$$

$$u_d(k_1) = R_s i_d(k_1) - L\omega_r(k_1) i_q(k_1) \tag{4.2c}$$

$$u_q(k_1) = R_s i_q(k_1) + \omega_r(k_1)\left[L i_d(k_1) + \psi_{pm}\right] \tag{4.2d}$$

Therefore, based on the above reference model, the multi-parameters, i.e., the stator winding resistance R_s, the dq-axis inductances L_d, L_q, and the rotor PM flux linkage ψ_{pm}, can be estimated as follows.

First, the q-axis inductance L_{q0} can be obtained from Eq. (4.2a).

Second, R_s and ψ_{pm} can be estimated from Eqs. (4.3) and (4.4). In addition, Eq. (4.3) is obtained by multiplying $i_d(k_1)/i_q(k_1)$ on both sides in Eq. (4.2c) and then adding into Eq. (4.2d).

$$u_d(k_1)\frac{i_d(k_1)}{i_q(k_1)} + u_q(k_1) = R_s\left[\frac{i_d^2(k_1)}{i_q(k_1)} + i_q(k_1)\right] + \omega_r(k_1)\psi_{pm} \tag{4.3}$$

$$\begin{cases} U(k_1) = R_s I(k_1) + \omega_r(k_1)\psi_{pm} \\ u_{q0}(k_0) = R_s i_{q0}(k_0) + \omega_r(k_0)\psi_{pm} \end{cases} \tag{4.4a}$$

with

$$U(k_1) = u_d(k_1)\frac{i_d(k_1)}{i_q(k_1)} + u_q(k_1) \text{ and } I(k_1) = \frac{i_d^2(k_1)}{i_q(k_1)} + i_q(k_1) \tag{4.4b}$$

Finally, the d-axis inductance L_d can be further estimated with the aid of the estimated R_s and ψ_{pm} directly from Eq. (4.2d) or from Eq. (4.5) = (4.2d)–(4.2b), which is expressed as follows:

$$\Delta u_q(k) = R_s \Delta i_q(k) + \omega_r(k_1)\left[L\, di_d(k_1) + \underbrace{\psi_{pm1} - \psi_{pm0}}_{\approx 0}\right] \tag{4.5a}$$

$$\text{with } \Delta u_q(k) = u_q(k_1) - u_{q0}(k_0) \text{ and } \Delta i_q(k) = i_q(k_1) - i_{q0}(k_0) \tag{4.5b}$$

where the variation of the rotor PM flux linkage is ignored, i.e., $\psi_{pm1} = \psi_{pm0}$.

As above, the iterative calculation of estimated parameters is presented in Figure 4.3. As can be seen, the estimation will be started after the completion of measuring and memorizing Data 0 and Data 1, which are saved in RAM, and the numbering will be reset to $k = k_0 = k_1 = 0$ at the beginning of the estimation. Thus, two sets of data are used for the estimation process for iterative computation from 0 to $N - 1$ in sequence, and k_0 will be kept always equal to k_1 in the whole estimation process for the convenience of the estimator design. In addition, all designed ANN estimators are described in Chapter 10.

4.2.2 Error Analysis

As aforementioned, the winding resistance and the rotor PM flux linkage are estimated under the assumptions of $L = L_d = L_q$ and $\psi_{pm} = \psi_{pm0} = \psi_{pm1}$, whereas the variations of parameters may cause the ill-conditioned problem, as discussed in Chapter 2. Therefore, an error analysis will be performed.

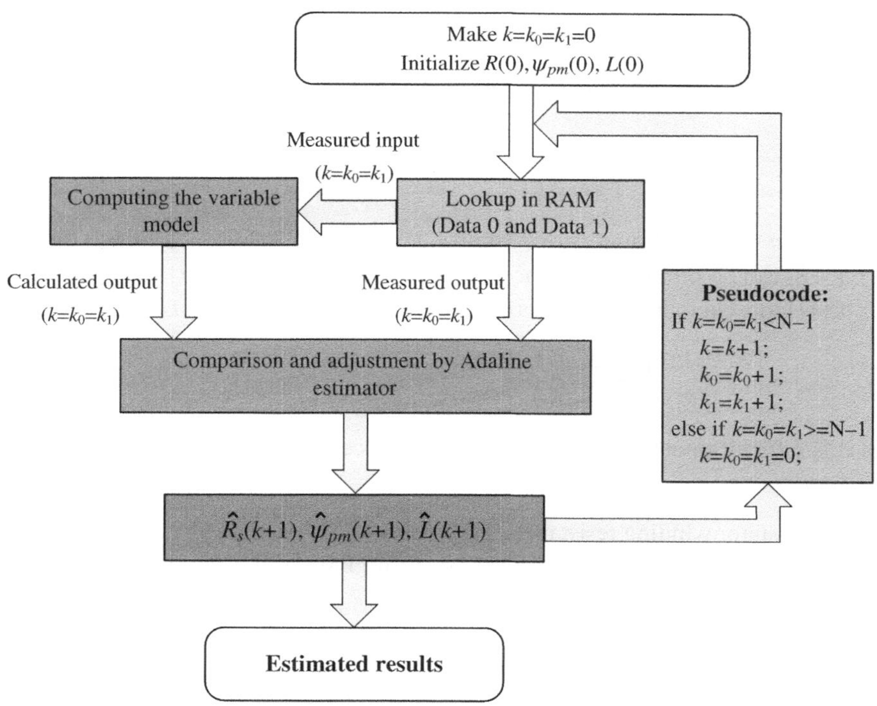

Figure 4.3 Iterative calculation of estimating R_s, ψ_{pm}, and L based on measured quantities.

First, in terms of the estimation error of the d-axis inductance L_d caused by the variation of PM flux linkage, i.e., $\Delta\psi_{pm} = \psi_{pm,0} - \psi_{pm}$, and the VSI non-linearity, Eq. (4.1) is rewritten as follows:

$$u_{d0}^*(k_0) = L_{q0}\,\omega_r(k_0)i_{q0}(k_0) - V_{\text{dead},0}Dd_0(k_0) \tag{4.6a}$$

$$u_{q0}^*(k_0) = R_s i_{q0}(k_0) + \omega_r(k_0)\psi_{pm0} - V_{\text{dead},0}Dq_0(k_0) \tag{4.6b}$$

$$u_d^*(k_1) = R_s i_d(k_1) - L_q\,\omega_r(k_1)i_q(k_1) - V_{\text{dead},1}Dd_1(k_1) \tag{4.6c}$$

$$u_q^*(k_1) = R_s i_q(k_1) + \omega_r(k_1)\left[L_d i_d(k_1) + \psi_{pm1}\right] - V_{\text{dead},1}Dq_1(k_1) \tag{4.6d}$$

where "*" denotes the measured dq-axis reference voltages from proportional–integral (PI) regulators. $V_{\text{dead},i}Dd_i(k_i)$ and $V_{\text{dead},i}Dq_i(k_i)$ with $\{i = 0 \text{ and } 1\}$ are the distorted voltage terms in the dq-axis caused by the VSI non-linearity at $i_d = 0$ and $i_d \neq 0$, respectively.

Subsequently, the result of Eq. (4.6d) subtracted by Eq. (4.6b) can be written as follows:

$$\begin{aligned}\Delta u_q^*(k) &= u_q^*(k_1) - u_{q0}^*(k_0) \\ &= R_s\Delta i_q(k) + \omega_r(k_1)\left[L_d i_d(k_1) - \Delta\psi_{pm}\right] - \left[V_{\text{dead},1}Dq_1(k_1) - V_{\text{dead},0}Dq_0(k_0)\right]\end{aligned} \tag{4.7}$$

It is evident that the cancellation of $V_{\text{dead},1}Dq_1(k_1) - V_{\text{dead},0}Dq_0(k_0)$ can improve the estimation accuracy compared with estimating L_d directly by Eq. (4.2d). However, the non-zero $\Delta\psi_{\text{pm}}$ caused by the current injection may also introduce additional estimation error.

Second, the estimation error of the PM flux linkage $\psi_{pm,\text{error}}$ between the estimated value $\hat{\psi}_{pm}$ and the actual value ψ_{pm0} can be calculated as follows:

$$\begin{aligned}\psi_{pm,\text{error}} &= \hat{\psi}_{pm} - \psi_{pm0} \\ &= -\frac{\Delta\psi_{pm}\,i_q(k_1)i_{q0}(k_0) + \Delta L i_d(k_1)i_{q0}(k_0)i_q(k_1)}{i_{q0}(k_0)i_q(k_1) - i_q^2(k_1) - i_d^2(k_1)}\end{aligned} \tag{4.8a}$$

$$\text{with}\begin{cases}\Delta\psi_{pm} = \psi_{pm0}(i_d = 0) - \psi_{\text{pm}1}(i_d \neq 0) \\ \Delta L = L_q(i_d \neq 0) - L_d(i_d \neq 0)\end{cases} \tag{4.8b}$$

Afterwards, under constant torque/speed control, i.e., $i_q(i_d \neq 0) \approx i_{q0}(i_d = 0)$, the estimation error $\psi_{pm,\text{error}}$ (Eq. 4.8) can be simplified as a monotone decreasing function of i_d and can be minimized as long as $|i_d|$ is large enough but irrespective of the values of $\Delta\psi_{pm}$ and ΔL.

$$\left|\psi_{pm,\text{error}}\right| \approx \left|\frac{\Delta\psi_{pm}\,i_q^2(k_1)}{i_d^2(k_1)} + \frac{\Delta L i_q^2(k_1)}{i_d(k_1)}\right| \tag{4.9}$$

Although it is difficult to obtain $\Delta\psi_{pm}$ and ΔL in practice, $\psi_{pm,\text{error}}$ may also be measured by the following steps.

1. Measure the stator winding resistance at standstill.
2. Using the measured resistance at Step 1, start the PMSM and obtain the PM flux linkage $\psi_{pm0}(\vartheta_0)$ under $i_d = 0$ A at the room temperature ϑ_r.
3. Calculate $\hat{\psi}_{pm}$ by using the ANN estimator.
4. Determine the rotor PM flux linkage error by $\psi_{pm,\text{error}} = \hat{\psi}_{pm} - \psi_{pm0}$.

Based on Prototype Machine 1 (see Appendix B), Figures 4.4 and 4.5 present the calculated and measured variations of $\left|\psi_{pm,\text{error}}/\psi_{pm0}(\vartheta_0)\right|$ with different amplitudes of the injected i_d. As can be

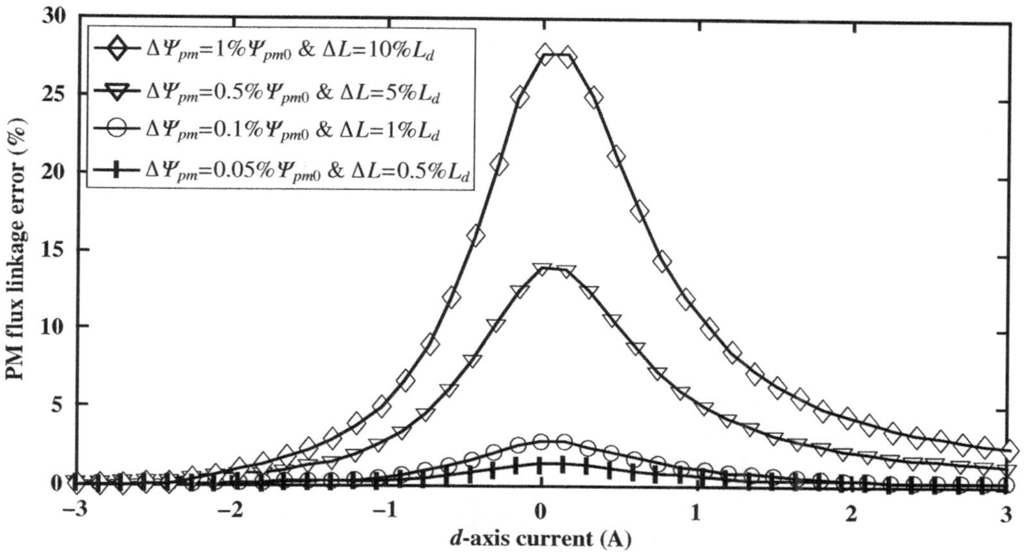

Figure 4.4 Calculated variation of $|\psi_{pm,\text{error}}/\psi_{pm0}(\vartheta_0)|$ with d-axis currents under various ΔL and $\Delta\psi_{pm}$ ($i_q = 3.46$ A, $i_{q0} = 3.44$ A, $\psi_{pm0}(\vartheta_0) = 77.6$ mWb, and $L_d = 3.24$ mH) [15], IEEE.

Figure 4.5 Measured variation of $|\psi_{pm,\text{error}}/\psi_{pm0}(\vartheta_0)|$ with d-axis currents ($i_q = 4$ A) [15], IEEE.

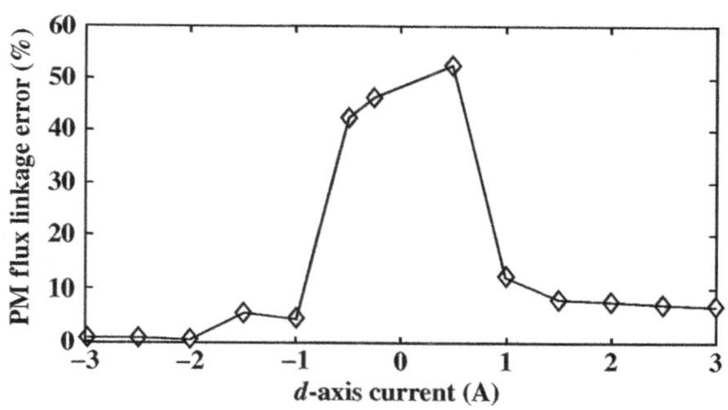

observed, the estimation error of PM flux linkage decreases with the amplitude of i_d. Meanwhile, it is preferable to employ a negative pulse of i_d in this case, since it gives rise to smaller errors due to error cancellation (note that $\Delta\psi_{pm} > 0$ and for majority PMSM machines $\Delta L > 0$). Meanwhile, it has been demonstrated in Figure 4.5 that the percentage error is kept within 5% under a negative pulse of $i_d = -2$ A. In addition, it is also important to ensure that there will be no irreversible demagnetization of PMs.

4.2.3 Experimental Results

Prototype Machine 1 is further employed for experimental verification, where a short pulse (<50 ms) of d-axis current of −2 A is injected. The sampled signals are presented in Figure 4.6, which have been filtered by the first-order Butterworth low-pass filters. It can be easily found that the injected current has a minor influence on the output torque and the rotor speed due to large mechanical inertia and fast response of the current-loop PI controller.

Figure 4.7 presents the estimated stator winding resistance, rotor PM flux linkage, and d-axis inductance. As observed, the estimated winding resistance 0.370 Ω is almost the same as the

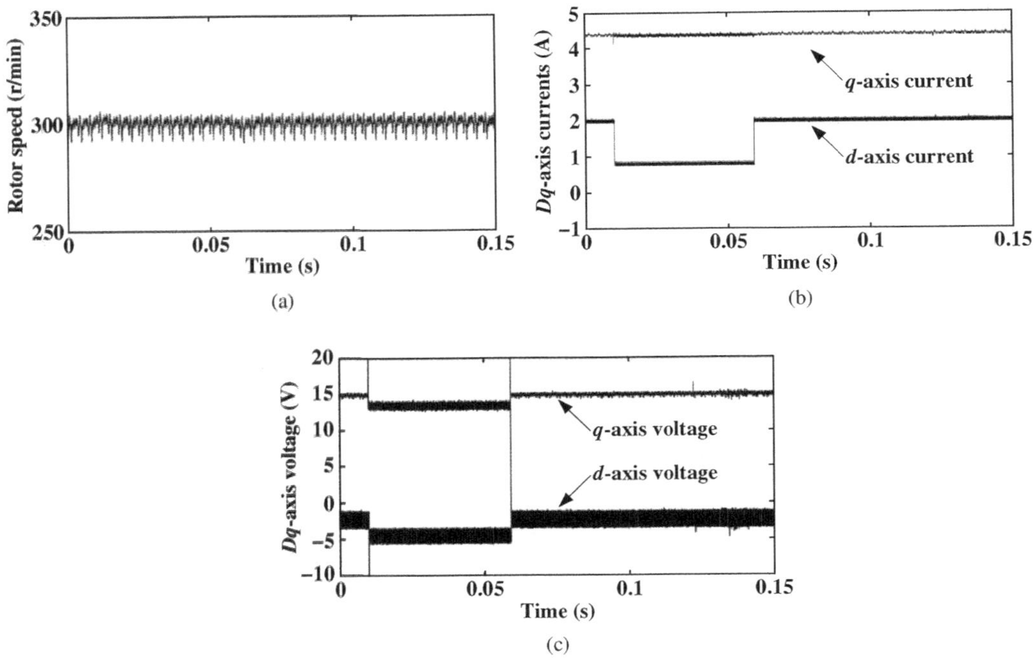

Figure 4.6 Sampled electrical angular speed and dq-axis currents and voltages ($i_d = -2$ A) [4]: (a) rotor speed, (b) Dq-axis currents, and (c) Dq-axis voltages. [4], John Wiley & Sons.

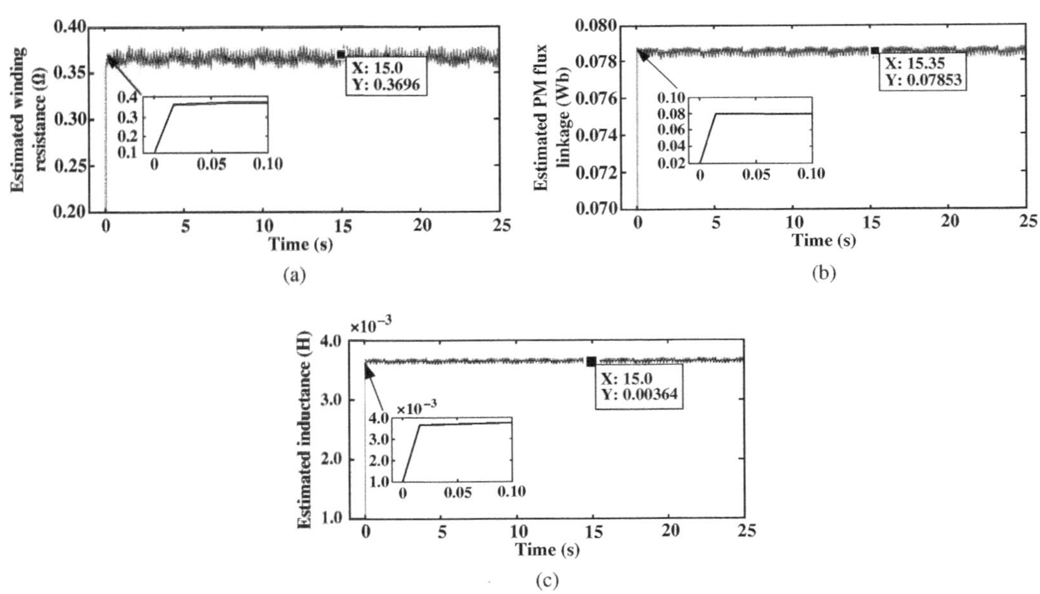

Figure 4.7 Estimated parameters based on d-axis current injection ($i_d = -2$ A) [15], IEEE: (a) stator winding resistance, (b) PM flux linkage, and (c) d-axis inductance.

measured value of 0.373 Ω, but larger than the nominal resistance of 0.320 Ω. It may be attributed to the loss due to the on-state resistance of VSI and the core-loss resistance, which will lead to an increase in the estimated winding resistance. Meanwhile, the estimated rotor PM flux linkage is 78.5 mWb, which is slightly larger than the nominal value of 77.6 mWb.

Meanwhile, the temperature tracking capability by the estimated parameters is also verified experimentally, where the test is designed as three steps.

1. Estimate the winding resistance, the rotor PM flux linkage, and the inductance at room temperature ϑ_0 ($t = 0$).
2. Keep on heating the prototype machine for 20 min and then estimate the above three parameters ($t = 20$ min).
3. Switch off the power source of the heater and PMSM. After natural cooling for 9 min, the parameters are estimated again ($t = 29$ min).

The experimental results are shown in Figure 4.8. First, the winding resistance R_s varies from 0.366 Ω ($t = 0$) to 0.464 Ω ($t = 20$ min) and then drops to 0.402 Ω ($t = 29$ min) after 9 –min of cooling down. Second, the PM flux linkage ψ_{pm0} decreases from 78.6 mWb ($t = 0$) to 76.9 mWb ($t = 20$ min) and rises to 78.2 mWb after 9 min of cooling ($t = 29$ min). In addition, as can be seen from Figure 4.8c, the estimated d-axis inductance L_d also varies with temperature, which is mainly caused by $\Delta\psi_{pm}$ in the designed estimator in Eq. (4.5). However, since the estimations of R_s, L_{q0}, and ψ_{pm0} are independent of the estimated L_d, the variation of L_d will not affect the estimation accuracy of the other three parameters.

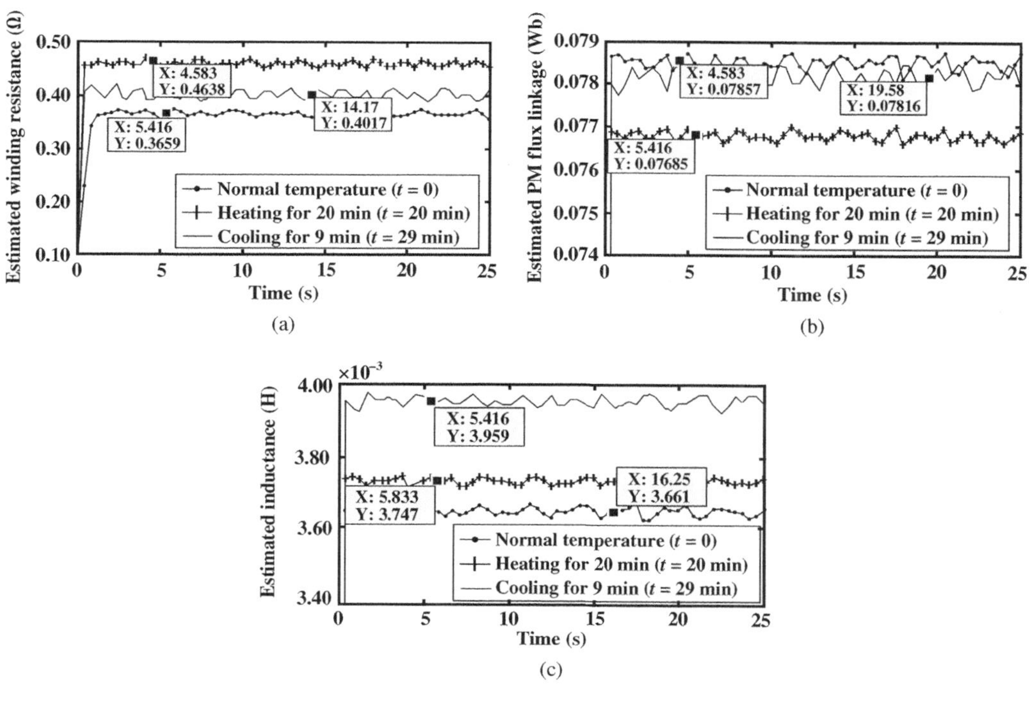

Figure 4.8 Estimated parameters for tracking temperature variations of Prototype Machine 1 (injected $i_d = -2$ A) [15], IEEE: (a) stator winding resistance, (b) PM flux linkage, and (c) d-axis inductance.

As above, it has been verified that a short pulse of d-axis current can overcome the rank-deficient issue. Hence, the winding resistance, the rotor PM flux linkage, and the dq-axis inductances can be estimated simultaneously based on two sets of dq-axis voltage equations, i.e., $i_d = 0$ and $i_d \neq 0$. Besides, the minimum amplitude of injected i_d can be determined by an error analysis to avoid the ill-conditioned problem (see Chapter 2). It has been proved that the nominal values of any parameters are not required, and the estimated values can fast converge to the actual values to reflect the temperature variations of the stator winding and the rotor PM.

4.3 Winding Resistance and Rotor Flux Linkage Estimation Based on Current Injection under Constant Torque/Speed Control

4.3.1 Designed Parameter Estimation Scheme

In this section, the d-axis current injection method is further applied to the machine under constant load torque T_e to estimate the stator winding resistance and the rotor PM flux linkage, where the rotor speed is kept constant under $i_d = 0$ and $i_d \neq 0$. In addition, Prototype Machine 1 is employed for following analysis:

$$\begin{aligned} T_e(k_0) &= T_e(k_1) \\ 1.5p\psi_{pm0}\, i_{q0}(k_0) &= 1.5p\left[\psi_{pm}\, i_q(k_1) + (L_d - L_q)\, i_d(k_1) i_q(k_1)\right] \end{aligned} \tag{4.10}$$

where p is the number of pole pairs. The subscripts "0" and "1" denote the electrical quantities/sampling indexes under $i_d = 0$ and $i_d \neq 0$, respectively.

By solving Eq. (4.10), the rotor PM flux linkage after the current injection can be derived as follows:

$$\psi_{pm} = \psi_{pm0}\frac{i_{q0}(k_0)}{i_q(k_1)} + (L_q - L_d)\, i_d(k_1)\, i_q(k_1) \tag{4.11}$$

Substituting Eq. (4.11) into Eq. (4.1d) and then multiplying it by $i_q(k_1)$ on both sides, it becomes as follows:

$$\begin{aligned} u_q(k_1) &= R_s i_q(k_1) + \left[\psi_{pm0}\frac{i_{q0}(k_0)}{i_q(k_1)} + L_q\, i_d(k_1)\right]\omega_r(k_1) \\ u_q(k_1) i_q(k_1) &= R_s i_q^2(k_1) + \left[\psi_{pm0}\, i_{q0}(k_0) + L_q\, i_d(k_1) i_q(k_1)\right]\omega_r(k_1) \end{aligned} \tag{4.12}$$

Afterwards, multiplying Eq. (4.1c) by $i_d(k_1)$, the equation can be rewritten as follows:

$$u_d(k_1)\, i_d(k_1) = R_s i_d^2(k_1) - L_q\, \omega_r(k_1)\, i_q(k_1)\, i_d(k_1) \tag{4.13}$$

Finally, combining Eqs. (4.12) + (4.13) and (4.1b), the employed estimation model becomes as follows:

$$u_q(k_1)\, i_q(k_1) + u_d(k_1)\, i_d(k_1) = R_s\left[i_d^2(k_1) + i_q^2(k_1)\right] + \psi_{pm0}\, i_{q0}(k_0)\, \omega_r(k_1) \tag{4.14a}$$

$$u_{q0}(k_0) = R_s\, i_{q0}(k_0) + \psi_{pm0}\, \omega_r(k_0) \tag{4.14b}$$

Hence, the rank of Eq. (4.14) is two, i.e., the winding resistance R_s and the rotor PM flux linkage ψ_{pm0} can be estimated simultaneously. Meanwhile, since the inductance terms are cancelled in Eq. (4.14), the variations in dq-axis inductances and rotor PM flux linkage due to $i_d \neq 0$ will not affect the estimated results. The measuring and recording processes under $i_d = 0$ and $i_d \neq 0$ and the

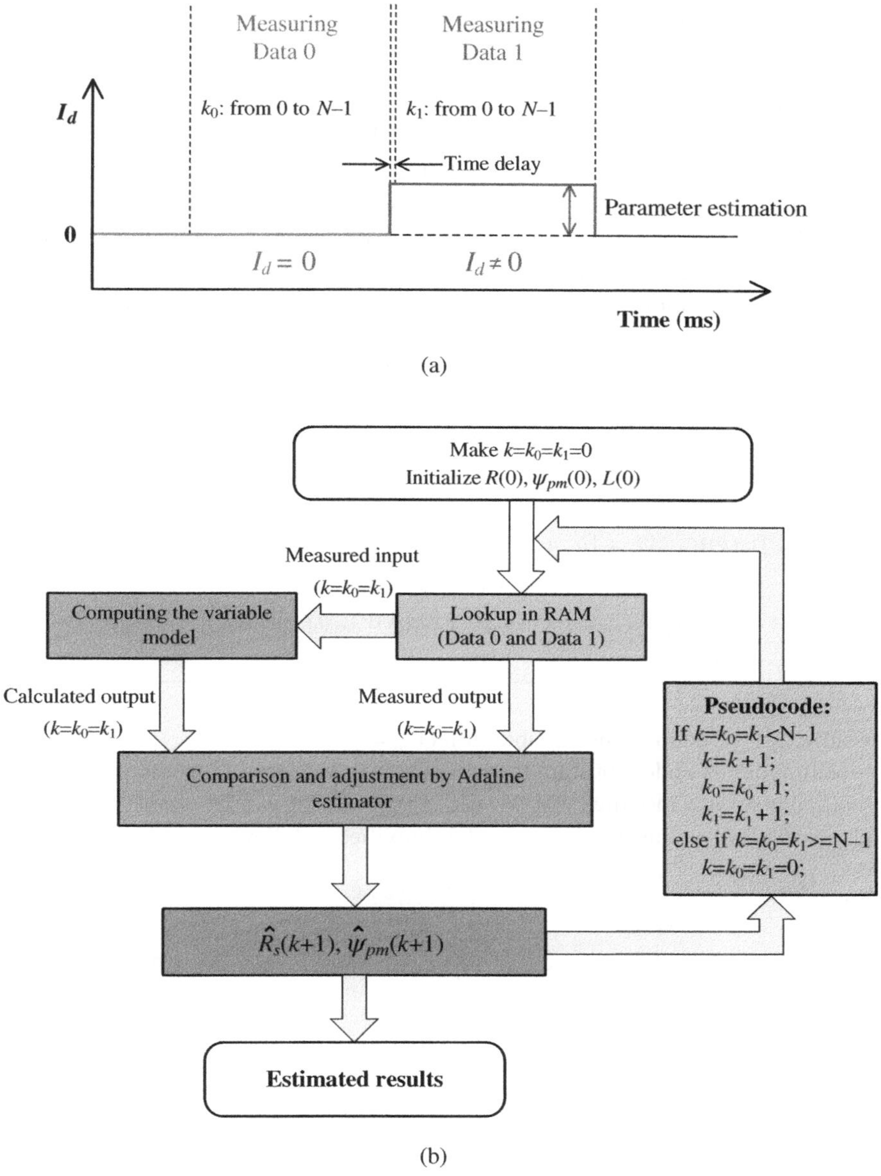

Figure 4.9 Estimation of R_s and ψ_{pm} based on measured inputs and outputs under constant torque/speed control: (a) measuring and recording data and (b) iterative calculation of parameter estimation.

iterative computation process are shown in Figures 4.9a,b, respectively. In addition, the design of ANN estimators can be found in Chapter 10.

4.3.2 Error Analysis and Experimental Validation

Although the designed reference model (Eq. 4.14) is full rank, the estimation still easily suffers from the ill-conditioned problem caused by the VSI non-linearity and the non-ideal voltage measurement. Thus, an error analysis is required to minimize the error.

First, the steady-state equations of a PMSM including the VSI non-linearity can be expressed as follows:

$$\begin{bmatrix} u_d(k) \\ u_q(k) \end{bmatrix} = \begin{bmatrix} u_d^*(k) + Dd(k)V_{\text{dead}} \\ u_q^*(k) + Dq(k)V_{\text{dead}} \end{bmatrix} = \begin{bmatrix} -L_q i_q(k) \\ L_d i_d(k) + \psi_{pm} \end{bmatrix} \omega_r(k) + R_s \begin{bmatrix} i_d(k) \\ i_q(k) \end{bmatrix} \tag{4.15}$$

Obviously, the voltage terms of $Dd(k)V_{\text{dead}}$ and $Dq(k)V_{\text{dead}}$ introduce voltage distortions in the dq-axis, which may give rise to the parameter misestimation. Therefore, the dq-axis reference voltages from the PI regulators should be compensated prior to the estimation, as presented in Section 2.2. However, the compensation cannot be 100% accurate, and thus, the estimation error due to non-ideal compensation should be analysed and minimized by a software approach, e.g., an error analysis.

By substituting Eq. (4.15) into Eq. (4.16), the estimated stator winding resistance R_s, including the real winding resistance R_a and the estimation error R_{err} caused by the distorted voltages of $Dd(k)V_{\text{dead}}$ and $Dq(k)V_{\text{dead}}$, can be expressed in Eq. (4.17):

$$u_q(k_1)\, i_q(k_1) + u_d(k_1)\, i_d(k_1) - u_{q0}(k_0)\, i_{q0}(k_0) = R\left(i_q^2(k_1) + i_d^2(k_1) - i_{q0}^2(k_0)\right) \tag{4.16}$$

$$\begin{aligned} R_s &= \frac{\left[u_q + Dq(k)V_{\text{dead}}\right] i_q + \left[u_d + Dd(k)V_{\text{dead}}\right] i_d - \left(u_{q0} + Dq_0(k)V_{\text{dead},0}\right) i_{q0}}{i_q^2 + i_d^2 - i_{q0}^2} \\ &= \underbrace{\frac{u_q i_q + u_d i_d - u_{q0} i_{q0}}{i_q^2 + i_d^2 - i_{q0}^2}}_{R_a} + \underbrace{\frac{Dq(k)V_{\text{dead}} i_q + Dd(k)V_{\text{dead}} i_d - Dd_0(k)V_{\text{dead},0}\, i_{q0}}{i_q^2 + i_d^2 - i_{q0}^2}}_{R_{\text{err}}} \end{aligned} \tag{4.17}$$

where the subscript "0" denotes the variables under $i_d = 0$.

Since the estimation is under constant torque control and the reluctance torque is negligible for a non-salient-pole surface-mounted PMSM, the q-axis current will be almost constant ($i_q = i_{q0}$) during the injection of i_d. Thus, Eq. (4.17) can be simplified as follows:

$$R_{\text{err}} \approx \frac{\left[Dq(k)V_{\text{dead}} - Dd_0(k)V_{\text{dead},0}\right] i_q}{i_d^2} + \frac{Dd(k)V_{\text{dead}}}{i_d} \tag{4.18}$$

Since the average value of $Dd_0(k)V_{\text{dead}0}$ is zero, Eq. (4.18) can be further simplified to be as follows:

$$R_{\text{err}} \approx \frac{Dq(k)V_{\text{dead}} i_q}{i_d^2} + \frac{Dd(k)V_{\text{dead}}}{i_d} \tag{4.19}$$

From Eq. (4.19), it is evident that R_{err} can be minimized by increasing i_d, $Dd(k)V_{dead}$, or $Dq(k)V_{dead}$. However, since $Dd(k)V_{\text{dead}}$ or $Dq(k)V_{\text{dead}}$ is more difficult to be minimized to zero in practice due to the non-ideal compensation, the amplitude of the injected i_d needs to be optimized for the minimization of estimation error R_{err}.

Figure 4.10 shows the error analysis results of the estimated winding resistance under different amplitudes of injected d-axis current i_d, where the nominal phase resistance and the terminal wire resistance at 25°C are 0.330 and 0.043 Ω, respectively. As can be seen, the estimation error decreases with the amplitude of i_d and becomes negligible when $i_d \geq 2$ A.

Therefore, the following experimental results are obtained based on the injection of $i_d = 2.5$ A with a pulse width of 50 ms, as shown in Figure 4.11. It is obvious that the influence of injected i_d on the speed is negligible due to the large moment of system inertia.

Figure 4.12 shows the estimated stator winding resistance and rotor PM flux linkage at normal condition, where the estimated parameter values are close to the nominal values of 0.373 Ω and 0.0776 Wb, respectively.

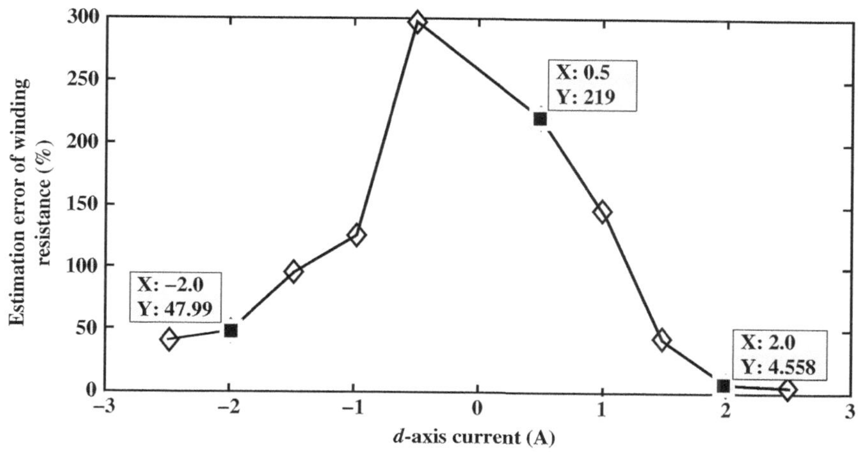

Figure 4.10 Measured variation of $|R_{err}/R_a|$ with different amplitudes of injected i_d [16], IEEE.

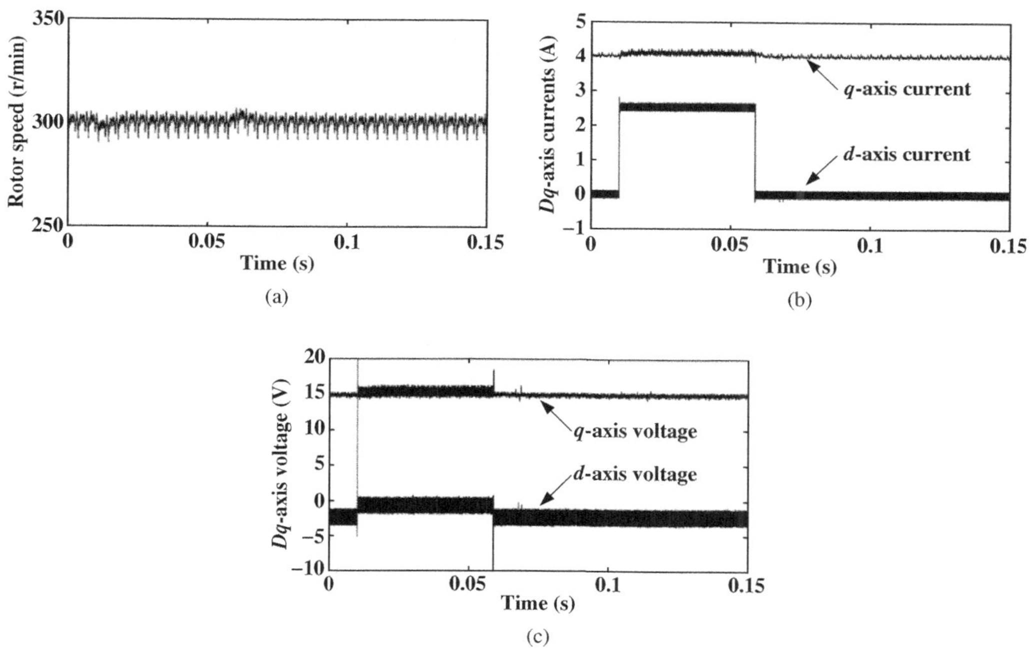

Figure 4.11 Measured electrical angular speed and dq-axis currents and voltages [4]: (a) rotor speed, (b) Dq-axis currents, and (c) Dq-axis voltages.

Furthermore, a heater is used to heat the PMSM for 20 min (from $t = 0$ to $t = 20$ min), whose temperature is measured by using an infrared thermometer. After 20 min of heating, the housing temperature is measured as 80°C. Afterwards, the PMSM is cooled down via natural convection (from $t = 20$ min to $t = 60$ min). The stator winding resistance and the rotor PM linkage are estimated at $t = 0$, 20, 35, and 60 min, as shown in Figure 4.13. It is evident that the estimated winding resistance increases from 0.388 to 0.473 Ω after 20 min of heating, and the winding

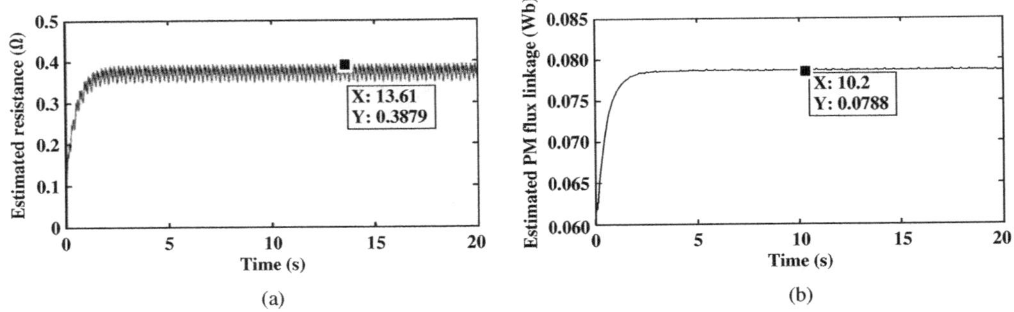

Figure 4.12 Estimated (a) stator winding resistance and (b) rotor PM flux linkage by using injection of $i_d = 2.5$ A under constant torque/speed control [16], IEEE.

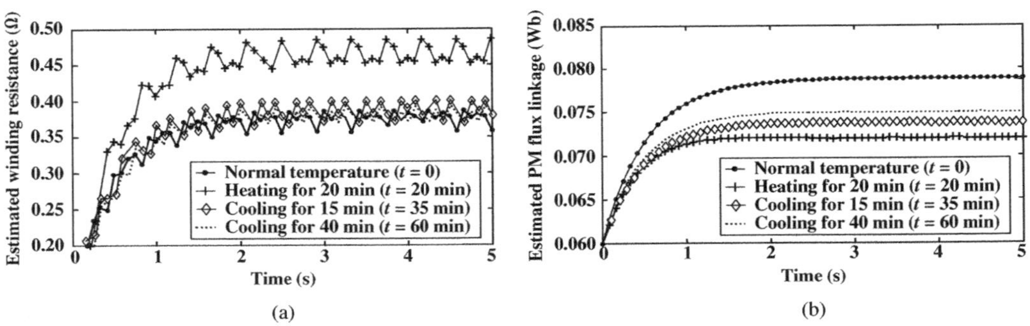

Figure 4.13 Estimated parameters for tracking temperature variations of prototype machine by using injection of $i_d = 2.5$ A under constant torque/speed control [16], IEEE: (a) estimated stator winding resistance and (b) estimated rotor PM flux linkage.

resistance value decreases to its normal value soon after the heating is removed. Besides, different from the variation in winding resistance, the value of rotor PM flux linkage decreases from 78.8 to 72.0 mWb after 20 min of heating. In addition, due to the poor heat dissipation of the rotor, the estimated rotor PM flux linkage increases back to its normal value and is much slower than the speed of winding resistance after 20 min.

4.4 Summary

This chapter has described two online parameter estimation methods based on the *d*-axis current pulse injections for PMSM, which can overcome the rank-deficient issue by introducing another set of PMSM equations when $i_d \neq 0$. More specifically, the first method can simultaneously estimate four electrical parameters for PMSMs, i.e., the stator winding resistance, the *dq*-axis stator inductance, and the rotor PM flux linkage. In comparison, the second method focuses on the estimation of stator winding resistance and the rotor PM flux linkage for thermal condition monitoring under constant torque/speed control. In addition, the experimental results have demonstrated that both methods exhibit good estimation accuracy and are capable of tracking temperature variations of PMSMs, while the injected current pulses have relatively minor influences on the system performance.

Besides, the error analyses have also been performed in the respective two methods to address the potential ill-conditioned problem, which may be caused by the VSI non-linearity, the non-ideal voltage measurement, and the parameter variation resulting from the current injection. It has been shown that the ill-conditioned problem can be minimized by the injection of a relatively large amplitude of $i_d \neq 0$. However, it is not preferred in real applications to inject a large $i_d \neq 0$ considering its negative impacts on the system stability and risk of irreversible PM demagnetization. Thus, the position-offset-based parameter estimation will be introduced in the following chapters, which do not rely on the current injection to establish a full-rank estimation system.

References

1 B. N. Mobarakeh, F. Meibody-Tabar, and F. M. Sargos, "Mechanical sensorless control of PMSM with online estimation of stator resistance," *IEEE Trans. Ind. Appl.*, vol. 40, no. 2, pp. 457–471, Mar./Apr. 2004.

2 L. Liu and D. A. Cartes, "Synchronisation based adaptive parameter identification for permanent magnet synchronous motors," *IET Control Theory Appl.*, vol. 1, no. 4, pp. 1015–1022, Jul. 2007.

3 M. Rashed, P. Macconnell, A. Stronach, and P. Acarnley, "Sensorless indirect-rotor-field-orientation speed control of a permanent-magnet synchronous motor with stator-resistance estimation," *IEEE Trans. Ind. Electron.*, vol. 54, no. 3, pp. 1664–1675, Jun. 2007.

4 S. Morimoto, M. Sanada, and Y. Yakeda, "Mechanical sensorless drives of IPMSM with online parameter identification," *IEEE Trans. Ind. Appl.*, vol. 42, no. 5, pp. 1241–1248, Sep./Oct. 2006.

5 A. Piippo, M. Hinkkanen, and J. Luomi, "Adaptation of motor parameters in sensorless PMSM drives," *IEEE Trans. Ind. Appl.*, vol. 45, no. 1, pp. 203–212, Jan./Feb. 2009.

6 K. Liu and Z.Q. Zhu, "Online estimation of the rotor flux linkage and voltage-source inverter nonlinearity in permanent magnet synchronous machine drives," *IEEE Trans. Power Electron.*, vol. 29, no. 1, pp. 418–427, Jan. 2014.

7 B. Stumberger, B. Kreca, and B. Hribernik, "Determination of parameters of synchronous motor with permanent magnets from measurement of load conditions," *IEEE Trans. Energy Convers.*, vol. 14, no. 4, pp. 1413–1416, Dec. 1999.

8 M. Uddin and M. Chy, "Online parameter-estimation-based speed control of PM AC motor drive in flux-weakening region," *IEEE Trans. Ind. Appl.*, vol. 44, no. 5, pp. 1486–1494, Sep./Oct. 2010.

9 S. Farag, R. Bartheld, and T. Habetler, "An integrated on-line motor protection system," *IEEE Trans. Ind. Appl.*, vol. 2, no. 2, pp. 21–6, Mar./Apr. 1996.

10 Z. Q. Zhu, D. Liang, and K. Liu, "Online parameter estimation for permanent magnet synchronous machines: an overview," *IEEE Access*, vol. 9, pp. 59059–59084, 2021.

11 E. Ifeachor and B. Jervis, *Digital Signal Processing*, 2th ed. Harlow: Addison Wesley, 2002.

12 Z. Q. Zhu, X. Zhu, and P. D. Sun, "Estimation of winding resistance and PM flux-linkage in brushless AC machines by reduced-order extended Kalman filter," in *Proc. IEEE Int. Conf. Netw., Sens. Control*, 2007, pp. 740–745.

13 K. W. Lee, D. H. Jung, and I. J. Ha, "An online identification method for both stator resistance and back-EMF coefficient of PMSMs without rotational transducers," *IEEE Trans. Ind. Electron.*, vol. 51, no. 2, pp. 507–510, Apr. 2004.

14 S. Wilson, P. Stewart, and B. Taylor, "Methods of resistance estimation in permanent magnet synchronous motors for real-time thermal management," *IEEE Trans. Energy Convers.*, vol. 25, no. 3, pp. 698–707, Sept. 2010.

15 K. Liu, Q. Zhang, J. Chen, Z.Q. Zhu, and J. Zhang, "Online multiparameter estimation of nonsalient-pole PM synchronous machines with temperature variation tracking," *IEEE Trans. Ind. Electron.*, vol. 58, no. 5, pp. 1776–1788, May 2011.

16 K. Liu, Z.Q. Zhu, and D. Stone, "Parameter estimation for condition monitoring of PMSM stator winding and rotor permanent magnets," *IEEE Trans. Ind. Electron.*, vol. 60, no. 12, pp. 5902–5913, Apr. 2013.

17 K. Liu, "Novel techniques for parameter estimation of permanent magnet synchronous machines," PhD dissertation, Uni. Sheffield, U.K., 2013.

18 Z. Liu, X. Li, L. Wu, S. Zhou, and K. Liu, "GPU-accelerated parallel coevolutionary algorithm for parameters identification and temperature monitoring in permanent magnet synchronous machines," *IEEE Trans. Ind. Informat.*, vol. 11, no. 5, pp. 1220–1230, Oct. 2015.

19 G. Feng, C. Lai, K. Mukherjee, and N. C. Kar, "Current injection-based online parameter and VSI nonlinearity estimation for PMSM drives using current and voltage DC components," *IEEE Trans. Transport. Electrific.*, vol. 2, no. 2, pp. 119–128, Jun. 2016.

5

Online Parameter Estimation Based on Position Offset Injection

5.1 Introduction

To solve the rank-deficient issue, Chapter 4 has described several current injection methods, which can estimate multi-parameters for PMSMs without relying on additional measurement instruments. However, it has been highlighted that during the estimation, the amplitude of the injected current should be as large as possible to achieve an adequate ratio of signal to noise and overcome the ill-condition problem. Nevertheless, the large current injection may cause significant disturbance and instability risk in the drive system. Meanwhile, the estimated parameters, e.g., VSI non-linearity, *dq*-axis inductances, and rotor PM flux linkage, are usually sensitive to the injected current. Consequently, there is a trade-off between the amplitude of the injected current and the variation of parameters.

From another new perspective, the rotor position is also one of the fundamental variables in a PMSM and essential for the control system [1]. The rotor position can be measured by a resolver/encoder, or estimated by various sensorless strategies [2]. Thus, if small rotor position offsets are added in the control system intentionally, the resultant variations of electrical quantities, e.g., *dq*-axis voltages, can reflect the inherent characteristics of the machine, i.e., the rotor PM flux linkage. Furthermore, when the *dq*-axis currents can remain constant during the estimation, the estimated rotor PM flux linkage is irrespective of the influences of VSI non-linearity and variations of other machine parameters. Besides, the addition of small position offsets has negligible influence on the control performance and the output torque capability.

Following this concept, this chapter introduces the principle of the position offset-based parameter estimation (POPE) method [3–8], where two small positive and negative position offsets are added to estimate the rotor PM flux linkage. Afterwards, based on the estimated rotor PM flux linkage ψ_{pm} by the POPE, other electrical parameters, i.e., stator winding resistance R_s, *dq*-axis inductances L_d, L_q, and/or VSI non-linearity V_{dead} can also be estimated subsequently. More specifically, the phasor analysis of rotor position offset in the *dq*-axis equation of PMSM is first analysed in Section 5.2, and four POPE methods under different control strategies will be introduced from Sections 5.3 to 5.6, i.e., $i_d = 0$ under constant torque/speed and variable speed controls and $i_d \neq 0$ under constant/variable speed control. Their features are summarized in Table 5.1.

It is worth mentioning that the *d*-axis inductance term L_d is inactive when $i_d = 0$, and thus, its estimation relies on introducing an additional excitation, e.g., a position offset or an injected *d*-axis current, which will be described in Sections 5.3 and 5.4. On the other hand, when $i_d \neq 0$, the described estimation method in Sections 5.5 and 5.6 could fully eliminate the influence of VSI

Parameter Estimation of Permanent Magnet Synchronous Machines, First Edition. Zi Qiang Zhu, Kan Liu, and Dawei Liang.

Table 5.1 Features of POPE methods under different control strategies.

	Section 5.3	Section 5.4	Section 5.5	Section 5.6
Estimated parameters	$i_d = 0$		$i_d \neq 0$	$i_d = 0$ and $i_d \neq 0$
	Constant torque/speed	Variable speed	Constant speed	Constant and variable speeds
R_s	q-axis equation without position offset (VSI non-linearity needs to be compensated)		*(not involved)*	*(not involved)*
ψ_{pm}	Addition of positive and negative position offsets			
L_q	d-axis equation without position offset		Required two different rotor speeds	Required positive, negative, and zero position offsets
L_d	Adding position offset *or* Injecting d-axis current	*(not involved)*		
V_{dead}	Priorly compensated	Online estimated	Eliminated	Eliminated

non-linearity, which can be used under constant and constant/variable speed control, respectively, and the uncertainties of R_s and distorted voltage due to VSI non-linearity can be eliminated. In contrast, the estimation equation in Section 5.5 relies on the dq-axis voltage equations of positive and negative position injections, while Section 5.6 introduced the machine state at zero position offset to build the full-rank model. Therefore, the POPE method described in Section 5.6 can be generally applied to both constant and variable speed control and is independent of whether the d-axis current is zero or not. Moreover, in terms of the amplitude selection of injected position offset, Section 5.6 analyses the upper and lower limits of position offset injection based on stability and signal-to-noise ratio (SNR) analysis, which can be applied to the machine in different voltage classes.

The effectiveness of all estimation methods is demonstrated by FEA and experimental results based on three prototype machines.

5.2 Phasor Analysis of Rotor Position Offset in PMSMs

As aforementioned, the POPE methods described in this chapter introduce the position offsets to estimate rotor PM flux linkages under both $i_d = 0$ and $i_d \neq 0$. Therefore, the influence of a position offset $\Delta\theta$ on the dq-axis PMSM equation needs to be analysed first. The phasor diagram with/without considering a position offset addition is shown in Figure 5.1, where d_1q_1-axes are the new reference frame caused by $\Delta\theta$. As can be seen, the addition of position offsets will result in the same change in the current vector and also cause a change in back-EMF terms, i.e., $\psi_{pm}\omega_r\sin(\Delta\theta)$ and $\psi_{pm}\omega_r\cos(\Delta\theta)$ in the d_1q_1-axes. Thus, the resulting voltage variations in the d_1q_1-axes can be used to indicate the rotor PM flux linkage value. In addition, a small position offset has negligible influences on the parameter variations.

Considering the distorted voltages V_{dead} caused by the VSI non-linearity, the dq- and d_1q_1-voltage equations are presented in Eqs. (5.1) and (5.2), respectively. The relationship of currents in the dq- and d_1q_1-axes is shown in Eq. (5.3).

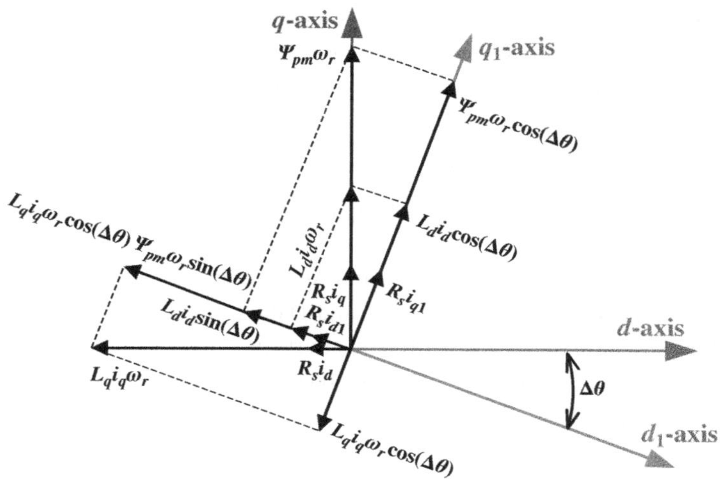

Figure 5.1 Phasor diagram of PMSM with and without position offset $\Delta\theta$.

1. *Voltage equations in the dq-axes:*

$$u_d^* + DdV_{dead} = R_s i_d - \omega_r L_q i_q \tag{5.1a}$$

$$u_q^* + DqV_{dead} = R_s i_q + \omega_r (L_d i_d + \psi_{pm}) \tag{5.1b}$$

2. *Voltage equations in the d_1q_1-axes:*

$$u_{d1} = u_{d1}^* + DdV_{dead1} = R_s i_{d1} - \omega_{r1}[\psi_{pm}\sin(\Delta\theta) + L_d i_d \sin(\Delta\theta) + L_q i_q \cos(\Delta\theta)] \tag{5.2a}$$

$$u_{q1} = u_{q1}^* + DqV_{dead1} = R_s i_{q1} + \omega_{r1}[\psi_{pm}\cos(\Delta\theta) + L_d i_d \cos(\Delta\theta) - L_q i_q \sin(\Delta\theta)] \tag{5.2b}$$

3. *Correlation of currents in the dq- and d_1q_1-axes:*

$$\begin{cases} i_d = i_{q1}\sin(\Delta\theta) + i_{d1}\cos(\Delta\theta) \\ i_q = i_{q1}\cos(\Delta\theta) - i_{d1}\sin(\Delta\theta) \end{cases} \tag{5.3}$$

where u_d^*, u_q^*, i_d, and i_q are the dq-axis reference voltages and currents. ω_r is the electrical angular speed. R_s L_d, L_q, and ψ_{pm} are the stator winding resistance, the dq-axis inductances, and the rotor PM flux linkage, respectively. The subscript "1" denotes the variables with the addition of $\Delta\theta$ in the d_1q_1-axes.

5.3 Position Offset-based Estimation with $i_d = 0$ Under Constant Torque/Speed Control

In this section, the POPE method with small positive and negative position offsets will be described with $i_d = 0$ under constant torque/speed control. First, the rotor PM flux linkage is estimated by utilizing the resulting dq-axis voltage variations caused by the position offsets. Afterwards, the stator winding resistance and the dq-axis inductances can be estimated subsequently with the aid of the priorly determined rotor PM flux linkage and/or constant torque equation, where the VSI non-linearity has been compensated in a prior. Meanwhile, this section will also describe a practical method to compensate for the non-ideal rotor position measurement, which has a significant effect on the estimation accuracy of dq-axis inductances. In addition, the experimental results based on a non-salient-pole and a salient-pole PMSMs will also be presented for illustration.

5.3.1 Designed Estimation Method

Under constant torque/speed control, the principle of POPE and its associated FOC system is shown in Figure 5.2.

As can be seen, two position offsets, i.e., $\Delta\theta_p$ and $\Delta\theta_n$, are introduced subsequently. The corresponding machine states are recorded as Data 1 and Data 2, including the dq-axis currents and voltages, as well as the rotor speed. Meanwhile, a short-time delay is added between two sets of measurements to ensure the settling of step transient correlated to the addition of $\Delta\theta_p$ and $\Delta\theta_n$.

First, the dq-axis voltage equations with $i_d = 0$ can be written as follows:

$$u_d^* + DdV_{\text{dead}} = -\omega_r L_q i_q \tag{5.4a}$$

$$u_q^* + DqV_{\text{dead}} = R_s i_q + \omega_r \psi_{pm}(L_d i_d + \psi_{pm}) \tag{5.4b}$$

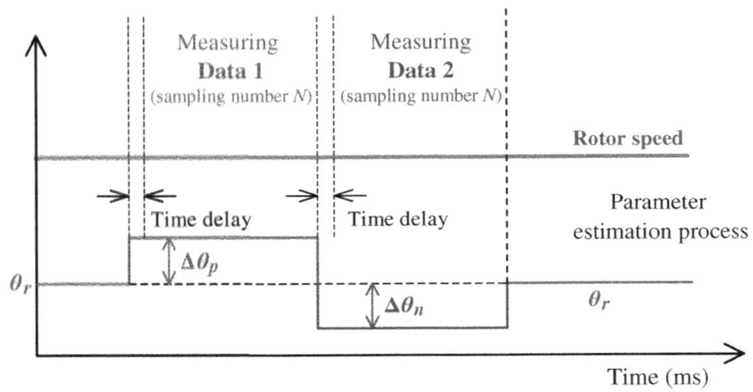

(a)

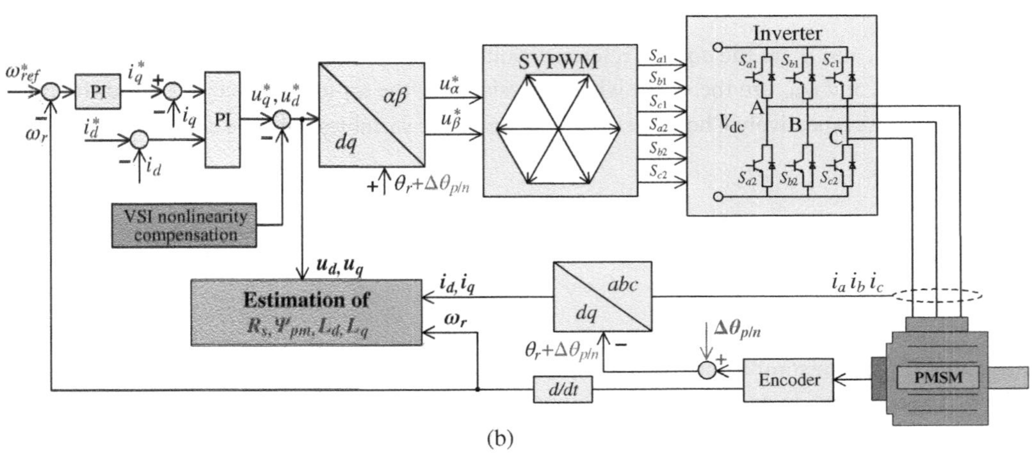

(b)

Figure 5.2 Principle of POPE with associated FOC system under constant torque/speed control: (a) principle and measurement process of POPE and (b) POPE under FOC system.

When there is a position offset $\Delta\theta$ in the drive system, whether it is added intentionally or caused by the non-ideal position measurement [9], the time-discrete d-axis voltage equations with $i_{d1} = 0$ can be simplified as follows:

$$u_{d1}^*(k) + Dd(k)V_{\text{dead1}} = -\omega_{r1}(k)\left[\psi_{pm}\sin(\Delta\theta) + L_q i_{q1}(k)\cos(\Delta\theta)\right] \tag{5.5}$$

where "k" is the index of the discrete sampling instant. The subscript "1" denotes the electrical quantities in the d_1q_1-axes with a position offset $\Delta\theta$.

When there are positive and negative position offsets with the same amplitudes, i.e., $\Delta\theta_p$ and $\Delta\theta_n$, the corresponding d-axis voltage equations under $i_d = 0$ can be expressed as follows:

$$u_{d1}^*(k) + Dd(k)V_{\text{dead}} = -\omega_r(k)\left[L_q i_q(k) + \psi_{pm}\sin(\Delta\theta + \Delta\theta_p)\right] \tag{5.6a}$$

$$u_{d2}^*(k) + Dd(k)V_{\text{dead}} = -\omega_r(k)\left[L_q i_q(k) + \psi_{pm}\sin(\Delta\theta + \Delta\theta_n)\right] \tag{5.6b}$$

where u_{d1}^* and u_{d2}^* are the reference d-axis voltages recorded after the PI regulator, which correspond to the additions of $\Delta\theta_p$ and θ_n, respectively.

5.3.1.1 Estimation of Rotor PM Flux Linkage ψ_{pm} and Stator Winding Resistance R_s

According to the reference estimation Eq. (5.6), the rotor PM flux linkage ψ_{pm} and the stator winding resistance R_s can be estimated subsequently, and the POPE-based iterative computation process is shown in Figure 5.3, where k_1 and k_2 are the indices of recorded Data 1 and Data 2, respectively, and $k_1 = k_2 = k$ in this study for convenience of calculations. Furthermore, ψ_{pm} and R_s can be estimated by minimizing the measured and calculated output quantities, e.g., ANN estimators used in this section.

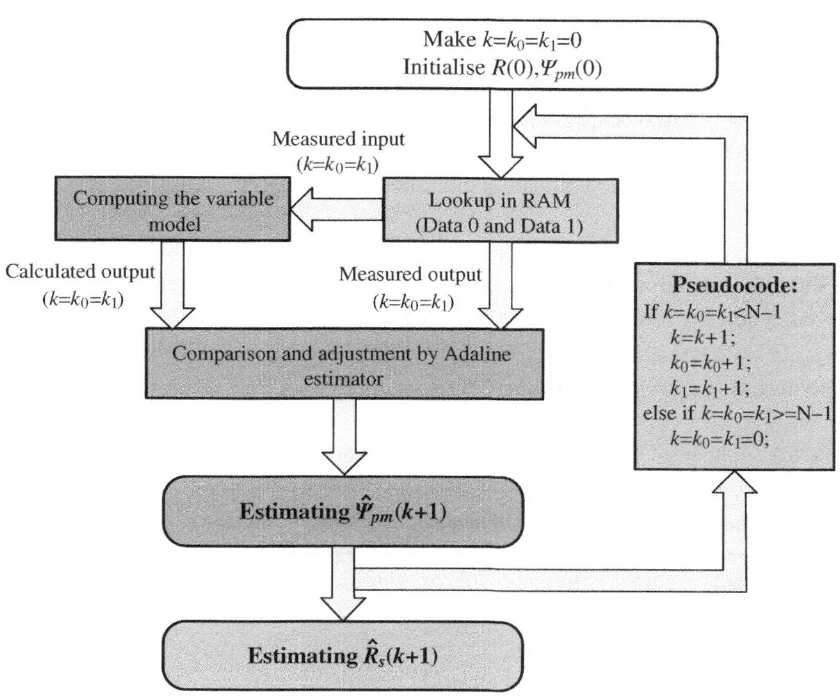

Figure 5.3 Iterative computation process of POPE for estimating $\hat{\psi}_{pm}$ and $\hat{R}_s$ [3], IEEE.

First, the rotor PM flux linkage $\hat{\psi}_{pm}$ can be calculated as follows:

$$\begin{aligned} u_{d1}^*(k) - u_{d2}^*(k) &= \psi_{pm}\omega_r(k)\left[\sin(\Delta\theta + \Delta\theta_p) - \sin(\Delta\theta + \Delta\theta_n)\right] \\ &= 2\psi_{pm}\omega_r(k)\cos(\Delta\theta)\sin(\Delta\theta_n) \end{aligned} \tag{5.7a}$$

$$\hat{\psi}_{pm} = \frac{u_{d1}^*(k) - u_{d2}^*(k)}{2\omega_r(k)\sin(\Delta\theta_n)\cos(\Delta\theta)} \tag{5.7b}$$

Obviously, the voltage term of $Dd(k)V_{\text{dead}}$ is cancelled in Eq. (5.7), since $Dd(k)$ only depends on the dq-axis currents, and its average value is zero (see Chapter 2). Meanwhile, $\cos(\Delta\theta) \approx 1$ for a small addition of position offset.

Second, in terms of the estimation of the stator winding resistance $\hat{R}_s$, it can be obtained from the q-axis voltage equation and expressed as follows:

$$\hat{R}_s = \frac{u_q^*(k) + Dq(k)V_{\text{dead}} - \frac{u_{d1}^*(k) - u_{d2}^*(k)}{2\sin(\Delta\theta_n)}}{i_q(k)} \tag{5.8}$$

As can be found in Eqs. (5.7b) and (5.8), the term $\sin(\Delta\theta_n)$ should be small enough to prevent the detrimental influence of position offsets on the control performance, i.e., the torque reduction should be smaller than 1% in this study. Meanwhile, to ensure accurate estimation performance under small position offsets, the main challenge is accurately recording the d-axis voltages, i.e., u_{d1}^* and u_{d2}^*, resulting from the addition of position offsets. However, they may vary with the cycle time of the control system and the precision of the analog-to-digital conversion. Thus, the estimation accuracies in Eqs. (5.7) and (5.8) can be guaranteed by a high ratio of signal to noise. In practice, it is suggested that the lower limit of the d-axis voltage change, i.e., $u_{d1}^*(k) - u_{d2}^*(k)$, should be set equal to or greater than 0.2 V.

$$u_{d1}^*(k) - u_{d2}^*(k) = 2\psi_{pm}\omega_r(k)\cos(\Delta\theta)\sin(\Delta\theta_n) \geq 0.2\ \text{V} \tag{5.9}$$

In addition, to analyse the influence of the position offsets on the dq-axis voltages, a unit electrical angle θ_u is defined in Eq. (5.10), and the inherent position error (or offset) $\Delta\theta$ due to non-ideal rotor position measurement can be expressed as a multiple of θ_u, i.e., $\Delta\theta = n\theta_u$. By way of example, for Prototype Machine 1 ($2p = 10$ and $M = 5000$), its unit electrical angle θ_u is 0.36°, and the influence of position offsets on the dq-axis voltages is tabulated in Table 5.2:

$$\theta_u = \frac{2\pi p}{M} \tag{5.10}$$

where p is the pole pair number and M is the line number of the incremental rotary encoder.

As observed, when the rotor speed is 300 r/min and the addition of $|\Delta\theta_n| = 20\theta_u$ (7.2°), the value of $u_{d1}^*(k) - u_{d2}^*(k) = 2 \times 0.1253\omega_r(k)\psi_{\text{pm}} \approx 2.76\ \text{V} > 0.2\ \text{V}$. In this case, the influence of the position offset on the output torque (or q-axis current) is calculated by Eq. (5.11), which can be neglected

Table 5.2 Influence of position offsets on dq-axis back-EMFs for Prototype Machine 1.

Position offset: $\Delta\theta = n\theta_u$	$\sin(\Delta\theta)$ in d-axis back EMF	$\cos(\Delta\theta)$ in q-axis back EMF
$n = 2$	0.0126	0.9999
$n = 5$	0.0314	0.9995
$n = 10$	0.0628	0.9980
$n = 20$	0.1253	0.9921

and immediately compensated by the PI regulator. Hence, the upper limit of the value of the position offset can be set to $|\Delta\theta_p| = |\Delta\theta_n| \leq 20\,\theta_u\,(7.2°)$.

$$\frac{1}{\cos(20\,\theta_u)} - 1 \approx 0.8\% < 1\% \tag{5.11}$$

In addition, the influence of the addition of position offsets on the dq-axis back-EMF is different. When $|\Delta\theta_n| = 20\,\theta_u\,(7.2°)$, it will result in $[\sin(20\,\theta_u)] \approx 12.5\%$ and $[1 - \cos(20\,\theta_u)] \approx 0.8\%$ changes in the dq-axis back-EMFs, respectively. Thus, it can be concluded that the influence of a relatively small position offset on the d-axis voltage (d-axis back EMF) will be significant, whereas it is negligible on the q-axis voltage (q-axis back EMF).

5.3.1.2 Estimation of q-axis Inductance L_q

Theoretically, with the aid of the earlier estimated rotor PM flux linkage ψ_{pm}, the q-axis inductances can be estimated from the d-axis voltage Eq. (5.12) straightforwardly when $i_d = 0$, whereas the non-ideal position measurement $\Delta\theta$ needs to be considered.

$$u_d^*(k) + Dd(k)\,V_{\text{dead}} = -\omega_r(k)\left[L_q\,i_q(k) + \psi_{pm}\sin(\Delta\theta)\right] \tag{5.12a}$$

$$\widehat{L}_q = \frac{u_d^*(k)}{\omega_r(k)\,i_q(k)} + \frac{\psi_{pm}\sin(\Delta\theta)}{i_q(k)} \tag{5.12b}$$

It can be seen from Eq. (5.12b), although $Dd(k)\,V_{\text{dead}}$ has a negligible effect on the estimation of L_q due to its zero-mean value, the potential measurement error of rotor position $\Delta\theta$ could give rise to a significant effect on the estimation accuracy. By way of example, for Prototype Machine 1, whose nominal values of L_q and ψ_{pm} are 3.24 mH and 70.7 mWb, under the assumptions of $i_q = 1$ A and $\Delta\theta = 1°$, the percentage error of the q-axis inductance ε_{L_q} can be calculated as follows:

$$\varepsilon_{L_q} = \frac{L_q - \widehat{L}_q}{L_q} = \frac{\psi_{pm}\sin(\Delta\theta)}{i_q(k)\,L_q} = \frac{0.0707 \times \sin(1°)}{0.00324 \times 1} \approx 38.8\% \tag{5.13}$$

It is evident that a one-degree error in electrical angle will result in more than 38% error. It has been mentioned beforehand that one pulse of the encoder is $\theta_u = 0.36°$ for Prototype Machine 1, and consequently, three pulses will cause the error of 1.08° and more than one electrical angle.

Therefore, the position measurement error must be compensated before the estimation. A simple compensation method will be described, and its flowchart is shown in Figure 5.4.

Firstly, based on a slight change of the q-axis current under a light load condition, i.e., from i_{q1} to i_{q2}, the corresponding d-axis voltage equations are written as:

$$u_{d1}^*(k) + Dd(k)V_{\text{dead}} = R_s i_d(k) - \omega_r(k)\left[L_q\,i_{q1}(k) + \psi_{pm}\sin(\Delta\theta)\right] \tag{5.14a}$$

$$u_{d2}^*(k) + Dd(k)V_{\text{dead}} = R_s i_d(k) - \omega_r(k)\left[L_q\,i_{q2}(k) + \psi_{pm}\sin(\Delta\theta)\right] \tag{5.14b}$$

Afterwards, after subtracting Eq. (5.14a) by Eq. (5.14b), the estimated q-axis incremental inductance L_q^{inc} can be calculated as follows:

$$L_q^{\text{inc}} = \frac{\psi_{q1} - \psi_{q2}}{i_{q1}(k) - i_{q2}(k)} = \frac{u_{d2}^*(k) - u_{d1}^*(k)}{\omega_r(k)\left[i_{q1}(k) - i_{q2}(k)\right]} \tag{5.15}$$

Finally, the measurement error of rotor position $\Delta\theta$ can be determined and then compensated with the aid of L_q^{inc} and expressed as follows:

$$\Delta\theta = -\arcsin\left[\frac{L_q^{\text{inc}}\,i_q(k)}{\psi_{pm}} + \frac{u_d^*(k)}{\omega_r(k)}\right] \tag{5.16}$$

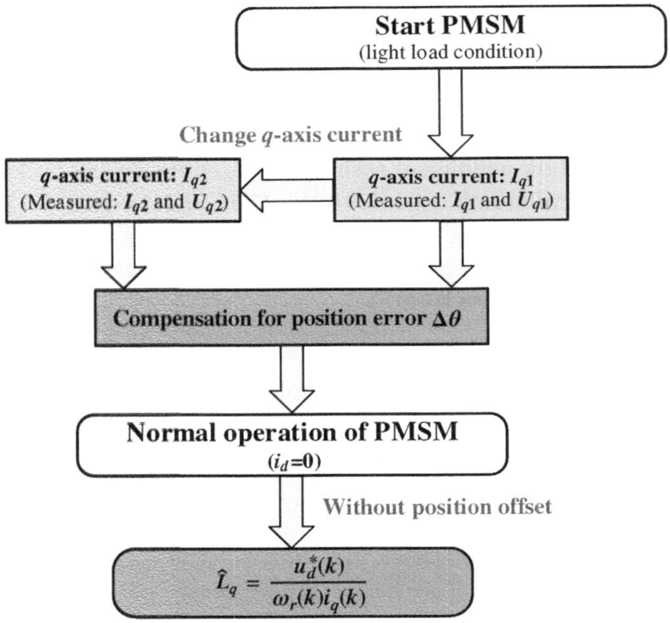

Figure 5.4 Flowchart of compensation for position measurement error $\Delta\theta$ and subsequent estimation of q-axis inductance $\hat{L}_q$.

It needs to be highlighted that L_q^{inc} is the same as the apparent inductance L_q when the magnetic circuit is unsaturated [10], whose q-axis inductance can be assumed as a constant under two different q-axis currents. However, once the magnetic circuit is saturated, the incremental inductance will be smaller, and the position measurement error may be misestimated.

For Prototype Machine 1, when its rotor speed is fixed to 300 r/min and i_q varies from 0.49 A (i_{q1}) to 0.60 A (i_{q2}), the measured electrical quantities are presented in Figure 5.5. As can be observed, since u_d varies from −0.117 to −0.051 V, the incremental q-axis inductance L_q^{inc} and the position measurement error $\Delta\theta$ can be calculated by Eqs. (5.15) and (5.16), respectively, which equal to 3.8 mH and 1.26°, respectively.

5.3.1.3 Estimation of d-axis Inductance L_d

Under $i_d = 0$, the voltage term of the d-axis inductance L_d is inactive. Thus, it can only be estimated based on the additional excitation, e.g., either a position offset or an injected d-axis current, which is illustrated in Figure 5.6.

First, the electromagnetic torque T_e with $i_d = 0$ can be calculated as follows:

$$T_e = 1.5p\psi_{pm} i_q \tag{5.17}$$

On the one hand, by utilizing an additional position offset $\Delta\theta_p$, the resultant electromagnetic torque T_e can be expressed as follows:

$$\begin{aligned} T_e &= 1.5p\left[\psi_{pm}\cos(\Delta\theta_p)i_{q1} - (L_d - L_q)i_d i_q\right] \\ &= 1.5p\left[\psi_{pm}\cos(\Delta\theta_p)i_{q1} - (L_d - L_q)i_{q1}^2\cos(\Delta\theta_p)\sin(\Delta\theta_p)\right] \end{aligned} \tag{5.18a}$$

$$\text{with } i_d = i_{q1}\cos(\Delta\theta_p) \text{ and } i_q = i_{q1}\sin(\Delta\theta_p) \tag{5.18b}$$

where the subscript "1" denotes the variables in d_1q_1-axes with the addition of $\Delta\theta_p$.

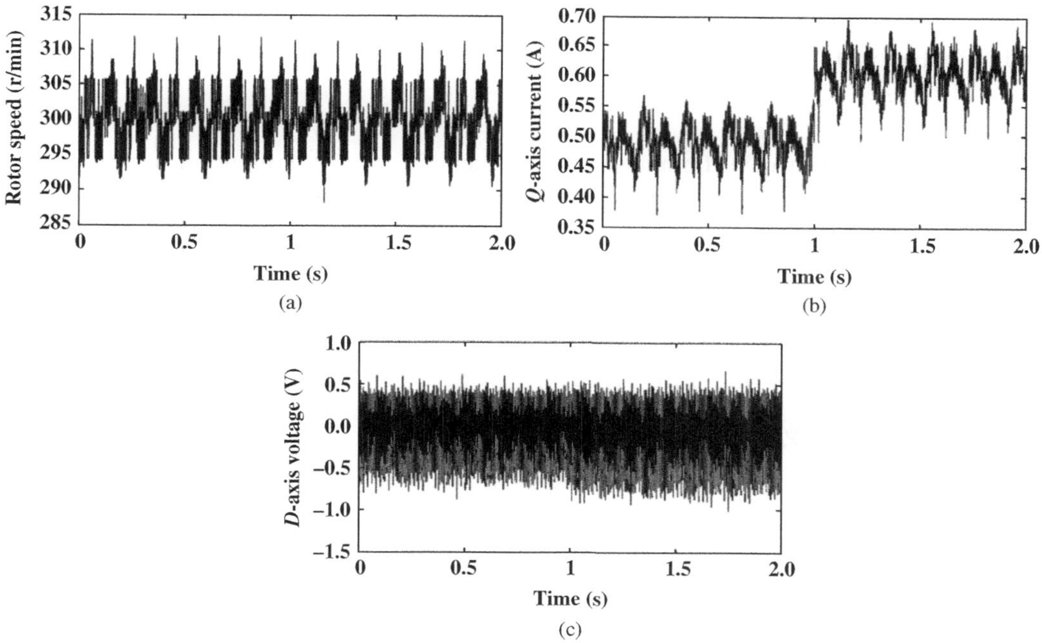

Figure 5.5 Measured (a) rotor speed, (b) q-axis current, and (c) d-axis voltage for Prototype Machine 1 [7].

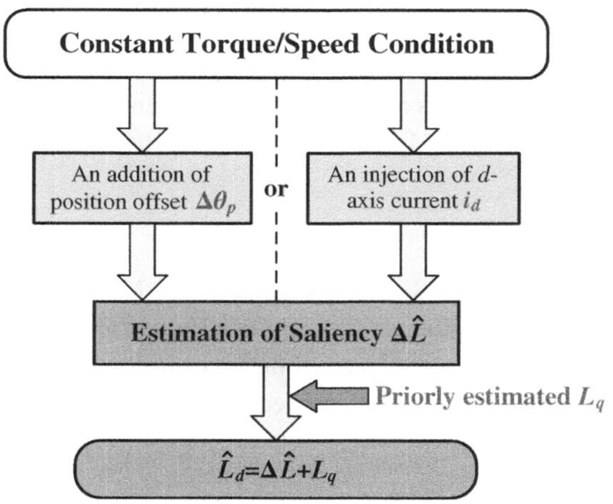

Figure 5.6 Flowchart for estimation of L_d under constant torque/speed control.

Afterwards, combining the torque Eqs. (5.17) and (5.18), the constant torque equation is as follows:

$$\begin{aligned} T_e &= 1.5p\psi_{pm}i_q \\ &= 1.5p\left[\psi_{pm}i_{q1} - (L_d - L_q)i_{q1}^2\sin(\Delta\theta_p)\right]\cos(\Delta\theta_p) \end{aligned} \tag{5.19}$$

Hence, the saliency of the dq-axis inductances $\Delta\hat{L}$ can be derived from Eq. (5.19) and written as follows:

$$\Delta\hat{L} = L_d - L_q = \frac{2\psi_{pm}\left[i_q - i_{q1}\cos(\Delta\theta_p)\right]}{i_{q1}^2\sin(2\Delta\theta_p)} \tag{5.20}$$

Finally, the d-axis inductance $\widehat{L}_d$ can be estimated with the aid of the estimated $\widehat{L}_q$ in Eq. (5.12b) and $\Delta\widehat{L}$ in Eq. (5.20).

$$\widehat{L}_d = \Delta\widehat{L} + \widehat{L}_q \tag{5.21}$$

On the other hand, with aiding from the estimated $\widehat{L}_q$ and $\widehat{\psi}_{pm}$, the d-axis inductance $\widehat{L}_d$ can also be estimated by a d-axis current injection ($i_d \neq 0$) but no longer introducing position offset. However, it is worth emphasizing that the amplitude of injected i_d should not be too large since it may cause significant parameter variations of L_q and ψ_{pm}. Thus, under constant torque control, the saliency term ΔL can be estimated by the constant torque condition.

$$\begin{aligned} T_e &= 1.5p\psi_{pm} i_q \\ &= 1.5p\left[\psi_{pm} i_{q2} - (L_d - L_q)\, i_{d2} i_{q2}\right] \end{aligned} \tag{5.22}$$

$$\Delta L = L_d - L_q = \frac{\psi_{pm}\left[i_q(k) - i_{q2}(k)\right]}{i_{d2}(k)\, i_{q2}(k)} \tag{5.23}$$

where the subscript "2" denotes the variables caused by the d-axis current injection.

It is worth mentioning that the above two saliency estimators, (Eq. 5.20) and (Eq. 5.23), are different from the current injection-based method presented in Section 4.3. It can be explained that since Eqs. (5.20) and (5.23) are derived based on the torque equation, the influences of inaccurate winding resistance and non-ideal compensation of VSI non-linearity can be eliminated.

5.3.2 Experimental and FEA Results

The described POPE method will be performed based on a non-salient-pole SPMSM (Prototype Machine 1) and a salient-pole IPMSM (Prototype Machine 2), whose main parameters are listed in Table 5.3. Their complete design parameters and specifications are presented in Appendix B. The amplitudes of introduced rotor position offsets for Prototype Machines 1 and 2 are $10\theta_u$ and $5\theta_u$, respectively, and the ANN estimators are employed.

5.3.2.1 Experimental Results of Estimated ψ_{pm} and R_s

First, during the estimation, the measured dq-axis voltages and currents of Prototype Machines 1 and 2 are shown in Figure 5.7, from which it is evident that the additions of position offsets $\pm\Delta\theta_p$ have negligible influence on the dq-axis currents and q-axis voltage, while its influence on the d-axis voltage is significant. Thus, the variation in d-axis voltage is measured and recorded for the

Table 5.3 Main parameters of Prototype Machines 1 and 2.

Parameters	Prototype Machine 1 (SPMSM)	Prototype Machine 2 (IPMSM)
Nominal rotor flux linkage	70.7 mWb	236.0 mWb
Nominal winding resistance	0.32 Ω	6.00 Ω
Nominal d-axis inductance	3.24 mH	31.20 mH
Nominal q-axis inductance	3.24 mH	55.00 mH
Unit electrical angle	0.36°	0.53°
Line number of encoder	5000	2048

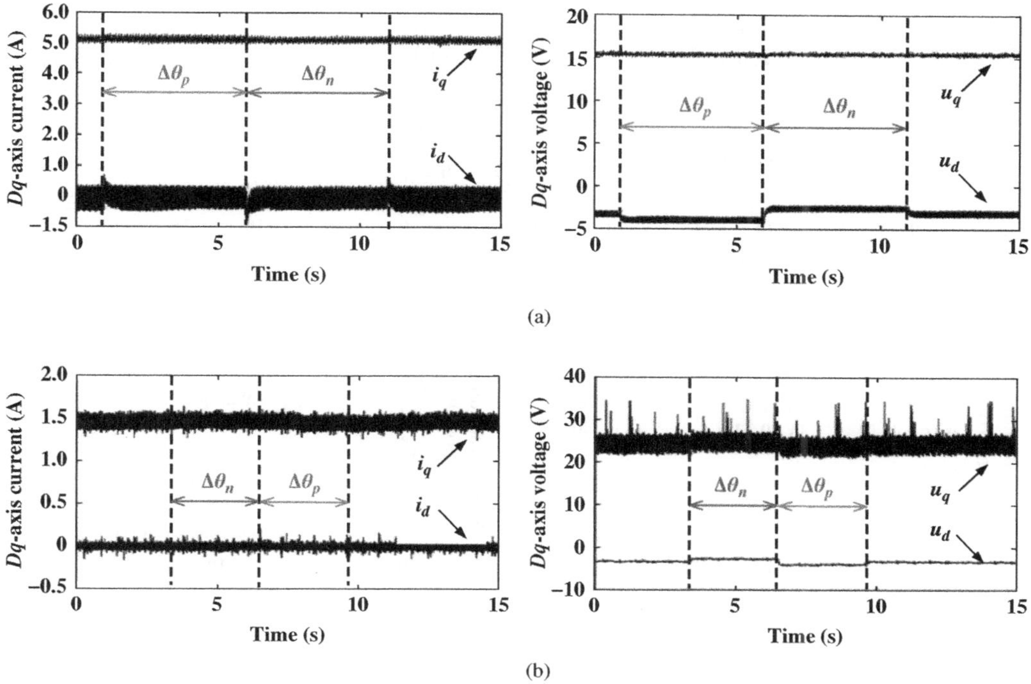

Figure 5.7 Measured *dq*-axis voltages and currents during estimation [3], IEEE: (a) Prototype Machine 1 ($i_q = 5.0$ A and $i_d = 0$ A at 300 r/min) and (b) Prototype Machine 2 ($i_q = 1.5$ A and $i_d = 0$ A at 200 r/min).

following estimation. In the tests, the durations of $\Delta\theta_p$ and $\Delta\theta_n$ are 5 s, and the measurements of electrical quantities are started after a 1 s delay to ensure the settling of measured signals.

Second, the estimated rotor PM flux linkage $\widehat{\psi}_{pm}$ and stator winding resistance $\widehat{R}_s$ with and without compensation of V_{dead} are shown in Figure 5.8, where the compensation method of VSI non-linearity has been described in Section 2.3. As can be observed, the estimation of $\widehat{\psi}_{pm}$ is independent of the distorted voltage V_{dead} due to VSI non-linearity, since it has been cancelled in Eq. (5.7). However, its influence on the estimation of $\widehat{R}_s$ is significant, as shown in Eq. (5.8). Thus, when VSI non-linearity is not compensated, the estimated stator winding resistance $\widehat{R}_s$ will lead to significant overestimation errors compared to the nominal values, i.e., 1.04 Ω versus 0.32 Ω (Prototype Machine 1) and 9.02 Ω versus 6.00 Ω (Prototype Machine 2). In contrast, after compensation, $\widehat{R}_s$ is only slightly larger than their nominal values caused by the iron losses and on-state resistances in IGBT and diode.

Third, for both PMSMs, the variations of the rotor PM flux linkage with the *q*-axis currents i_q are also compared with the FEA results [11] under $i_d = 0$ A, as shown in Figure 5.9. It can be easily found that the rotor PM flux linkage decreases with i_q. For Prototype Machines 1 and 2, the slopes of estimated variations in the rotor PM flux linkage are 0.47 and 5.2 mWb/A, respectively. In comparison, the respective slopes were calculated by FEA of 0.57 and 3.0 mWb/A. The errors may be attributed to the non-ideal manufacture of the PMSMs and the temperature variations of PMs.

Finally, to demonstrate the tracking capability of the POPE method when the values of winding resistance and the rotor PM flux linkage are changed, two experiments are performed based on Prototype Machine 1.

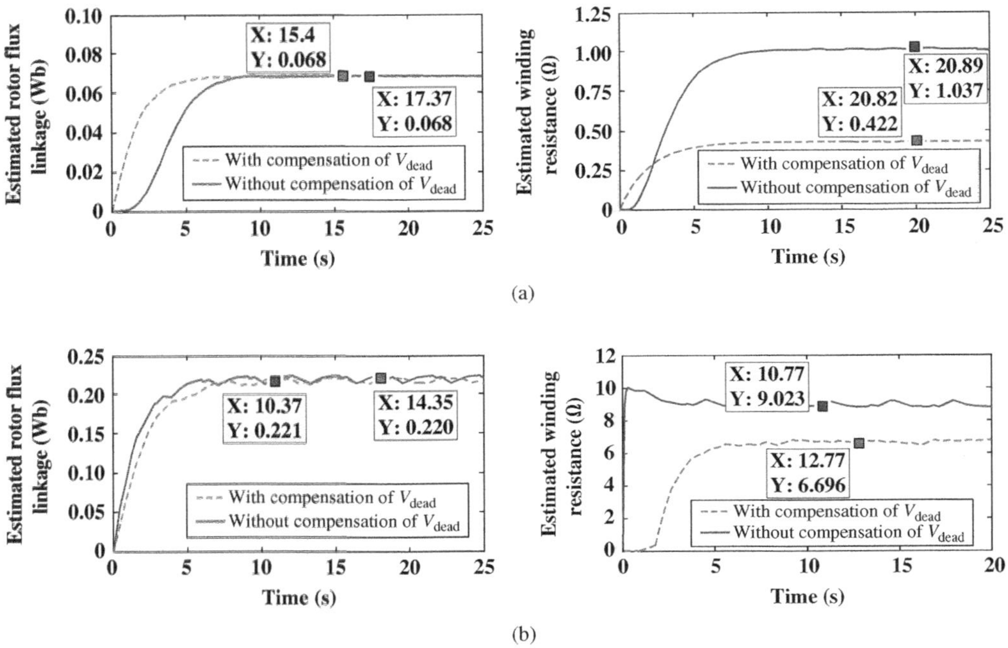

Figure 5.8 Estimated rotor PM flux linkage and winding resistance with and without compensation of VSI non-linearity (V_{dead}): (a) Prototype Machine 1 ($i_q = 5.0$ A and $i_d = 0$ A at 300 r/min) and (b) Prototype Machine 2 ($i_q = 1.5$ A and $i_d = 0$ A at 200 r/min).

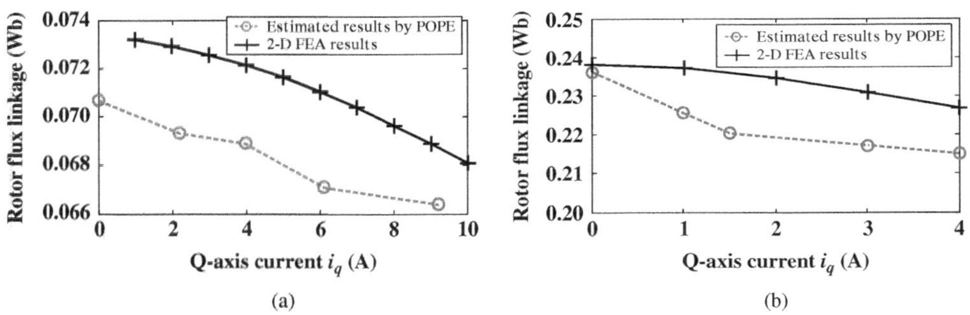

Figure 5.9 Variations of estimated rotor PM flux linkages with i_q against FEA results [3], IEEE: (a) Prototype Machine 1 (at 300 r/min) and (b) Prototype Machine 2 (at 200 r/min).

In the first test, three resistors (1.072 Ω each) are added in series with the three phases of PMSM, where i_q and i_d are 4 and 0 A, respectively, and the rotor speed is kept at 300 r/min. It can be found from Figures 5.10a and b that the estimated rotor PM flux linkage is also close to the nominal value of 0.07 Wb, and the estimated value of the total resistance is 1.431 Ω, which is close to the real value of 1.494 Ω (1.072 Ω + 0.422 Ω).

In the second test, the PM temperature tracking capability of the POPE method is shown. The experiment is designed as follows.

1. A heater is employed to heat the machine for 20 min and the temperature of the rotor shaft is measured as 105 °C at $t = 20$ min.

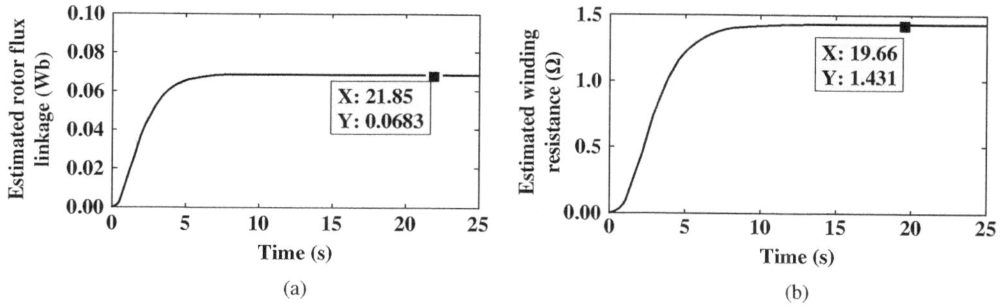

Figure 5.10 Estimation results after the addition of external resistors [3], IEEE: (a) rotor PM flux linkage and (b) stator winding resistance.

2. The rotor PM flux linkage at no load condition will be measured from the line back-EMF by using FFT at $t = 0$, 20, 35, 47, 67, 87, 107, 152, and 300 min.
3. The rotor PM flux linkage at the load condition ($i_d = 0$ A and $i_q = 4$ A at 300 r/min) will be estimated by the POPE method at the time points of $t = 0$, 20, 30, 50, 70, 90, 110, 132, 155, and 300 min.

The measured and estimated rotor PM flux linkages and their variations are shown in Figure 5.11. As can be seen, the measured and estimated rotor PM flux linkages and their variations exhibit similar tendencies. The estimated rotor PM flux linkages are slightly lower than the measured values. It can be explained that since the measured results are calculated from the no-load line back-EMFs, the POPE is conducted under the on-load condition. The rapid decrease in the rotor PM flux linkage at the first 20 min is caused by the decreased residual flux density of NdFeB magnets with the temperature increase. In addition, after the removal of the heater at $t = 20$ min, it is found that the value of rotor PM flux linkage continues to decrease for about 10 min, which can be explained by the heat transfer between the shaft and the PM being delayed. Afterwards, the estimated rotor PM flux linkage will slowly increase to its normal value after about 280-min natural cooling. Thus, the POPE method has the capability to track the variation in the rotor PM flux linkage accurately and is appropriate for the thermal condition monitoring of the rotor PM.

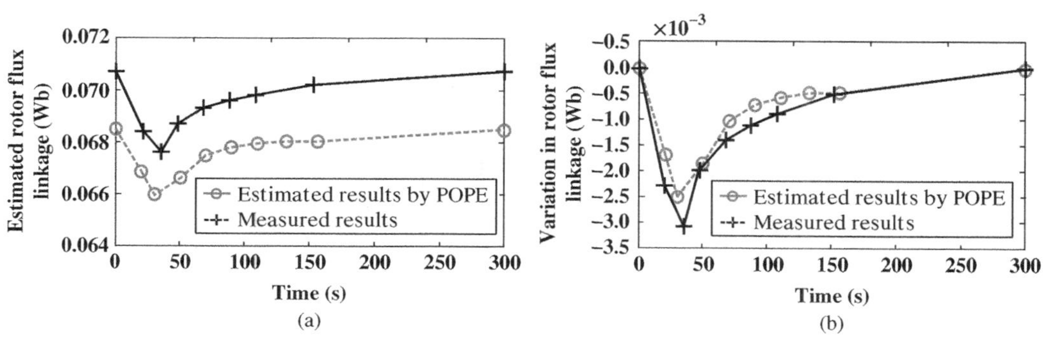

Figure 5.11 Measured and estimated rotor PM flux linkages and their variations with changed temperatures [3], IEEE: (a) measured and estimated rotor PM flux linkages and (b) variations of rotor PM flux linkages.

5.3.2.2 Experimental Results of Estimated L_d and L_q

In terms of the estimation of *dq*-axis inductances, it has been highlighted in this section that the position measurement error must be compensated appropriately prior to the estimation of the *q*-axis inductance L_q. Furthermore, the *d*-axis inductance L_d can be estimated subsequently by either adding position offsets or injecting *d*-axis current.

On the one hand, for the non-salient-pole SPMSM (Prototype Machine 1), the position measurement error is 1.26° and calculated by Eq. (5.16). After compensation, with the aid of the estimated rotor PM flux linkage $\hat{\psi}_{pm}$ of 68.2 mWb, the estimated *q*-axis inductance L_q is shown in Figure 5.12a, where $i_d = 0$ A and $i_q = 4$ A at 300 r/min. The estimated *d*-axis inductances are shown in Figure 5.12b–d.

As observed, when L_d is estimated by using the position offsets of $|\Delta\theta_n| = \Delta\theta_p = 20\,\theta_u$, $L_d = 3.27$ mH, which is slightly smaller than the estimated L_q of 3.80 mH, as shown in Figure 5.12b. In comparison, when L_d is estimated by using the injected *d*-axis current i_d, when $i_d = 1$ and 0.5 A, the estimated L_d decreases from 4.23 to 3.84 mH. It shows that the injection of i_d will introduce an estimation error in Eq. (5.23) due to non-negligible variations of ψ_{pm} and L_q. Thus, in real applications, the amplitude of injected i_d or position offset should be limited to avoid significant parameter variations.

On the other hand, for the salient-pole IPMSM (Prototype Machine 2), by adding the rotor position offsets, i.e., $|\Delta\theta_n| = \Delta\theta_p = 20\,\theta_u$, the estimated rotor PM flux linkage is 0.2224 Wb. Furthermore, the *dq*-axis inductances are shown in Figure 5.13, where $i_q = 2$A and $i_d = 0$A at 200 r/min. It is obvious that the saliency $L_q/L_d = 3.25/5.0 = 55.8\%$, which is close to the nominal saliency of 55.8% (see Table 5.3).

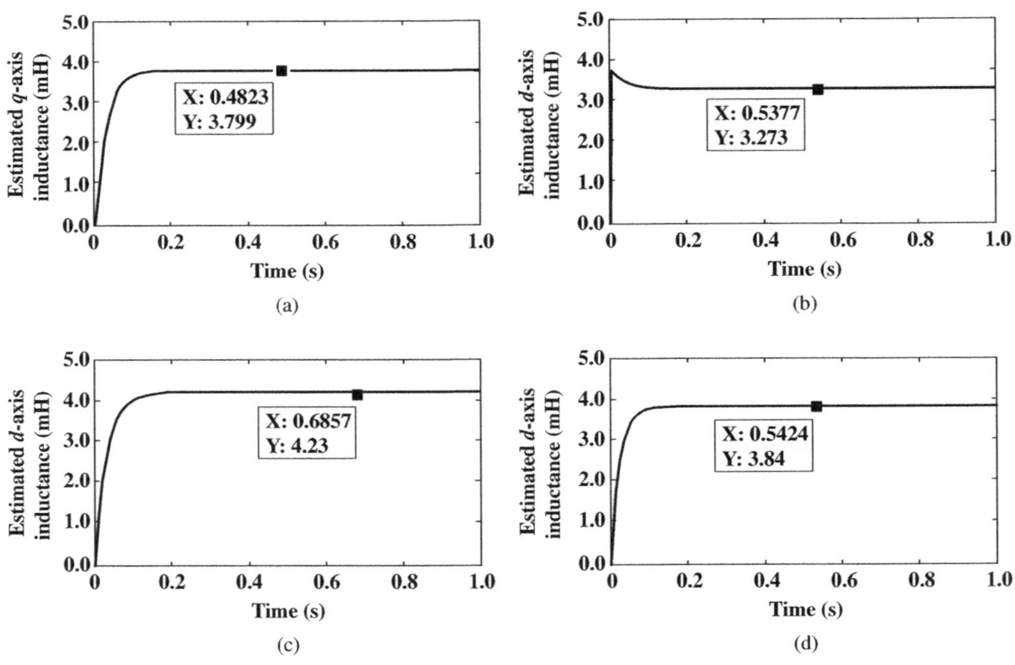

Figure 5.12 Estimation of *dq*-axis inductances at 300 r/min for non-salient-pole SPMSM (Prototype Machine 1) [4]: (a) estimated *q*-axis inductance, (b) estimated *d*-axis inductance by addition of $|\Delta\theta_n| = \Delta\theta_p = 20\,\theta_u$, (c) estimated *d*-axis inductance by injection of $i_d = -1$ A, and (d) estimated *d*-axis inductance by injection of $i_d = -0.5$ A.

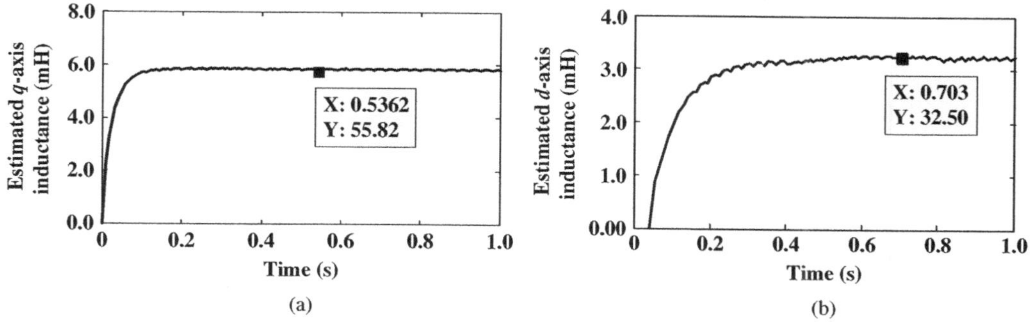

Figure 5.13 Estimation of *dq*-axis inductances at 200 r/min for salient-pole IPMSM [4]: (a) estimated *q*-axis inductance and (b) estimated *d*-axis inductance.

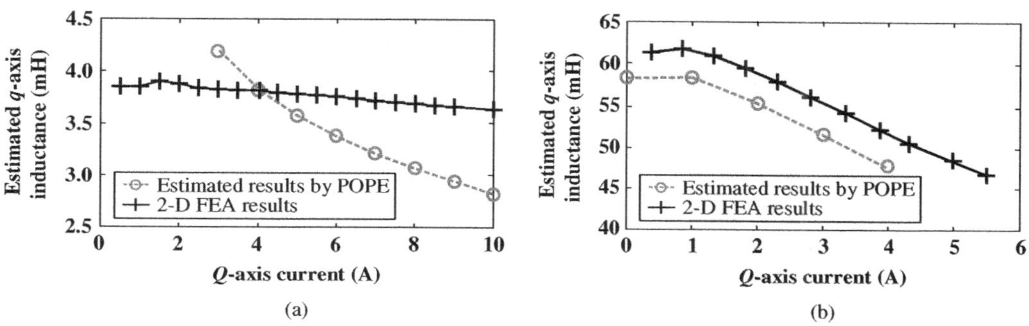

Figure 5.14 Comparison of *q*-axis inductances estimated by POPE and FEA with different *q*-axis currents [4]: (a) Prototype Machine 1 and (b) Prototype Machine 2.

In addition, the variation of the *q*-axis inductances predicted by the POPE and the FEA for both Prototype Machines 1 and 2 are compared in Figure 5.14, where the rotor speed is fixed to 200 r/min.

For Prototype Machine 1, the slope of the *q*-axis inductance estimated by the POPE is different from that predicted by the two-dimensional (2D) FEA result. It may be attributed to the fact that the *q*-axis inductance is very small, and the compensation of position measurement error cannot be 100% accurate. In comparison, for Prototype Machine 2, the estimated slope of the *q*-axis inductance is almost the same as that predicted by FEA, since it has more than 10 times *dq*-axis inductances than those in Prototype Machine 1, and thus, its *dq*-axis inductances will cause large voltage drops, i.e., high signal-to-noise ratios.

5.4 Position Offset-based Estimation with $i_d = 0$ Under Variable Speed Control

As described in the previous section (Section 5.3), the POPE method has been successfully applied to PMSMs under constant torque/speed control. In this sub-section, the POPE method is extended to variable speed control. As will be shown, the described POPE method can estimate the rotor PM flux linkage, the stator winding resistance, the distorted voltage, and the *q*-axis inductance, which can be used for condition monitoring.

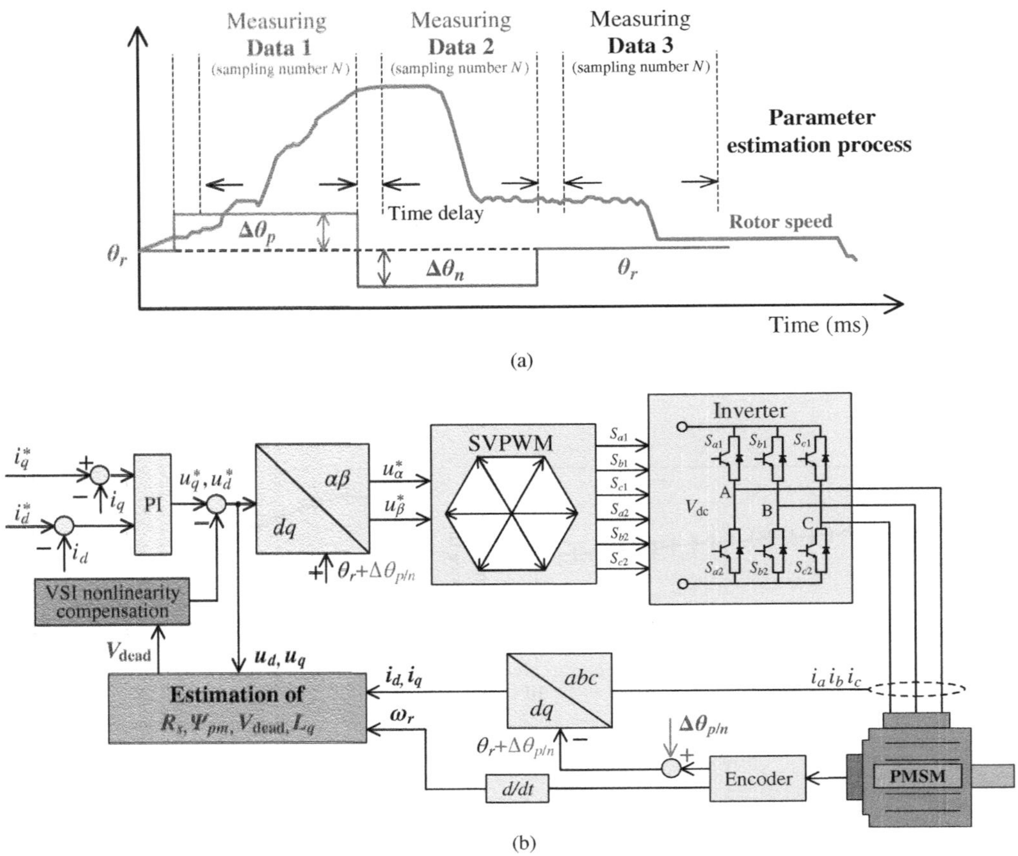

Figure 5.15 Measurement process of POPE under variable speed control: (a) principle and measurement process of POPE and (b) POPE under the FOC system.

Under the variable speed control, the principle of the POPE and its associated FOC system is shown in Figure 5.15, where three sets of data, i.e., Data 1, Data 2, and Data 3, are used for the estimation and correspond to positive offsets $\Delta\theta_p$, negative offsets $\Delta\theta_n$, and none offset θ_r, respectively.

5.4.1 Designed Estimation Method

The dq-axis voltage equations under positive $\Delta\theta_p$ and negative $\Delta\theta_n$ position offsets at variable rotor speeds can be derived from Eqs. (5.1) and (5.2) and written as follows:

$$u_{d1}^*(k) + Dd_1(k)V_{\text{dead1}} = R_s i_{d1}(k) - \omega_{r1}(k)\left[L_q i_{q1}(k) + \psi_{pm}\sin(\Delta\theta + \Delta\theta_p)\right] \tag{5.24a}$$

$$u_{q1}^*(k) + Dq_1(k)V_{\text{dead1}} = R_s i_{q1}(k) + \omega_{r1}(k)\left[L_d i_{d1}(k) + \psi_{pm}\cos(\Delta\theta + \Delta\theta_p)\right] \tag{5.24b}$$

$$u_{d2}^*(k) + Dd_2(k)V_{\text{dead2}} = R_s i_{d2}(k) - \omega_{r2}(k)\left[L_q i_{q2}(k) + \psi_{pm}\sin(\Delta\theta + \Delta\theta_n)\right] \tag{5.25a}$$

$$u_{q2}^*(k) + Dq_2(k)V_{\text{dead2}} = R_s i_{q2}(k) + \omega_{r2}(k)\left[L_d i_{d2}(k) + \psi_{pm}\cos(\Delta\theta + \Delta\theta_n)\right] \tag{5.25b}$$

where the subscripts "1" and "2" denote the variables corresponding to Data 1 and Data 2, respectively. Meanwhile, the dq-axis currents are set to constants during the data measurement to avoid parameter

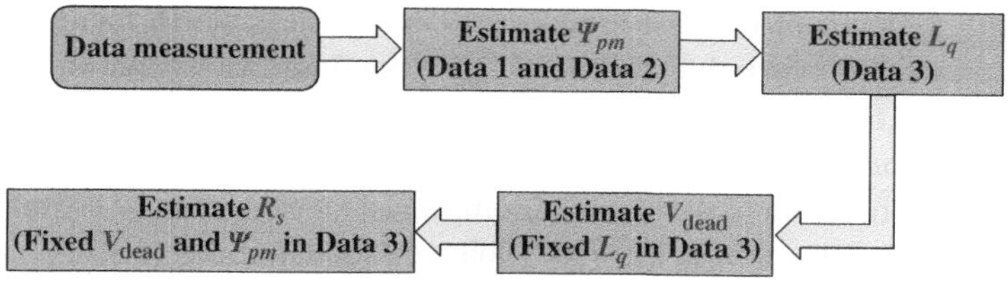

Figure 5.16 Complete estimation method of POPE under variable speed control.

variations due to current changes. Meanwhile, the bandwidth of the current-loop control is much higher than that of the speed loop. Thus, it is reasonable to assume that $i_{d1} = i_{d2}$ and $i_{q1} = i_{q2}$.

Based on the above three sets of measurement data, four unknown parameters, i.e., the rotor PM flux linkage ψ_{pm}, the q-axis inductance L_q, the distorted V_{dead}, and the winding resistance R_s, will be estimated subsequently. The complete estimation method is shown in Figure 5.16.

First, by subtracting Eq. (5.24a) × $\omega_{r2}(k)$ by Eq. (5.25a) × $\omega_{r1}(k)$, the term of winding resistance R_s can be cancelled in Eq. (5.26).

$$u_{d1}^*(k)\omega_{r2}(k) - u_{d2}^*(k)\omega_{r1}(k) = 2\psi_{pm}\omega_{r1}(k)\omega_{r2}(k)\left[\cos(\Delta\theta)\sin\left(\Delta\theta_p\right)\right] \\ -[Dd_1(k)V_{dead1}\omega_{r2}(k) - Dd_2(k)V_{dead2}] \tag{5.26}$$

Hence, the rotor PM flux linkage ψ_{pm} can be obtained as follows:

$$\psi_{pm} \approx \frac{u_{d1}^*(k)\,\omega_{r2}(k) - u_{d2}^*(k)\,\omega_{r1}(k)}{2\omega_{r1}(k)\,\omega_{r2}(k)\sin\left(\Delta\theta_p\right)} - \frac{[Dd_1(k)V_{dead1}\omega_{r2}(k) - Dd_2(k)V_{dead2}\omega_{r1}(k)]}{2\omega_{r1}(k)\,\omega_{r2}(k)\sin\left(\Delta\theta_p\right)} \tag{5.27}$$

As detailed in Section 5.3, the position measurement error $\Delta\theta$ is usually quite small and can be compensated, and thus, $\psi_{pm} \approx \psi_{pm}\cos(\Delta\theta)$. Meanwhile, the second term in Eq. (5.27) involves two VSI non-linearity voltages, i.e., $Dd_1(k)V_{dead1}$ and $Dd_2(k)V_{dead2}$, whose average values are 0 under $i_d = 0$ control.

Thus, under $i_d = 0$ control, Eq. (5.27) can be rewritten as follows:

$$\psi_{pm} = \frac{u_{d1}^*(k)\,\omega_{r2}(k) - u_{d2}^*(k)\,\omega_{r1}(k)}{2\omega_{r1}(k)\,\omega_{r2}(k)\sin\left(\Delta\theta_p\right)} \tag{5.28}$$

It can be easily found that the expression of Eq. (5.28) is the same as Eq. (5.7b) when $\omega_{r1}(k) = \omega_{r2}(k)$ under constant torque/speed control. Moreover, $\widehat{\psi}_{pm}$ can be estimated by either an ANN estimator or other optimization algorithms, e.g., genetic algorithm (GA), quantum genetic algorithm (QGA), and immune clonal algorithm, whose cost function P_1 can be designed as follows:

$$P_1\left(\widehat{\psi}_{pm}\right) = \sum_{k=1}^{n}\left[K_m\left|\,u_{d1}^*(k)\,\omega_{r2}(k) - u_{d2}^*(k)\,\omega_{r1}(k) - 2\,\omega_{r1}(k)\,\omega_{r2}(k)\sin\left(\Delta\theta_p\right)\right|\right] \tag{5.29}$$

where K_m is the gain of the cost function. n is the data length of Data 1–3. More importantly, as can be found in Eqs. (5.28) and (5.29), the influence of VSI non-linearity has been cancelled. In addition, the detailed instructions for GA optimisation will be described in Chapter 11.

Second, in terms of the estimation of the q-axis inductance L_q and the distorted voltage V_{dead}, they can be estimated from the d-axis voltage Eq. (5.4a) directly from Data 3, but without adding position offset ($\Delta\theta = 0$). The respective cost functions are designed as follows:

$$P_2\left(\widehat{L}_q, \widehat{V}_{dead}\right) = \sum_{k=1}^{n}\left[K_u\left|\,u_d^*(k) + Dd(k)\,\widehat{V}_{dead} + \omega_r(k)\,\widehat{L}_q\,i_q(k)\right|\right] \tag{5.30}$$

$$P_3\left(\widehat{V}_{dead}\right) = \sum_{k=1}^{n}\left[K_v\left|\,u_d^*(k) + Dd(k)\,\widehat{V}_{dead} + \omega_r(k)\,L_q\,i_q(k)\right|\right] \tag{5.31}$$

where K_u and K_v are the gains of the cost functions P_2 and P_3. It is worth mentioning that the voltage drop caused by the q-axis inductance ($\omega_r L_q i_q$) is dominant compared to the zero-mean DdV_{dead}. Therefore, once L_q is accurately estimated in P_2, it will be fixed to minimize the cost function of P_3 to determine V_{dead}.

Third, in terms of the estimation of the stator winding resistance R_s, it can be estimated by minimizing the cost function P_4 derived from the q-axis voltage equation in Eq. (5.4b):

$$P_4(\hat{R}_s) = \sum_{k=1}^{n} \left[K_w \middle| u_q^*(k) + Dq(k)V_{\text{dead}} - \omega_r(k)\psi_{pm}\cos(\Delta\theta) - \hat{R}_s i_q(k) \right| \tag{5.32}$$

where K_w is the gain of the cost function P_4. ψ_{pm} and V_{dead} used in Eq. (5.32) have been estimated by the cost functions P_2 and P_3, respectively. In addition, there is no d-axis inductance term under $i_d = 0$ control, and its estimation may rely on other methods, e.g., FEA, measurement instruments, and d-axis current injection methods.

5.4.2 Experimental and FEA Results

The salient-pole Prototype Machine 2 (see Table 5.3) is used for illustration, and the QGA estimator (Chapter 10) is employed to minimize the designed cost functions of $P_1 - P_4$. Under variable speed control ($i_d = 0$, $i_q = 3$ A, and $|\Delta\theta_p| = |\Delta\theta_n| = 10\theta_u$), the measured data are presented in Figure 5.17.

It is obvious that the VSI non-linearity function of Dd is a zero-mean disturbance when $i_d = 0$, which confirms the previous analysis that it does not affect the estimation of rotor PM flux linkage by minimizing P_1. Meanwhile, the measured dq-axis currents are kept as constants during the position offsets, which underpins that the current response is much faster than speed change.

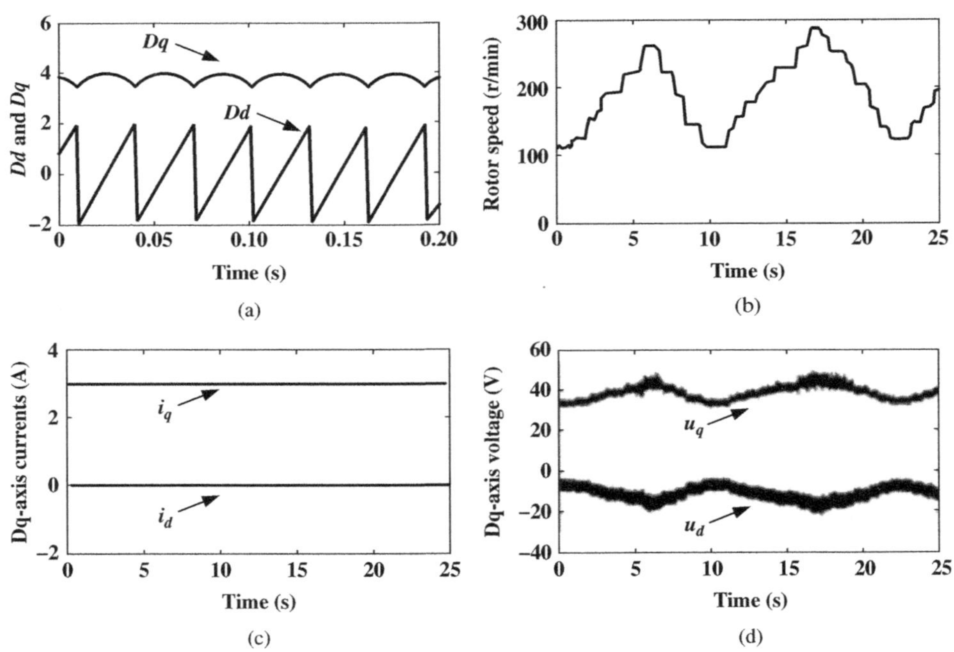

Figure 5.17 Measured electrical quantities for Prototype Machine 2 [4]: (a) *Dd* and *Dq*, (b) rotor speed, (c) *Dq*-axis currents, and (d) *Dq*-axis voltages.

Furthermore, the estimated rotor PM flux linkage, q-axis inductance, distorted voltage, and winding resistance for Prototype Machine 2 are shown in Figure 5.18. As can be seen, the estimated parameters are close to their nominal values (see Table 5.3), and all estimated parameters converge within 100 generations in QGA calculation. It is also worth mentioning that the estimated winding resistance will be much larger than its nominal value if VSI is not compensated, i.e., 8.95 Ω versus 6.20 Ω (nominal value).

Meanwhile, Figure 5.19 compares the variations in the estimated rotor PM flux linkage with q-axis currents against FEA results, where the amplitudes of the position offsets are $10\theta_u$ and $20\theta_u$. As is observed, the difference between the two sets of estimation results is small, and the estimated

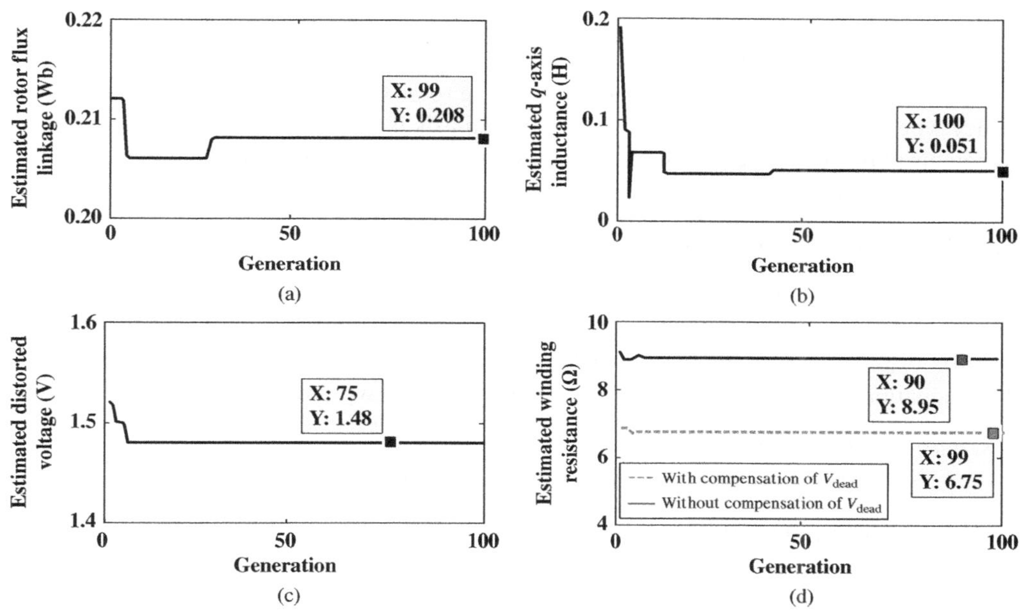

Figure 5.18 Estimated parameters based on POPE under variable speed control for Prototype Machine 2 [4]: (a) rotor PM flux linkage, (b) Q-axis inductance, (c) distorted voltage, and (d) stator winding resistance.

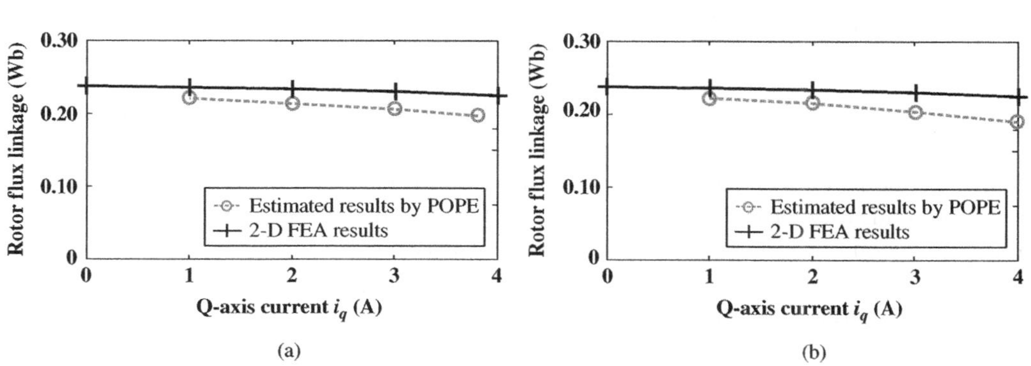

Figure 5.19 Comparison of estimated and FE-predicted variations of rotor PM flux linkage with different q-axis currents under variable speed control for Prototype Machine 2 [4]: (a) $|\Delta\theta_p| = |\Delta\theta_n| = 10\,\theta_u$ and (b) $|\Delta\theta_p| = |\Delta\theta_n| = 20\,\theta_u$.

slope is 8.6 mWb/A, which is larger than the FEA results of 3.2 mWb/A. The errors are attributed to the temperature rise in PM during the experiments and the mis-matches of material parameters used in the FEA.

5.5 Position Offset-based Estimation with $i_d \neq 0$ Under Constant Torque/Speed Control

As aforementioned, the POPE methods achieve excellent estimation performance with $i_d = 0$ under constant torque/speed and variable speed controls. However, for salient-pole PMSMs, i_d is usually set as a negative value to utilize the reluctance torque and/or to extend the speed range, i.e., MTPA/MTPV strategies [12–14]. When $i_d \neq 0$, the d-axis inductance is activated, and consequently, both the saliency term of the dq-axis inductance and current-dependent VSI non-linearity need to be considered during the estimation.

In this section, the POPE is applied to estimate the rotor PM flux linkage and the inductance saliency when $i_d \neq 0$ at constant rotor speed. The dq-axis inductances can be further estimated individually at two rotor speeds. The designed estimation method could fully eliminate the influence of current-dependent VSI non-linearity.

5.5.1 Designed Estimation Method

As analysed by the phasor diagram in Section 5.2, when the d-axis current $i_d \neq 0$, the influence of the additional rotor position $\Delta\theta$ on the d-axis current is significant. Thus, considering the addition of positive $\Delta\theta_p$ and negative $\Delta\theta_n$ position offsets, i.e., $|\Delta\theta_p| = |\Delta\theta_n| = \Delta\theta_\sigma$, the corresponding dq-axis voltage equations can be derived from Eqs. (5.1) and (5.2) and written as follows:

$$\begin{aligned} u'_{d1} &= u^*_{d1}(k) + Dd_1(k)V_{\text{dead1}} \\ &= R_s i'_{d1}(k) - \omega_{r1}(k)\left[\psi_{pm}\sin(\Delta\theta_\sigma) + L_d i_{d1}(k)\sin(\Delta\theta_\sigma) + L_q i_{q1}(k)\cos(\Delta\theta_\sigma)\right] \end{aligned} \tag{5.33a}$$

$$\begin{aligned} u'_{q1} &= u^*_{q1}(k) + Dq_1(k)V_{\text{dead1}} \\ &= R_s i'_{q1}(k) + \omega_{r1}(k)\left[\psi_{pm}\cos(\Delta\theta_\sigma) + L_d i_{d1}(k)\cos(\Delta\theta_\sigma) - L_q i_{q1}(k)\sin(\Delta\theta_\sigma)\right] \end{aligned} \tag{5.33b}$$

$$\begin{aligned} u'_{d2} &= u^*_{d2}(k) + Dd_2(k)V_{\text{dead2}} \\ &= R_s i'_{d2}(k) + \omega_{r2}(k)\left[\psi_{pm}\sin(\Delta\theta_\sigma) + L_d i_{d2}(k)\sin(\Delta\theta_\sigma) - L_q i_{q2}(k)\cos(\Delta\theta_\sigma)\right] \end{aligned} \tag{5.33c}$$

$$\begin{aligned} u'_{q2} &= u^*_{q2}(k) + Dq_2(k)V_{\text{dead2}} \\ &= R_s i'_{q2}(k) + \omega_{r2}(k)\left[\psi_{pm}\cos(\Delta\theta_\sigma) + L_d i_{d2}(k)\cos(\Delta\theta_\sigma) + L_q i_{q2}(k)\sin(\Delta\theta_\sigma)\right] \end{aligned} \tag{5.33d}$$

where the variables with subscripts "1" and "2" correspond to states with positive and negative position offsets. The superscript ′ denotes the variables in the d_1q_1- and d_2q_2-axes with positive and negative position offsets. Otherwise, they are variables in the dq-axes without position offset. Since the VSI non-linearity terms $Dd(k)V_{\text{dead}}$ and $Dq(k)V_{\text{dead}}$ are functions of currents, their average values depend on the current angle or the ratio of dq-axis currents. Thus, when the dq-axis currents are the same, i.e., $\overline{Dd(k)V_{\text{dead1}}} = \overline{Dd(k)V_{\text{dead2}}}$ and $\overline{Dq(k)V_{\text{dead1}}} = \overline{Dq(k)V_{\text{dead2}}}$. Thus, the following calculations are based on their average values.

Besides, the correlations between the dq-axis currents i_{d1} and i_{d2} and the d_1q_1-/d_2q_2-axis currents i'_{d1}, i'_{q1}, i'_{d2}, and i'_{q2} can be derived from Eq. (5.3) and written as follows:

$$\begin{cases} i_{d1} = +i'_{q1}\sin(\Delta\theta_\sigma) + i'_{d1}\cos(\Delta\theta_\sigma) \\ i_{q1} = +i'_{q1}\cos(\Delta\theta_\sigma) - i'_{d1}\sin(\Delta\theta_\sigma) \end{cases} \tag{5.34a}$$

$$\begin{cases} i_{d2} = -i'_{q2}\sin(\Delta\theta_\sigma) + i'_{d2}\cos(\Delta\theta_\sigma) \\ i_{q2} = +i'_{q2}\cos(\Delta\theta_\sigma) + i'_{d2}\sin(\Delta\theta_\sigma) \end{cases} \tag{5.34b}$$

5.5.1.1 Estimation of Rotor PM Flux Linkage ψ_{pm} and Inductances' Saliency ΔL

By utilizing the reference equation in Eq. (5.33), ψ_{pm} and ΔL will be estimated by using deviations of two sets of dq-axis voltages, where it is assumed that $i'_d = i'_{d1} = i'_{d2}$ and $i'_q = i'_{q1} = i'_{q2}$. The rotor speed is tuned to be $\omega_{r1} = \omega_{r2} = \omega_r$ to derive the inductance saliency term in the following calculation.

First, the inductance saliency $\Delta L = L_d - L_q$ can be obtained by using the deviation of the two q-axis equations, i.e., Eqs. (5.33b)–(5.33d), which is expressed as follows:

$$u^*_{q2} - u^*_{q1} = (L_q - L_d)\omega_r i'_q \sin(2\Delta\theta_\sigma) \tag{5.35}$$

$$\Delta\hat{L} = L_d - L_q = \frac{u^*_{q1} - u^*_{q2}}{i'_q\,\omega_r(k)\sin(2\Delta\theta_\sigma)} \tag{5.36}$$

Second, the deviation of the two d-axis equations, i.e., Eqs. (5.33c)–(5.33a), is derived as follows:

$$u^*_{d2} - u^*_{d1} = 2\psi_{pm}\sin(\Delta\theta_\sigma)\omega_r + L_d\omega_r\sin(\Delta\theta_\sigma)(i_{d1} + i_{d2}) + L_q\omega_r\cos(\Delta\theta_\sigma)(i_{q1} - i_{q2}) \tag{5.37}$$

Afterwards, by introducing Eq. (5.33) into Eq. (5.36), it is represented as follows:

$$u^*_{d2} - u^*_{d1} = 2\psi_{pm}\sin(\Delta\theta_\sigma)\omega_r + (L_d - L_q)\omega_r i'_d \sin(2\Delta\theta_\sigma) \tag{5.38}$$

By substituting Eq. (5.34) into Eq. (5.37), the rotor PM flux linkage ψ_{pm} can be estimated as follows:

$$2\psi_{pm}\sin(\Delta\theta_\sigma)\omega_r = \left(u^*_{d2} - u^*_{d1}\right) + \frac{\left(u^*_{q2} - u^*_{q1}\right) i'_d}{i'_q} \tag{5.39}$$

$$\psi_{pm} = \frac{u^*_{d2} - u^*_{d1}}{2\sin(\Delta\theta_\sigma)\omega_r} + \frac{\left(u^*_{q2} - u^*_{q1}\right) i'_d}{2\sin(\Delta\theta_\sigma) i'_q\,\omega_r} \tag{5.40}$$

Compared to the estimation of rotor PM flux linkage when $i_d = 0$ in Section 5.3, Eq. (5.40) includes the deviation of q-axis voltages (the second term). Thus, the associated inductance saliency (Eq. 5.35) will be introduced into the estimation results when $i_d \neq 0$. In addition, the VSI non-linearity has been cancelled in the above calculations, and thus, this section does not involve the estimation of winding resistance, which is strongly affected by the VSI non-linearity. Nevertheless, the winding resistance is easier to be obtained, e.g., by using the thermocouples buried in the stator winding in Chapter 3.

5.5.1.2 Estimation of dq-axis Inductances

When the inductance saliency ΔL is determined, the d- and q-axis inductances, L_d, L_q, can be estimated by a simple method by utilizing the dq-axis voltage equations at two different rotor speeds. The complete estimation flowchart is shown in Figure 5.20.

At two different rotor speeds of ω_3 and ω_4 without rotor position offset, the d-axis voltage equations at different rotor speeds can be expressed as follows:

$$u_{d3} = u^*_{d3} + DdV_{\text{dead3}} = R_s i_{d3} - \omega_{r3} L_q i_{q4} \tag{5.41a}$$

$$u_{d4} = u^*_{d4} + DdV_{\text{dead4}} = R_s i_{d4} - \omega_{r4} L_q i_{q4} \tag{5.41b}$$

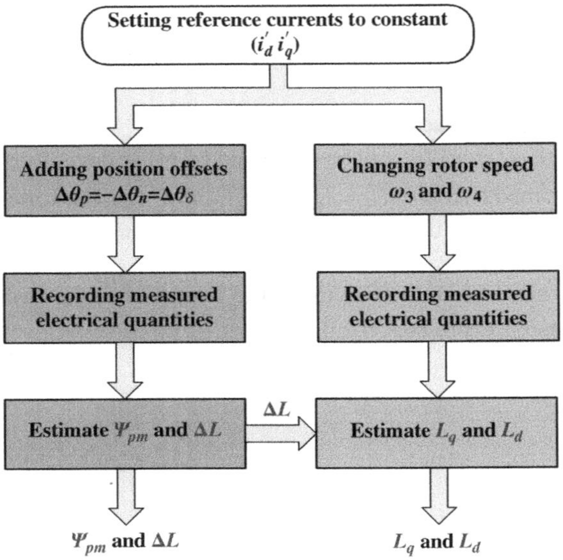

Figure 5.20 Complete estimation flowchart for rotor PM flux linkage, inductance saliency, and dq-axis inductances when $i_d \neq 0$.

In addition, since the currents are fixed, i.e., $i_{d3} = i_{d4} = i_d'$ and $i_{q3} = i_{q4} = i_q'$, it can be deduced that $DdV_{\text{dead3}} = DdV_{\text{dead4}}$. Furthermore, the q-axis inductance L_q can be estimated from subtracting Eq. (5.41a)–(5.41b) and expressed as follows:

$$\hat{L}_q = \frac{u_{d3}^* - u_{d4}^*}{i_q(\omega_{r4} - \omega_{r3})} \tag{5.42}$$

It needs to be mentioned that the speed difference between ω_{r3} and ω_{r4} needs to ensure the adequate ratio of signal to noise in the resulting d-axis voltage difference $u_{d3}^* - u_{d4}^*$.

With the earlier estimated saliency term ΔL in Eq. (5.36), the d-axis inductance L_d can be calculated as follows:

$$\hat{L}_d = \hat{L}_q + \Delta L \tag{5.43}$$

Obviously, the dq-axis flux linkages, i.e., ψ_d and ψ_q, can also be calculated as follows:

$$\psi_d = L_d i_d + \psi_{pm} \tag{5.44a}$$

$$\psi_q = L_q i_q \tag{5.44b}$$

5.5.2 Experimental and FEA Results

Prototype Machines 1 (which is a SPMSM) and Prototype Machines 2 (which is an IPMSM) under the FOC system are used for experimental illustration. At each load point of data measurement, the dq-axis currents should be set to constants by the dq-axis current regulator, while the rotor speed is tuned to be constant by the external load, e.g., a dc machine in this study. Besides, as reported in Ref. [15], the non-ideal signal measurement will usually result in a fluctuation in the estimation results due to the ill-conditioned issue. This issue could be mitigated by a surface fitting to the estimation results [16]. In this study, a two-variable quadratic polynomial function is employed and expressed as follows:

$$f(i_d, i_q) = a\,i_d + b\,i_q + c\,i_d^2 + d i_q^2 + e\,i_d i_q + g \tag{5.45}$$

where a, b, c, d, and e are the coefficients obtained by using the least square regression.

Figures 5.21 and 5.22 show the estimated dq-axis inductances and rotor PM flux linkage, as well as the estimation errors of the dq-axis flux linkages against FEA results.

First, the estimated rotor PM flux linkages in both PMSMs slightly increase with the amplitude $i_d < 0$ due to the demagnetization effect in the d-axis, while decreasing with the amplitude of i_q due to the increased saturation level.

Second, the estimated q-axis inductance for the Prototype Machine 1 decreases with the amplitude of i_q but slightly increases with the amplitude of i_d, which agrees well with the FEA results. However, the marginally opposite tendency of the q-axis inductance for Prototype Machine 2 in the d-axis may be caused by the non-ideal position measurement, since its mounted encoder only has a lower resolution of 2400 lines.

Third, the estimated d-axis inductances for both machines increase with the amplitude of i_d, whereas slightly decrease with i_q.

Besides, it is worth mentioning that when i_q is relatively small, the estimated q-axis flux linkages ψ_q will exhibit a significant error, as shown in Figures 5.21e and 5.22e. It can be explained that when there is a small non-ideal position measurement error $\Delta\theta$, the dq-axis flux linkages are rewritten as follows:

$$\psi_d = L_d i_d + \psi_{pm} \cos(\Delta\theta) \tag{5.46a}$$

$$\psi_q = L_q i_q + \psi_{pm} \sin(\Delta\theta) \tag{5.46b}$$

It can be easily found that when i_q approaches zero, the role of $\psi_{pm} \sin(\Delta\theta)$ will be more dominant. In contrast, the influence of $\Delta\theta$ is negligible in the calculation of ψ_d, since $\cos(\Delta\theta)$ is almost equal to zero.

Moreover, to demonstrate the online tracking performance of the POPE method considering the variations of rotor PM flux linkage, Prototype Machine 1 is employed and heated for 25 min. Due

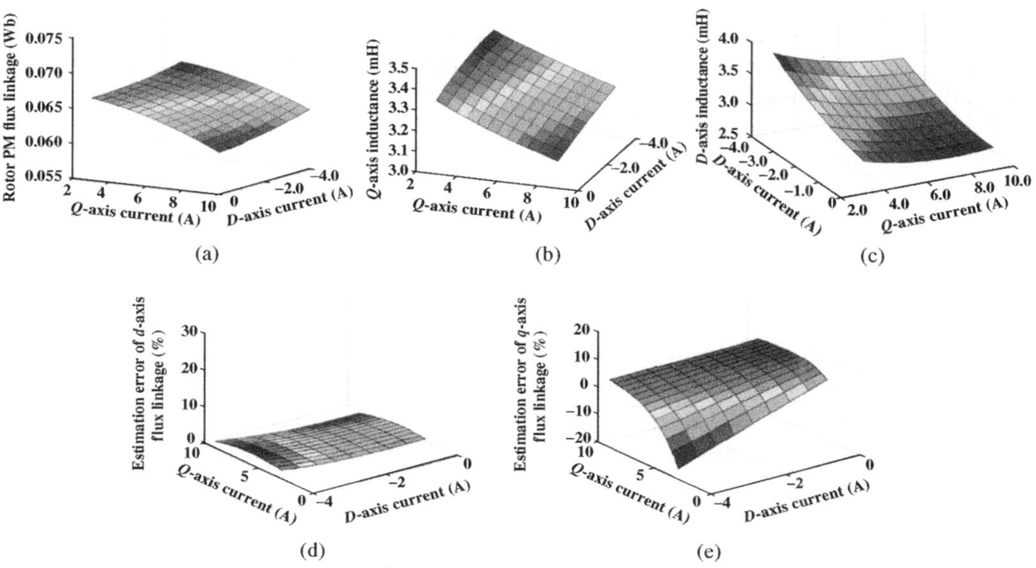

Figure 5.21 Estimated parameters and dq-axis flux linkages errors against FEA results for non-salient-pole SPMSM (Prototype Machine 1) [5], John Wiley & Sons: (a) rotor PM flux linkage, (b) Q-axis inductance, (c) D-axis inductance, (d) D-axis flux linkage error, and (e) Q-axis flux linkage error.

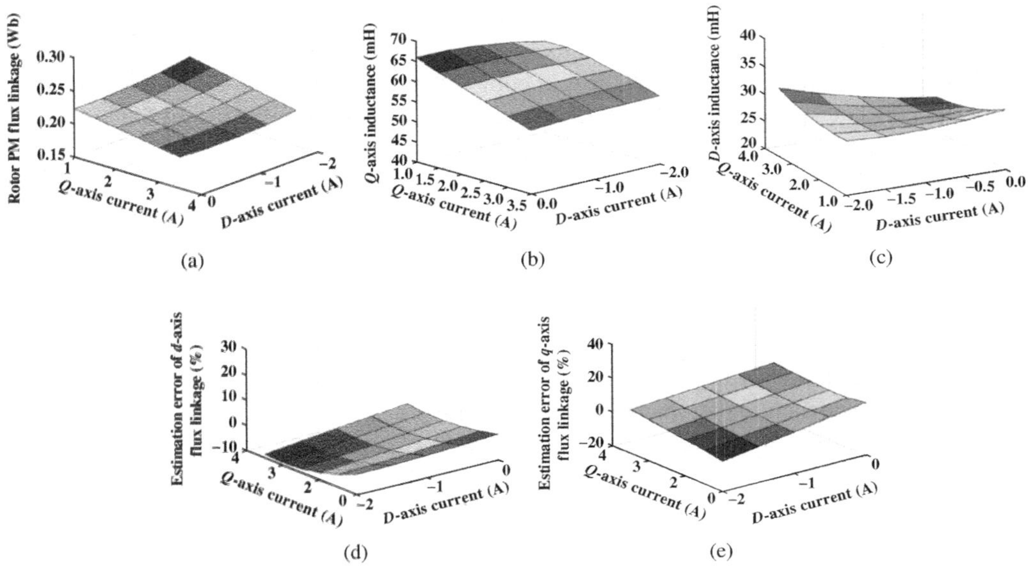

Figure 5.22 Estimated parameters and *dq*-axis flux linkage errors against FEA results for salient-pole IPMSM (Prototype Machine 2) [5], John Wiley & Sons: (a) rotor PM flux linkage, (b) *Q*-axis inductance, (c) *D*-axis inductance, (d) *D*-axis flux linkage error, and (e) *Q*-axis flux linkage error.

to the difficulty of direct measurement of PM temperatures, an infrared thermometer is used to measure the housing temperature to indicate the thermal behaviour of the machine, as shown in Figure 5.23a. The measured (from back-EMF) and estimated rotor PM flux linkages are compared in Figure 5.23b. As can be seen, the measured and estimated rotor PM flux linkages exhibit the same tendencies, and their differences are due to the saturation resulting from the current excitation. The peak housing temperature is at about $t = 25\,\text{min}$, while the minimum value of the PM flux linkage is at about $t = 40\,\text{min}$. This can be explained that the heat dissipation from the rotor is more difficult compared to that in the stator.

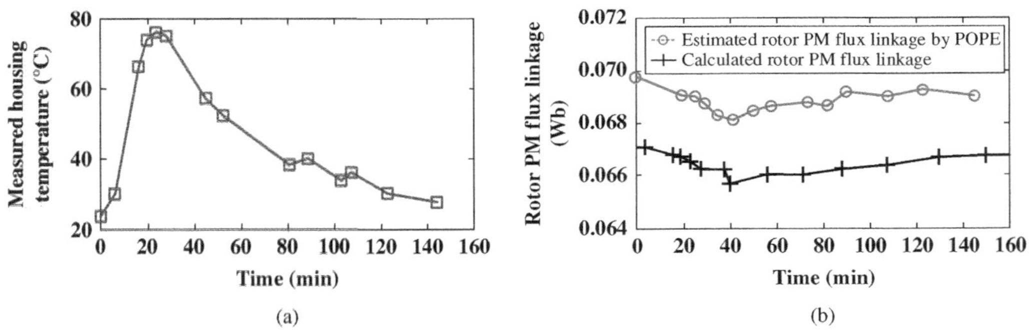

Figure 5.23 Variations of measured/estimated rotor PM flux linkages with measured housing temperature for non-salient-pole SPMSM (Prototype Machine 1) [5], John Wiley & Sons: (a) measured housing temperature and (b) measured and estimated rotor PM flux linkages.

5.6 Position Offset-based Estimation with $i_d = 0$ and $i_d \neq 0$ Under Constant and Variable Speed Control

5.6.1 Designed Estimation Method

The designed estimation method is illustrated in Figure 5.24, which is the same as those in the previous sections. The positive and negative position offsets are injected subsequently, and the dq-axis current, the rotor speed, as well as the reference dq-axis voltages are recorded and used for the estimation.

According to Eqs. (5.1)–(5.3), the dq-axis voltage equations with positive and negative position offsets can be expressed as follows:

Addition of positive position offset:

$$u_{d1}^{*}(k) + Dd_1(k)V_{\text{dead1}} = R_s i_{d1}(k) - \omega_{r1}(k)\left[L_q i_{q1}(k) + \psi_{pm}\sin\left(\Delta\theta + \Delta\theta_p\right)\right] \tag{5.47a}$$

$$u_{q1}^{*}(k) + Dq_1(k)V_{\text{dead1}} = R_s i_{q1}(k) + \omega_{r1}(k)\left[L_d i_{d1}(k) + \psi_{pm}\cos\left(\Delta\theta + \Delta\theta_p\right)\right] \tag{5.47b}$$

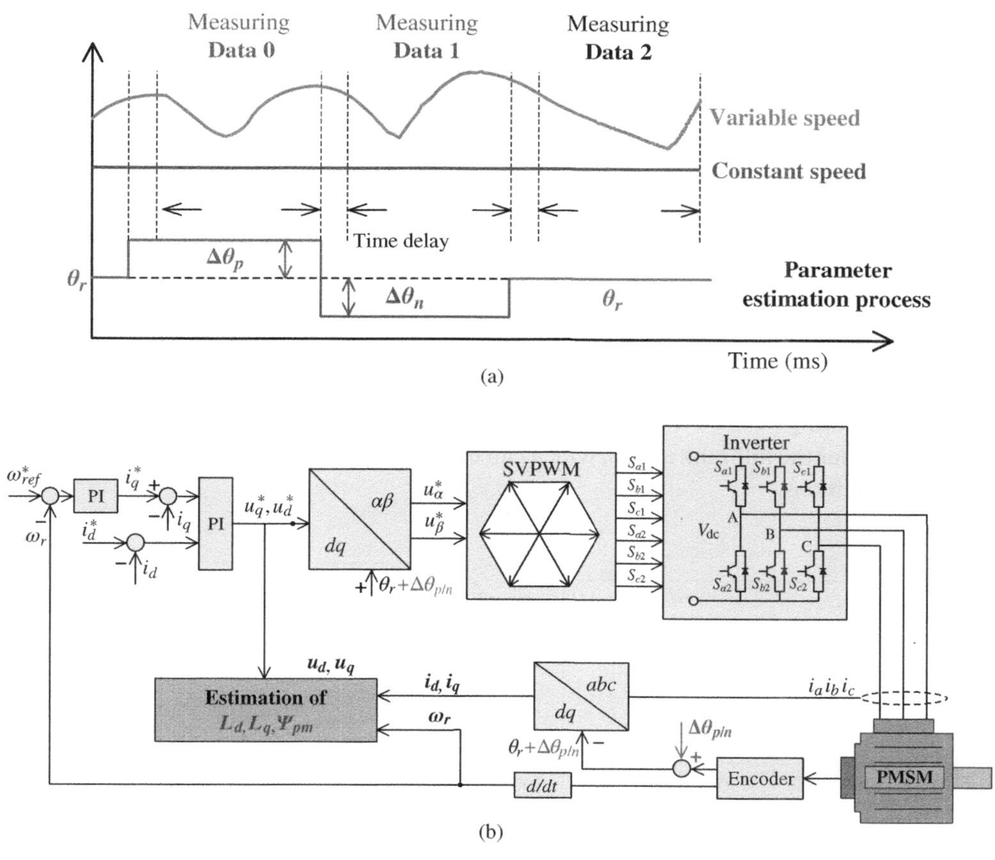

Figure 5.24 General POPE method with $i_d = 0$ and $i_d \neq 0$ under constant/variable speed control: (a) principle and measurement process and (b) POPE under FOC system.

Addition of negative position offset:

$$u_{d2}^{*}(k) + Dd_2(k)V_{\text{dead2}} = R_s i_{d2}(k) - \omega_{r2}(k)\left[L_q i_{q2}(k) + \psi_{pm}\sin(\Delta\theta + \Delta\theta_n)\right] \tag{5.48a}$$

$$u_{q2}^{*}(k) + Dq_2(k)V_{\text{dead2}} = R_s i_{q2}(k) + \omega_{r2}(k)\left[L_d i_{d2}(k) + \psi_{pm}\cos(\Delta\theta + \Delta\theta_n)\right] \tag{5.48b}$$

Since the injected position offset $\Delta\theta_{p(n)}$ is small, the resultant changes in the *dq*-axis currents can be neglected. Therefore, it is reasonable to assume that $V_{\text{dead1}} = V_{\text{dead2}}$. Besides, the gain functions of the VSI non-linearity, i.e., $Dd_{1,2}$ and $Dq_{1,2}$, depend on the current vector angle γ, as presented in Figure 5.25.

When a position offset is injected, the current vector angel γ remains constant, and thus, $\overline{Dd} = \overline{Dd}_1 = \overline{Dd}_2$ and $\overline{Dq} = \overline{Dq}_1 = \overline{Dq}_2$. Therefore, the *dq*-axis voltage deviations under positive and negative position offsets can be expressed as follows:

$$\begin{aligned}\overline{u}_{q1}^{*} - \overline{u}_{q2}^{*} = &+\left(\overline{\omega}_{r1}\overline{i}_{d1} - \overline{\omega}_{r2}\overline{i}_{d2}\right)\cos(\Delta\theta_p)L_d + (\overline{\omega}_{r1} - \overline{\omega}_{r2})\cos(\Delta\theta_p)\psi_{pm} \\ &+ \left[(\overline{\omega}_{r1} - \overline{\omega}_{r2})\overline{i}_q - \left(\overline{\omega}_{r1}\overline{i}_{d1} - \overline{\omega}_{r2}\overline{i}_{d2}\right)\cos(\Delta\theta_p)\right]L_q\end{aligned} \tag{5.49}$$

$$\begin{aligned}\overline{u}_{d1}^{*} - \overline{u}_{d2}^{*} = &+\left(\overline{\omega}_{r1}\overline{i}_{d1} - \overline{\omega}_{r2}\overline{i}_{d2}\right)\sin(\Delta\theta_p)L_d + (\overline{\omega}_{r1} + \overline{\omega}_{r2})\sin(\Delta\theta_p)\psi_{pm} \\ &- \left[(\overline{\omega}_{r1} - \overline{\omega}_{r2})\overline{i}_q + \left(\overline{\omega}_{r1}\overline{i}_{d1} + \overline{\omega}_{r2}\overline{i}_{d2}\right)\sin(\Delta\theta_p)\right]L_q\end{aligned} \tag{5.50}$$

It can be easily found that although the VSI non-linearity V_{dead} and the stator resistance R_s are cancelled in Eqs. (5.49) and (5.50), they are still not enough to estimate three unknown parameters, i.e., L_d, L_q, and ψ_{pm}, due to rank-deficient issue. Hence, the machine state without position offset injection is introduced to establish the full-rank estimation model.

Similarly, to eliminate the distorted voltage term V_{dead} in the *dq*-axis voltage without position offset, Eq. (5.1) can be rewritten as follows:

$$\overline{Dd}\,\overline{u}_{q0}^{*} - \overline{Dq}\,\overline{u}_{d0}^{*} = R_s\left(\overline{Dd}\,\overline{i}_q - \overline{Dq}\,\overline{i}_d\right) + \overline{Dd}\,\omega_{r0}\overline{i}_d L_d + \overline{Dq}\,\omega_{r0}\overline{i}_q L_q + \overline{Dd}\,\omega_{r0}\psi_{pm} \tag{5.51}$$

where the subscript "0" denotes the sampled data without position offset injection.

Furthermore, since $\overline{Dd}\,\overline{i}_q - \overline{Dq}\,\overline{i}_d = 0$, Eq. (5.51) can be simplified as follows:

$$\overline{Dd}\,\overline{u}_{q0}^{*} - \overline{Dq}\,\overline{u}_{d0}^{*} = \overline{Dd}\,\omega_{r0}\overline{i}_d L_d + \overline{Dq}\,\omega_{r0}\overline{i}_q L_q + \overline{Dd}\,\omega_{r0}\psi_{pm} \tag{5.52}$$

Finally, by combining three state equations, i.e., Eqs. (5.49), (5.50), and (5.52), L_d, L_q, and ψ_{pm} can be estimated under the full-rank model, which can be performed under both variable and constant rotor speed conditions.

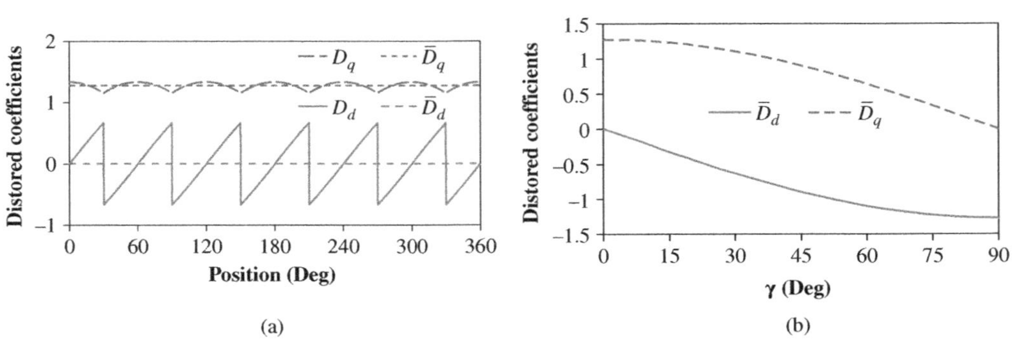

Figure 5.25 Gain functions of *Dd* and *Dq* [8], IEEE: (a) gain functions of *Dd* and *Dq* at $\gamma = 0$ and (b) variations of mean values of $\overline{Dd}$ and $\overline{Dq}$ with current vector angle γ.

Under variable speed control:

When $\omega_{r0} \neq \omega_{r1} \neq \omega_{r2}$, the complete estimation equation can be written as follows:

$$\begin{bmatrix} L_d \\ L_q \\ \psi_{pm} \end{bmatrix} = \begin{bmatrix} a_1 & b_1 & c_1 \\ a_2 & b_2 & c_2 \\ a_3 & b_3 & c_3 \end{bmatrix}^{-1} \begin{bmatrix} \overline{u}_{q1}^* - \overline{u}_{q2}^* \\ \overline{u}_{d1}^* - \overline{u}_{d2}^* \\ \overline{Dd}\,\overline{u}_{q0}^* - \overline{Dq}\,\overline{u}_{d0}^* \end{bmatrix} \tag{5.53a}$$

$$\text{with} \begin{cases} a_1 = \left(\overline{\omega}_{r1}\overline{i}_{d1} - \overline{\omega}_{r2}\overline{i}_{d2}\right)\cos\left(\Delta\theta_p\right) \\ a_2 = \left(\overline{\omega}_{r1}\overline{i}_{d1} + \overline{\omega}_{r2}\overline{i}_{d2}\right)\sin\left(\Delta\theta_p\right) \\ a_3 = \overline{Dd}\,\omega_{r0}\overline{i}_d \end{cases} \tag{5.53b}$$

$$\begin{cases} b_1 = \left(\overline{\omega}_{r1} - \overline{\omega}_{r2}\right)\overline{i}_q - \left(\overline{\omega}_{r1}\overline{i}_{d1} - \overline{\omega}_{r2}\overline{i}_{d2}\right)\cos\left(\Delta\theta_p\right) \\ b_2 = \left(\overline{\omega}_{r2} - \overline{\omega}_{r1}\right)\overline{i}_q - \left(\overline{\omega}_{r1}\overline{i}_{d1} + \overline{\omega}_{r2}\overline{i}_{d2}\right)\sin\left(\Delta\theta_p\right) \\ b_3 = \overline{Dq}\,\omega_{r0}\overline{i}_q \end{cases} \tag{5.53c}$$

$$\begin{cases} c_1 = \left(\overline{\omega}_{r1} - \overline{\omega}_{r2}\right)\cos\left(\Delta\theta_p\right) \\ c_2 = \left(\overline{\omega}_{r1} + \overline{\omega}_{r2}\right)\sin\left(\Delta\theta_p\right) \\ c_3 = \overline{Dd}\,\omega_{r0} \end{cases} \tag{5.53d}$$

Under constant speed control:

When the PMSM is operated at the constant speed, i.e., $\omega_{r0} = \omega_{r1} = \omega_{r2} = \omega_r$, the estimation equation has the same form as Eq. (5.53) and can be simplified as follows:

$$\begin{bmatrix} L_d \\ L_q \\ \psi_{pm} \end{bmatrix} = \begin{bmatrix} -\overline{\omega}_r\overline{i}_q\sin\left(2\Delta\theta_p\right) & \overline{\omega}_r\overline{i}_d\sin\left(2\Delta\theta_p\right) & \overline{Dd}\omega_r\overline{i}_d \\ \overline{\omega}_r\overline{i}_q\sin\left(2\Delta\theta_p\right) & -\overline{\omega}_r\overline{i}_d\sin\left(2\Delta\theta_p\right) & \overline{Dq}\omega_r\overline{i}_q \\ 0 & 2\overline{\omega}_r\sin\left(2\Delta\theta_p\right) & \overline{Dd}\,\omega_r \end{bmatrix}^{-1} \begin{bmatrix} \overline{u}_{q1}^* - \overline{u}_{q2}^* \\ \overline{u}_{d1}^* - \overline{u}_{d2}^* \\ \overline{Dd}\,\overline{u}_{q0}^* - \overline{Dq}\,\overline{u}_{d0}^* \end{bmatrix} \tag{5.54}$$

As can be found, when $i_d \neq 0$ control, L_d, L_q, and ψ_{pm} are involved in the estimation equations (Eq. 5.53). Otherwise, when $i_d = 0$, only L_q and ψ_{pm} can be also estimated except L_d, since the value of the voltage term $\omega_r L_d i_d$ becomes 0 and inactive.

Furthermore, the cost function is designed to find the global optimum in Eqs. (5.53) and (5.54), which is expressed as follows:

$$P\left(\widehat{L}_d, \widehat{L}_q, \widehat{\psi}_{pm}\right) = \sum_{i=1}^{3}\left[\left(a_i\widehat{L}_d + b_i\widehat{L}_q + c_i\widehat{\psi}_{pm}\right) - f_i\right] \tag{5.55}$$

where $f = \left[\overline{u}_{q1}^* - \overline{u}_{q2}^* \quad \overline{u}_{d1}^* - \overline{u}_{d2}^* \quad \overline{Dd}\,\overline{u}_{q0}^* - \overline{Dq}\,\overline{u}_{d0}^*\right]^T$. The optimization problem can be solved by any gradient-based or evolutionary algorithms, as introduced in Chapter 10.

5.6.2 Experimental and FEA Results

The experiments are carried out based on an IPMSM (Prototype Machine 2) drive platform with a dSPACE controller for illustration, as shown in Figure 5.26. The nominal values of L_d, L_d, and ψ_{pm} are 40 mH, 60 mH, and 250.5 mWb, respectively, and the detailed specifications are listed in Appendix B. In addition, the method in Ref. [17] is used to compensate for the control delay in the digital system.

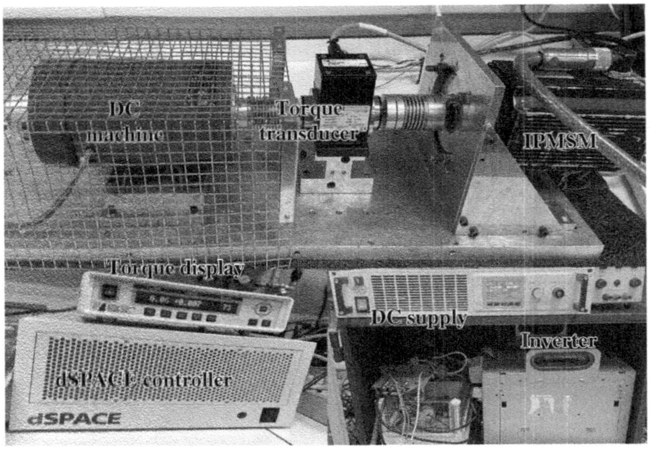

Figure 5.26 Experimental platform and Prototype Machine 2.

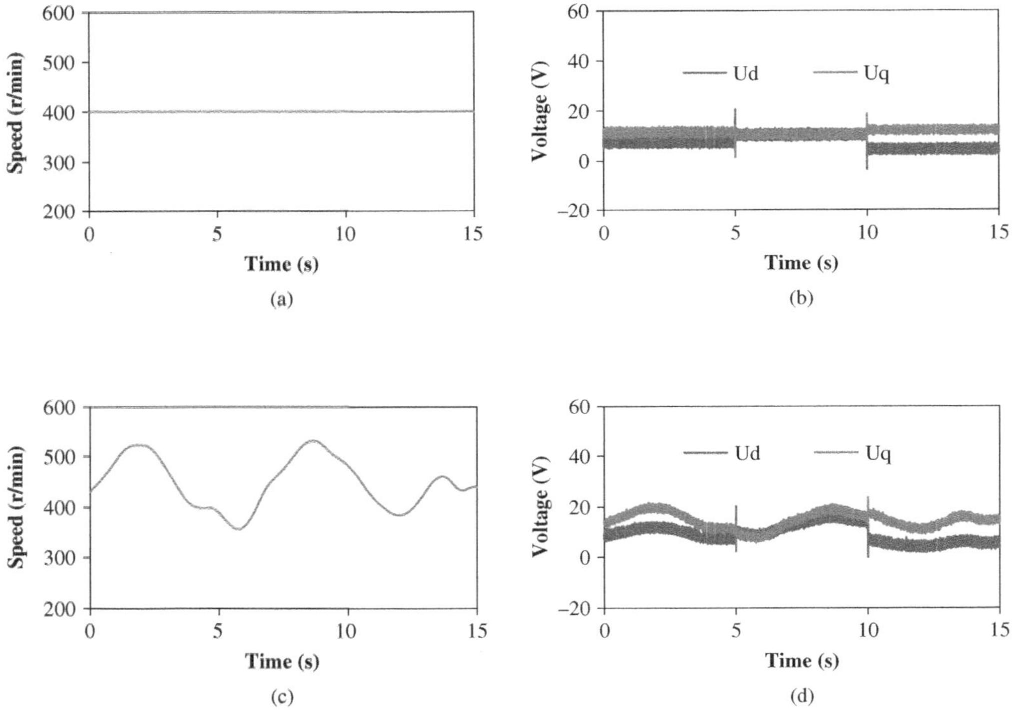

Figure 5.27 Measured rotor speeds and reference *dq*-axis voltages under constant and variable speed controls [8], IEEE: (a) constant speed of 400 r/min and (b) *Dq*-axis voltages at constant speed; (c) variable speed and (d) *Dq*-axis voltages at variable speed.

When the amplitude of positive and negative position offsets is set to 5°, the described POPE method is implemented under both constant and variable speed controls. The time durations of zero, positive, and negative position offsets are all 5 s. The measured rotor speeds and the reference *dq*-axis voltages under constant and variable rotor speeds are shown in Figure 5.27, where the

d-axis current $i_d = -1$ A. The estimated results are shown in Figure 5.28, where the estimated L_d, L_q, and ψ_{pm} are converged to 40.66 mH, 58.75 mH, and 252.97 mWb, and the errors are 5.4%, 2.75%, and 0.9%, respectively.

Afterwards, the dq-axis inductance maps estimated by POPE under constant and variable speed conditions are compared with the FEA results, as presented in Figures 5.29 and 5.30, and the estimation errors are shown in Figure 5.31.

As can be observed, the estimated dq-axis inductance maps match well with the FEA value under both constant and variable speed conditions. With increasing d-axis demagnetizing current, the estimated d-axis inductance increases first and then keeps constant generally. It can be explained that the d-axis demagnetizing current will counter the flux linkage of the d-axis magnetic circuit, which reduces magnetic saturation. Furthermore, with the increase in the q-axis load current, the estimated q-axis inductance decreases due to magnetic saturation.

Furthermore, Figure 5.32 presents the variations in estimation errors with different dead times of VSI. Clearly, with increasing the dead time from 1 to 4 μs, the errors of L_d, L_d, and ψ_{pm} are almost kept constant at 4%, 3.8%, and 0.8%, respectively. Therefore, it can be concluded that the influence of VSI non-linearity has been fully eliminated in the designed estimation method.

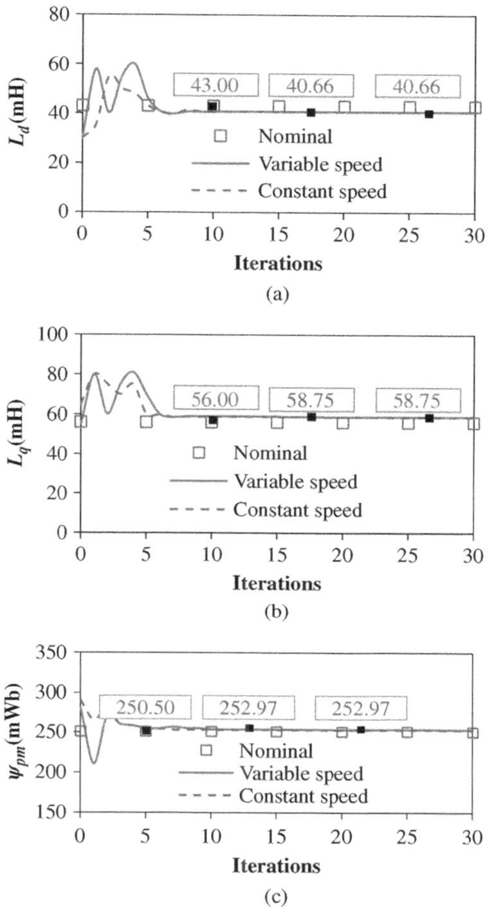

Figure 5.28 Estimated parameters under constant and variable speed operations [8], IEEE: (a) estimated d-axis inductance L_d, (b) estimated q-axis inductance L_q, and (c) estimated rotor PM flux linkage ψ_{pm}.

5.7 Analysis of Amplitude of Position Offset Injection

In the POPE-based methods, the amplitude of position offsets is critical. A small value would give rise to considerable estimation error due to the low SNR. In contrast, a large value would deteriorate system stability. Based on engineering experience, the amplitudes of position offset are set as 3.6° and 2.65° for Prototype Machines 1 and 2 in Sections 5.3–5.5, while it is 5° for Prototype Machine 2.

This section analyses the selection of the upper and lower limits of position offset injection in depth, which can be applied to the machine in different voltage classes.

On the one hand, the upper limit of position offset can be selected through the PI parameter setting of the current loop, as shown in Figure 5.33. More specifically, Figure 5.33a–c shows the typical current

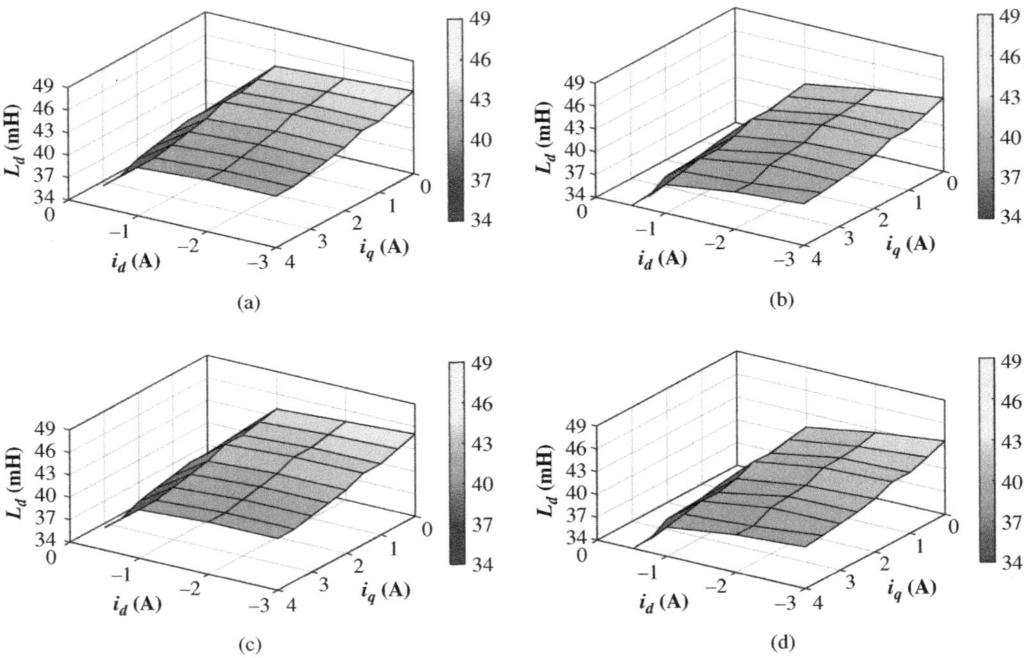

Figure 5.29 FEA and POPE estimated d-axis inductance maps with i_d and i_q at constant and variable speed operations [8], IEEE: (a) POPE estimated L_d at constant speed, (b) FEA-predicted L_d at constant speed, (c) POPE estimated L_d at variable speed, and (d) FEA-predicted L_d at variable speed.

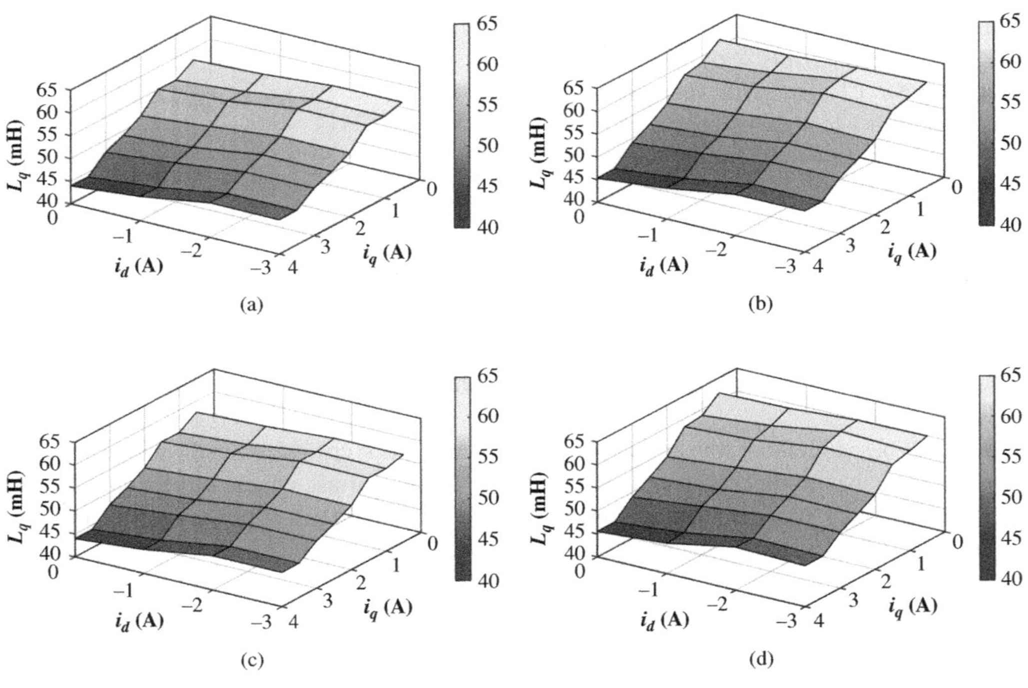

Figure 5.30 FEA and POPE estimated q-axis inductance maps with i_d and i_q at constant and variable speed operations [8], IEEE: (a) POPE estimated L_q at constant speed, (b) FEA-predicted L_q at constant speed, (c) POPE estimated L_q at variable speed, and (d) FEA-predicted L_q at variable speed.

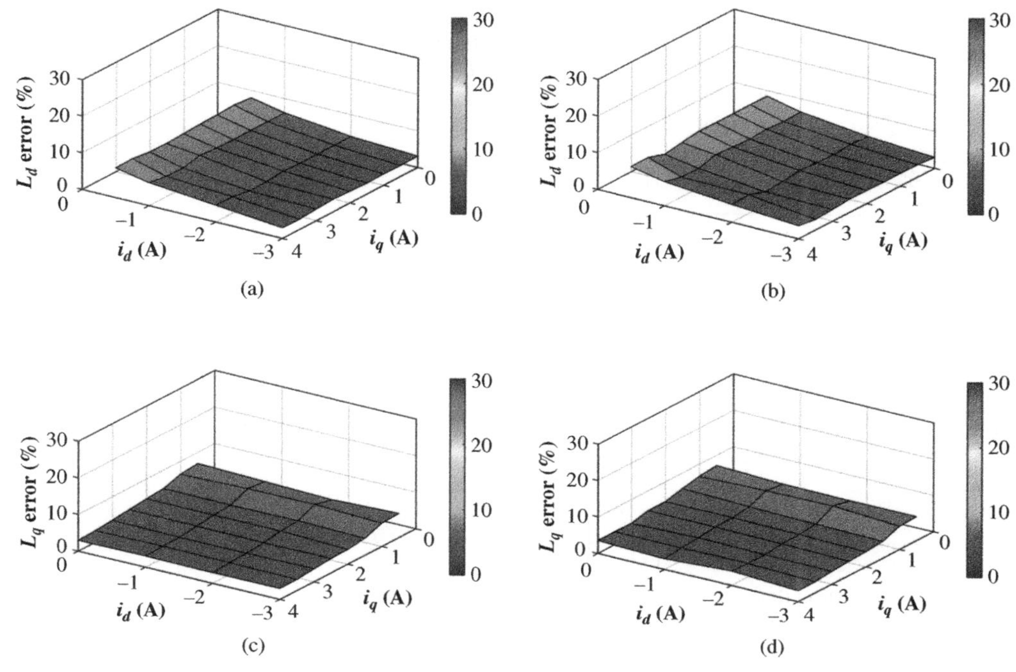

Figure 5.31 Estimation errors of dq-axis inductance maps with i_d and i_q by POPE against FEA results [8], IEEE: (a) estimation error of L_d at constant speed, (b) estimation error of L_d at variable speed, (c) estimation error of L_q at constant speed, and (d) estimation error of L_q at variable speed.

loop block, the current loop after position offset is injected (where the reference position of the q-axis voltage is changed), and its equivalent transformation, where $T_{\Sigma i} = T_{\mathrm{PWM}} + T_{ic}$ lumps two small time constants.

As can be seen from Figure 5.33c, the control object of the current loop is a dual inertia system and can be simplified as an I-type system by using a PI regulator (see Figure 5.34). Therefore, the transfer function of the PI regulator is written as follows:

$$W_{\mathrm{ACR}}(s) = \frac{K_p(\tau_i s + 1)}{\tau_i s} \tag{5.56}$$

Furthermore, the open-loop transfer function is represented as follows:

$$G(s)H(s) = \frac{K_I}{s(1 + T_{\Sigma i} s)} \tag{5.57a}$$

with

$$K_I = \frac{K_p K_{\mathrm{PWM}} \csc(\alpha) \sin\left(\alpha - \Delta\theta_p\right)}{R\tau_i} \tag{5.57b}$$

According to log magnitude–frequency characteristics in Eq. (5.57), to keep the stability of I-type system, the condition should be satisfied as follows:

$$K_I \leq \frac{1}{T_{\Sigma i}} \tag{5.58}$$

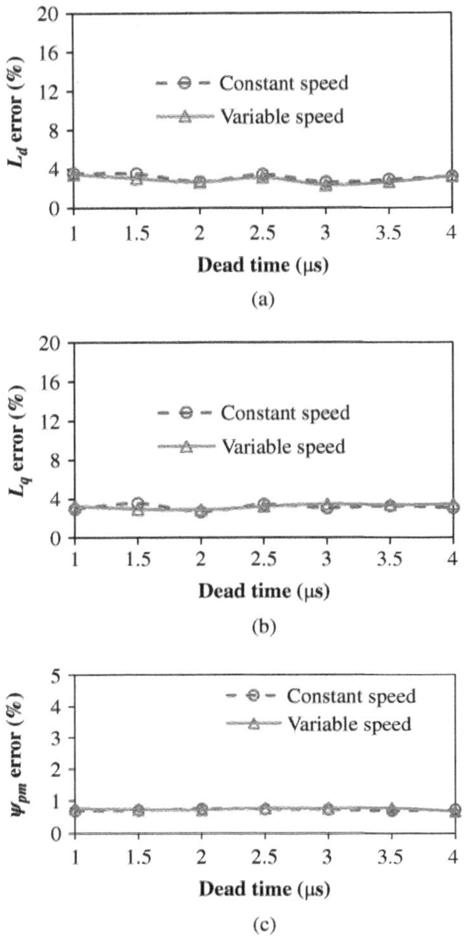

Figure 5.32 Variations of parameter estimation error with different dead time of VSI [8], IEEE: (a) estimation error of L_d, (b) estimation error of L_q, and (c) estimation error of ψ_{pm}.

According to Eq. (5.58), the larger the K_I is, the larger the cut-off frequency ω_c is, and thus, the faster the system response is, but the smaller the phase margin is. Therefore, to maintain the stability of the drive system, $\Delta\theta_p$ should be smaller than the maximum value $\Delta\theta_{p,\max}$.

On the other hand, the injected position offset should be large enough to make a large voltage fluctuation Δu to ensure a higher SNR. Most drive systems can guarantee that SNR $\geq$ 48 dB [3, 18].

According to Eqs. (5.49) and (5.50), when the position offsets are injected, the fluctuation between u_{d1}, u_{d2} and u_{q1}, u_{q2} are described as follows:

$$u_{q1}^* - u_{q1}^* = (L_q - L_q)\omega_r i_q \sin(2\Delta\theta_p) \tag{5.59}$$

$$\begin{aligned} u_{d1}^* - u_{d2}^* &= \omega_r i_d \sin(2\Delta\theta_p)(L_d - L_q) + 2\omega_r \sin(\Delta\theta_p)\psi_{pm} \\ &= 2\omega_r \sin(\Delta\theta_p)\psi_{pm} - \frac{\left(u_{q1}^* - u_{q2}^*\right) i_d}{i_q} \\ &\geq 2\omega_r \sin(\Delta\theta_p)\psi_{pm} \end{aligned} \tag{5.60}$$

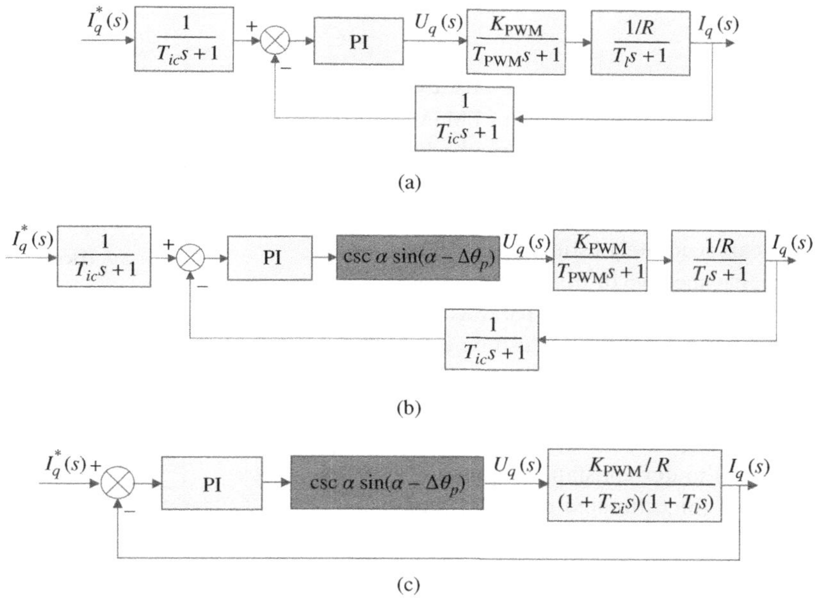

Figure 5.33 Schematic diagrams of current loop before and after position offset injection [8], IEEE: (a) before position offset injection, (b) after position offset injection, and (c) equivalent transformation (after position offset injection).

Therefore, the voltage fluctuation Δu can be expressed as follows:

$$\Delta u = \min\{(L_q - L_d)\omega_r i_q \sin(2\Delta\theta_p), 2\omega_r \sin(\Delta\theta_p)\psi_{pm}\} \tag{5.61}$$

The SNR of the drive system in different voltage classes can be calculated by [18] the following:

$$\text{SNR}_{\text{dB}} = 20\log_{10}\left(\frac{V_{\text{dc}}/\sqrt{3}}{u_N}\right) \tag{5.62}$$

where V_{dc} is the dc supply voltage and u_N is the system noise.

Since the voltage fluctuation Δu should be larger than the noise u_N, the lower limit of position offset is expressed as follows:

$$\Delta\theta_p \geq \max\left\{\frac{1}{2(L_q - L_d)\omega_r i_q}\frac{V_{\text{dc}}/\sqrt{3}}{10^{\text{SNR}_{\text{dB}}/20}}, \frac{1}{2\omega_r\psi_{\text{pm}}}\frac{V_{\text{dc}}/\sqrt{3}}{10^{\text{SNR}_{\text{dB}}/20}}\right\} \tag{5.63}$$

The estimation errors, including L_d, L_q, and ψ_{pm}, at different amplitudes of position offset injection are shown in Figure 5.35 based on Prototype Machine 2 for illustration. The parameter estimation errors decrease with increasing amplitudes of position offset injection and then keep constant when $\Delta\theta_p$ is larger than 2°, which is the lower limits of position offset due to SNR issue.

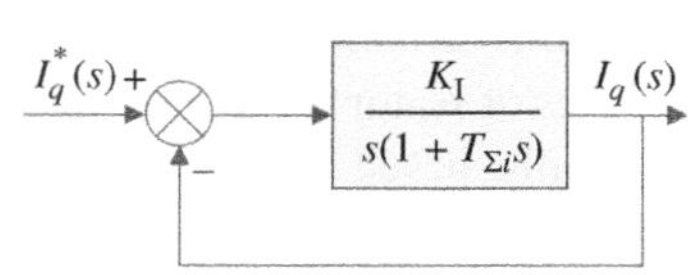

Figure 5.34 Typical I-type system [8], IEEE.

5.8 Summary

This chapter has analysed the influence of rotor position offsets on the dq-axis equation of PMSM first and then described three POPE methods, i.e., $i_d = 0$ and $i_d \neq 0$ under constant torque/speed and variable speed controls. It shows that the POPE methods can accurately estimate the rotor PM flux linkage, irrespective of the influences of VSI non-linearity and other machine parameters. The other electrical parameters can also be estimated subsequently, and any nominal parameter values of PMSMs and additional measurement instruments are not required.

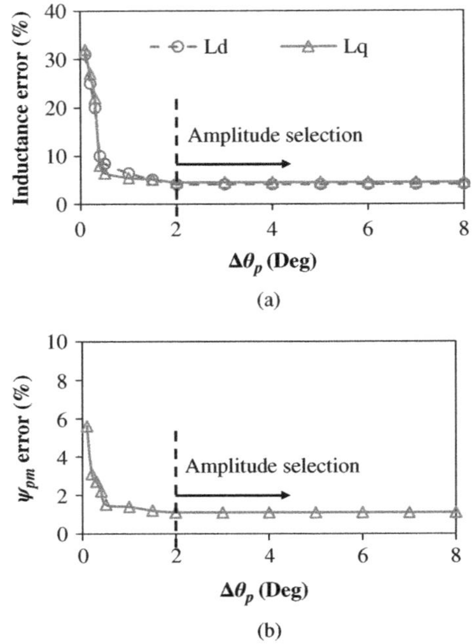

Figure 5.35 Variations of estimation errors with different amplitudes of injected position offset [8], IEEE: (a) estimation error of dq-axis inductances and (b) estimation error of rotor PM flux linkage.

First, when $i_d = 0$ under constant torque control, with the aid of the estimated rotor PM flux linkage by POPE, the stator winding resistance and the q-axis inductance can be estimated from the q- and d-axis voltage equations, respectively. However, the estimation accuracies of stator winding resistance and q-axis inductance are severely affected by the VSI non-linearity and the non-ideal rotor position measurement, which need to be compensated beforehand. On the other hand, since the d-axis inductance is inactive when $i_d = 0$, its estimation relies on introducing another excitation, e.g., position offset addition or d-axis current injection.

Second, when $i_d = 0$ under variable speed control, the estimation of rotor PM flux linkage is similar to the POPE under constant torque/speed control. Afterwards, the distorted voltage caused by the VSI non-linearity, the q-axis inductance, and the winding resistance can be estimated separately. It has been proved that the described estimation method is appropriate for condition monitoring of PMSM.

Third, when $i_d \neq 0$ at constant rotor speed, the described estimation method could estimate the rotor PM flux linkage and saliency term of dq-axis inductances, where the influence of current-dependent VSI non-linearity can be fully eliminated. Furthermore, the dq-axis inductances can be estimated at two different rotor speeds.

Fourth, when three machine states, including zero, positive, and negative position offsets, are introduced in the designed estimation method, three unknown parameters, including dq-axis inductances and rotor PM flux linkage, can be estimated simultaneously, which can be used under constant/variable speed controls. Meanwhile, the built estimation method eliminates the uncertainties of stator resistance and distorted voltage due to VSI non-linearity.

Finally, based on the stability and SNR analyses, the upper and lower limits of position offset injection have also been analysed. Overall, the POPE methods do not require current changes to build a full-rank model, and the addition of small position offsets has negligible influence on the control performance and the output torque capability. It has proved that all three estimation methods exhibit good agreements with the FEA and experimental results for both salient-pole and non-salient-pole PMSMs.

References

1 D. W. Novotny and T. A. Lipo, *Vector Control and Dynamics of AC Drives*. Oxford: Clarendon Press, 1997.

2 Z.Q. Zhu and X. M. Wu, *Sensorless Control of Permanent Magnet Synchronous Machine Drives*. New York: Wiley and IEEE Press, 2023.

3 K. Liu and Z. Q. Zhu, “Position-offset-based parameter estimation using the Adaline NN for condition monitoring of permanent-magnet synchronous machines,” *IEEE Trans. Ind. Electron.*, vol. 62, no. 4, pp. 2372–2383, April 2015.

4 K. Liu and Z. Q. Zhu, “Position offset-based parameter estimation for permanent magnet synchronous machines under variable speed control,” *IEEE Trans. Power Electron.*, vol. 30, no. 6, pp. 3438–3446, June 2015.

5 K. Liu, Jianghua Feng, Shuying Guo, Lei Xiao, Z.Q. Zhu, “Improved position offset based parameter determination of permanent magnet synchronous machines under different load conditions,” *IET Electric Power Appl.*, vol. 11, no. 4, pp. 603–612, April 2017.

6 K. Yu, Z. Wang, X. Wang and Z. Zou, “An online flux estimation for dual three-phase SPMSM drives using position-offset injection,” *IEEE Trans. Power Electron.*, vol. 36, no. 10, pp. 11606–11617, Oct. 2021.

7 K. Liu, “Novel techniques for parameter estimation of permanent magnet synchronous machines,” PhD dissertation, Uni. Sheffield, U.K., 2013.

8 P. Wang, Z. Q. Zhu, and D. Liang, “Improved position-offset based online parameter estimation of PMSMs under constant and variable speed operations,” *IEEE Trans. Energy Convers.*, vol. 39, no. 2, pp. 1325–1340, June 2024.

9 K. Lee, D. Jung, and I. Ha, “An online identification method for both stator resistance and back-EMF coefficient of PMSMs without rotational transducers,” *IEEE Trans. Ind. Electron.*, vol. 51, no. 2, pp. 507–510, April 2004.

10 Y.S. Chen, “Motor topologies and control strategies for permanent magnet brushless AC drives,” Ph.D dissertation, Univ. Sheffield, Sheffield, U.K., 1999.

11 *ANSYS*. Ansys, inc. [Online]. Available: https://www.ansys.com.

12 T.M. Jahns, G.B. Kliman, and T.W. Neumann, “Interior permanent-magnet synchronous motors for adjustable-speed drives,” *IEEE Trans. Ind. Appl.*, vol. IA-22, no. 4, pp. 738–747, Jul./Aug. 1986.

13 S. Morimoto, M. Sanada, and Y. Takeda, “Wide-speed operation of interior permanent magnet synchronous motors with high-performance current regulator,” *IEEE Trans. Ind. Appl.*, vol. 30, no. 4, pp. 920–926, Jul./Aug. 1994.

14 C.T. Pan, and S.M. Sue, “A linear maximum torque per ampere control for IPMSM drives over full-speed range,” *IEEE Trans. Energy Conver.*, vol. 20, no. 2, pp. 359–366, Jun. 2005.

15 W. Xu and R. D. Lorenz, “High-frequency injection-based stator flux linkage and torque estimation for DB-DTFC implementation on IPMSMs considering cross-saturation effects,” *IEEE Trans. Ind. Appl.*, vol. 50, no. 6, pp. 3805–3815, Nov./Dec. 2014.

16 B. Stumberger, G. Stumberger, D. Dolinar, A. Hamler, and M. Trlep, “Evaluation of saturation and cross-magnetization effects in interior permanent-magnet synchronous motor,” *IEEE Trans. Ind. Appl.*, vol. 39, no. 5, pp. 1264–1271, Sept./Oct. 2003.

17 B. H. Bae and S. K. Sul, “A compensation method for time delay of full-digital synchronous frame current regulator of PWM AC drives,” *IEEE Trans. Ind. Appl.*, vol. 39, no. 3, pp. 802–810, May 2003.

18 R. D. Lorenz, T. A. Lipo, and D. W. Novotny, “Motion control with induction motors,” *Proc. IEEE*, vol. 82, no. 8, pp. 1215–1240, Aug. 1994.

6

Online Parameter Estimation Under Variable Speed Control

6.1 Introduction

To improve the control performance and monitor the thermal conditions of permanent magnet synchronous machines (PMSMs), accurate knowledge of electrical parameters under different load conditions is critical. Although the finite element (FE) method can provide sufficient information on electromagnetic characteristics, accurate geometric dimensions and magnetic properties of different materials are not always available. Therefore, by using the measured quantities, i.e., rotor speed, three-phase currents, and voltages, several online parameter estimation methods have been described in the previous chapters with the aid of current/position-offset injections or additional measurement devices to estimate the electrical parameters of PMSMs.

Unlike the constant speed control, when the PMSM operates at variable speeds, the full-rank estimation model can be easily established by using the recorded PMSM states at different speeds [1–3], but without introducing additional disturbance signals.

This chapter describes two online parameter estimation methods under variable speed control, and the principle is illustrated in Figure 6.1.

In Section 6.2, an electrical parameter estimation method will be described. It has the following features: (a) It can estimate the rotor PM flux linkage and the stator resistance separately irrespective of the value of the *d*-axis inductance when $i_d = 0$. (b) The identifiability of all estimated parameters is considered, and the nominal machine parameter values are not required. (c) The influence of VSI non-linearity on the estimated rotor PM flux linkage is cancelled, whereas its influence on the estimated winding resistance can be well compensated. (d) To estimate the *d*-axis inductance, the output torque is fixed, while a small direct current (dc) offset $\Delta i_d \neq 0$ is also added to the *d*-axis current for a short period.

In Section 6.3, a *dq*-axis flux linkage map estimation method will be introduced, which was originally developed to aid the derivation of maximum torque per ampere (MTPA) trajectory. To fulfil the requirement, the *dq*-axis flux linkages and their whole maps under different load conditions can be identified quickly by minimizing the designed cost functions. Meanwhile, it is worth mentioning that in the estimation process, the influence of the uncertain inverter non-linearity and circuit resistance, consisting of winding resistance, on-state resistance of power devices, and equivalent core-loss resistance, can be fully eliminated.

In addition, both estimation methods are implemented by evolutionary algorithms (EA) and applied to an IPMSM-based drive system. The experimental results are also presented for illustration.

Parameter Estimation of Permanent Magnet Synchronous Machines, First Edition. Zi Qiang Zhu, Kan Liu, and Dawei Liang.

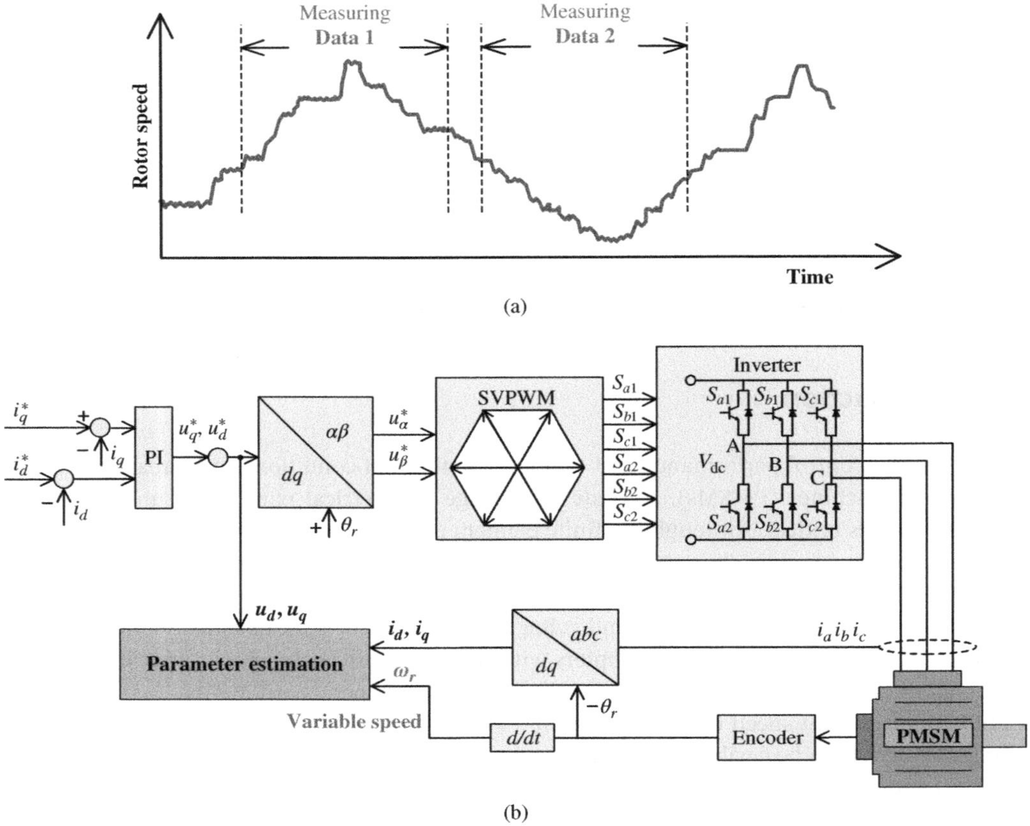

Figure 6.1 Online parameter estimation of PMSM under variable speed control: (a) variable rotor speed and data measurements and (b) schematic diagram of FOC-based parameter estimation.

6.2 Estimation of Stator Resistance, Inductances, and Rotor PM Flux Linkage

The PMSM model in the dq-axis reference frame can be expressed as:

$$u_d = R_s i_d + L_d \frac{di_d}{dt} - \omega_r L_q i_q \tag{6.1a}$$

$$u_q = R_s i_q + L_q \frac{di_q}{dt} + \omega_r (L_d i_d + \psi_{pm}) \tag{6.1b}$$

where u_d, u_q, i_d, and i_q are the dq-axis voltages and currents, respectively. L_d, L_q, and ψ_{pm} are the dq-axis inductances and rotor PM flux linkage. ω_r is the electrical angular speed.

At steady state, the dq-axis equations can be discretized by Euler's method and simplified as follows:

$$u_d(k) = R_s i_d(k) - \omega_r(k) L_q i_q \tag{6.2a}$$

$$u_q(k) = R_s i_q(k) + \omega_r(k) \big(L_d i_d(k) + \psi_{pm}\big) \tag{6.2b}$$

where "k" is the index of the discrete sampling instants in the drive system. It is evident that the steady-state PMSM model is a linear system and consists of four electrical parameters, i.e., R_s, L_d, L_q, and ψ_{pm}. Since the rank of Eq. (6.2) is two, there are infinite solutions when the above four parameters are intended to be estimated simultaneously, which will result in an ill-convergence/rank-deficient issue, as discussed in Chapter 2.

6.2.1 Identifiability Analysis and Influence of VSI Non-linearity

Under variable speed control, the rank-deficient issue can be fully overcome with the aid of independent machine states at different rotor speeds. Hence, all four electrical parameters can be accurately identified from the conventional simple cost function P, which has been used for induction machine [1] and PMSM [4]:

$$P = \sum_{k=1}^{n} \left[K_d \left| i_d(k) - \hat{i}_d(k) \right| + K_q \left| i_q(k) - \hat{i}_q(k) \right| \right] \tag{6.3}$$

where K_d and K_q are the gains of the cost function. i_d, $\hat{i}_d$, i_q, and $\hat{i}_q$ are the measured and calculated dq-axis currents. n is the sampling number.

However, the conventional cost function (Eq. 6.3) easily suffers from the VSI non-linearity, which is dependent upon different working conditions. The reason will be illustrated as follows.

When the influence of VSI non-linearity is considered, Eq. (6.2) can be rewritten as follows:

$$u_d^*(k) + Dd(k)V_{\text{dead}} = R_s i_d(k) - \omega_r(k) L_q i_q \tag{6.4a}$$

$$u_q^*(k) + Dq(k)V_{\text{dead}} = R_s i_q(k) + \omega_r(k)\left(L_d i_d(k) + \psi_{pm}\right) \tag{6.4b}$$

where Dd and Dq are the gains of the distorted voltage V_{dead} due to VSI non-linearity, and their detailed description can be found in Chapter 2. It is obvious that V_{dead} will be introduced into the PMSM model, and consequently, when the VSI non-linearity is not well compensated, it will give rise to non-negligible errors.

Furthermore, when $i_d = 0$, the steady-state PMSM model (6.4) can be further simplified as:

$$u_d^*(k) + Dd(k)V_{\text{dead}} = -\omega_r(k) L_q i_q \tag{6.5a}$$

$$u_q^*(k) + Dq(k)V_{\text{dead}} = R_s i_q(k) + \omega_r(k)\psi_{pm} \tag{6.5b}$$

As can be observed, although the q-axis inductance L_q can be estimated irrespective of the influence of V_{dead} due to the zero-mean Dd, the stator resistance R_s and the rotor PM flux linage ψ_{pm} in the q-axis equation are severely affected by V_{dead} since Dq has a dc harmonic [5]. Therefore, the identifiability of R_s and ψ_{pm} cannot be ensured by only one q-axis Eq. (6.5b). Moreover, even if the estimation model is full rank, e.g., under variable speed control, the estimation of R_s and ψ_{pm} based on Eq. (6.3) will still suffer from the misestimation issue due to the VSI non-linearity.

6.2.2 Improved Estimation Scheme

To address the above issue, this section describes an improved estimation scheme to estimate R_s, ψ_{pm}, L_d, and L_q under variable speed control. The complete estimation procedure consists of six steps, as shown in Figure 6.2.

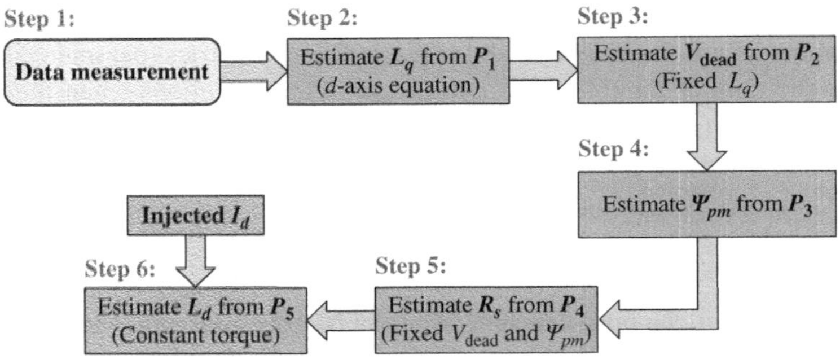

Figure 6.2 Flowchart of estimation scheme under variable speed control.

Step 1. Data Measurement:
A set of measurement quantities, i.e., dq-axis currents, reference voltages, and rotor speed/position, are recorded when $i_d = 0$, where the currents are set to constant to ensure that L_q and ψ_{pm} will not vary during the estimation.

Step 2. Estimation of L_q:
The d-axis equation is used to estimate L_q by minimizing the cost function P_1.

Step 3. Estimation of V_{dead}:
L_q is fixed to the estimated value at Step 2, and V_{dead} is estimated by minimizing the cost function P_2.

Step 4. Estimation of ψ_{pm}:
ψ_{pm} is estimated by minimizing the cost function P_3.

Step 5. Estimation of R_s:
ψ_{pm} and V_{dead} are fixed to their estimated values at Steps 3 and 4, respectively, and R_s is estimated by minimizing the cost function P_4.

Step 6. Estimation of L_d:
A small d-axis current Δi_d is injected to activate the d-axis inductance term under $i_d = 0$ control. Under the assumption of constant electromagnetic torque T_e with and without the addition of Δi_d in Eq. (6.6), L_d can be estimated based on the cost function P_5:

$$T_e = 1.5p\psi_{pm} i_q = 1.5p\left[\psi_{pm} i_{q2} + (L_d - L_q) i_{q2} i_{d2}\right] \tag{6.6}$$

where p is the pole pair number. i_{d2} and i_{q2} are the dq-axis currents and $i_{d2} = \Delta i_d$.

The employed five cost functions, $P_1 - P_5$, are expressed as follows, which are derived from the dq-axis model in Eq. (6.5) and the torque equation in Eq. (6.6):

$$P_1\left(\hat{L}_q, \hat{V}_{\text{dead}}\right) = \sum_{k=1}^{n} \left[K_u\left|u_d^*(k) + \hat{V}_{\text{dead}} Dd(k) + \hat{L}_q i_q(k)\,\omega_r(k)\right|\right] \tag{6.7}$$

$$P_2\left(\hat{V}_{\text{dead}}\right) = \sum_{k=1}^{n} \left[K_v\left|u_d^*(k) + \hat{V}_{\text{dead}} Dd(k) + L_q i_q(k)\,\omega_r(k)\right|\right] \tag{6.8}$$

$$P_3(\hat{\psi}_{pm}) = \sum_{k=1}^{0.5n} \left[K_m\left|u_q^*(k) - u_q^*(k+0.5n) - \hat{\psi}_{pm}[\omega_r(k) - \omega_r(k+0.5n)]\right|\right] \tag{6.9}$$

$$P_4(\hat{R}_s) = \sum_{k=1}^{n}\left[K_w\left|u_q^*(k) - \hat{R}_s i_q(k) - \psi_{pm}\omega_r(k) + V_{\text{dead}}Dq(k)\right|\right] \tag{6.10}$$

$$P_5(\hat{L}_d) = \sum_{k=1}^{n}\left[K_d\left|\left(\hat{L}_d - L_q\right)i_{q2}(k)i_{d2}(k) - \psi_{pm}\left(i_q(k) - i_{q2}(k)\right)\right|\right] \tag{6.11}$$

where n is the length of the measured data. K_u, K_v, K_m, K_w, and K_d are the gains of the cost functions.

The principle of the designed cost functions is analysed hereinafter in detail.

In Step 2, since $Dd(k)V_{\text{dead}}$ is a zero-mean disturbance voltage when $i_d = 0$, the contribution of the distorted voltage V_{dead} to the cost function P_1 is minor compared with that of the term $\hat{L}_q i_q \omega_r$. Consequently, the minimization of P_1 can ensure the estimation accuracy of L_q, but may converge to a local optimum of V_{dead}.

In Step 3, to ensure V_{dead} is converged correctly, L_q is set to the estimated value in P_2, and thus, the term $Dd(k)V_{\text{dead}}$ becomes dominant.

In Step 4, the measured data will be divided into two parts, of which the first part is indexed from 1 to $0.5n$ and the second part is indexed from $0.5n + 1$ to n. Since i_d and i_q are kept constant, V_{dead} remains unchanged during the measurement. Thus, the cost function P_3 can be analyzed as follows:

$$\begin{aligned}\sum_{k=1}^{0.5n}\left[u_q^*(k) - u_q^*(k+0.5n)\right] \approx & \sum_{k=1}^{0.5n}\left\{\psi_{pm}(k)\left[\omega_r(k) - \omega_r(k+0.5n)\right]\right\} \\ & + V_{\text{dead}}\left\{\sum_{k=1}^{0.5n}\left[Dq(k+0.5n) - Dq(k)\right]\right\}\end{aligned} \tag{6.12}$$

Since Dq is not changed if i_d and i_q are kept constant, Eq. (6.12) can be simplified as follows:

$$\sum_{k=1}^{0.5n}\left[u_q^*(k) - u_q^*(k+0.5n)\right] \approx \sum_{k=1}^{0.5n}\left\{\psi_{pm}(k)\left[\omega_r(k) - \omega_r(k+0.5n)\right]\right\} \tag{6.13}$$

Thus, it is obvious that the influence of VSI non-linearity has been cancelled in Eq. (6.13), where the estimation of rotor PM flux linkage ψ_{pm} will achieve high estimation accuracy by minimizing P_3.

In Step 5, with the aid of the estimated V_{dead} and ψ_{pm}, R_s can be estimated with high accuracy by the minimization of cost function P_4.

In Step 6, a disturbance signal $\Delta i_d \neq 0$ is injected to estimate L_d. Since the VSI non-linearity terms in dq-axis voltages, DdV_{dead} and $Dq\,V_{\text{dead}}$, vary with the ratio of i_q to i_d considering the IGBT-fed drive system [6, 7], the dq-axis equation is inappropriate to be employed and the estimated V_{dead} in Step 3 is also changed due to addition of $\Delta i_d \neq 0$. Therefore, when the constant torque (Eq. 6.6) is employed for the estimation of L_d, the influence of VSI non-linearity can be eliminated.

6.2.3 Experimental Results

The designed estimation scheme is applied to an IPMSM (Prototype Machine 2 in Appendix 1), which is controlled by FOC based on a dSPACE DS1006 processing board. The quantum genetic algorithm (QGA) [8] is employed for estimation, which is introduced in Appendix 2. In addition, the main parameters of the IPMSM are tabulated in Table 6.1.

The measured rotor speed, the Dd and Dq, the dq-axis currents, and voltages are presented in Figure 6.3, where the IPMSM operates at the steady state of constant torque/speed control before

Table 6.1 Specification and measured parameters of Prototype Machine 2 – IPMSM.

Parameters	Value
Rated current/rotor speed	4 A/1000 r/min
dc-link voltage	158 V
Measured phase resistance (20 °C)	6.00 Ω
Measured terminal wire resistance	0.20 Ω
Measured *dq*-axis inductance	38.1 mH/55.00 mH
Measured rotor PM flux linkage	230.0 mWb

$t = 2.5$ s and will be switched to work under variable speed control after $t = 2.5$ s. The measured data under variable speed controls at low and medium-/high-speed regions are designated as S1 and S2, respectively.

By using the measured data in the low-speed region (S1) under variable speed control, the estimated R_s and ψ_{pm} without and with the compensation of VSI non-linearity are shown in Figures 6.4 and 6.5, respectively. It is evident that the estimated rotor PM flux linkages are always converged to the same value of 0.224 Wb since the influence of VSI non-linearity has been cancelled in the cost function P_3. In addition, the estimated PM flux linkages are slightly smaller than the nominal value (0.236 Wb). It can be explained that both magnetic saturation and temperature rise may reduce the PM flux linkage.

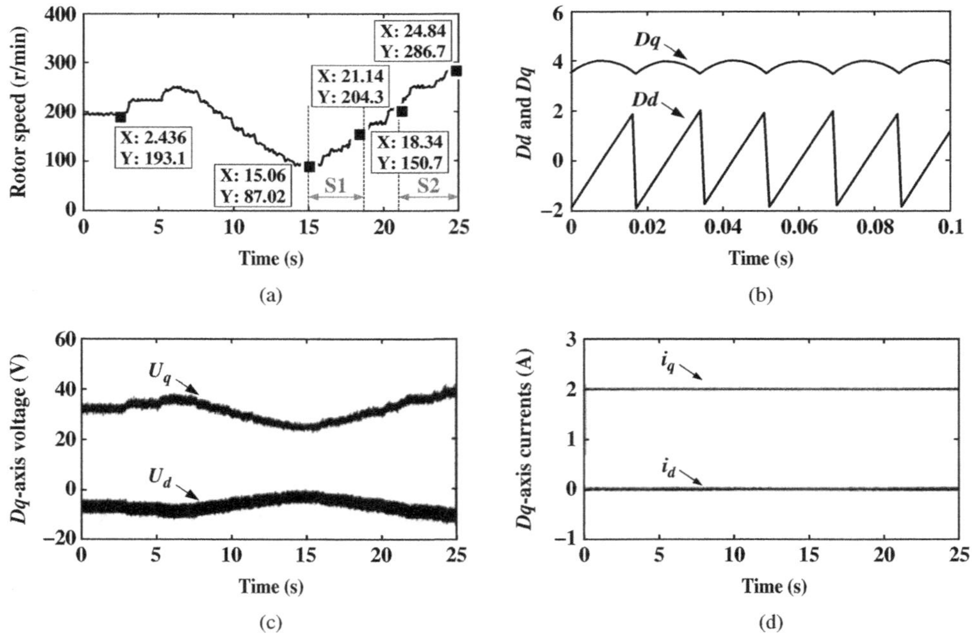

Figure 6.3 Measured data under variable speed control [1], IEEE: (a) rotor speed, (b) *Dd* and *Dq*, (c) *Dq*-axis voltages, and (d) *Dq*-axis currents.

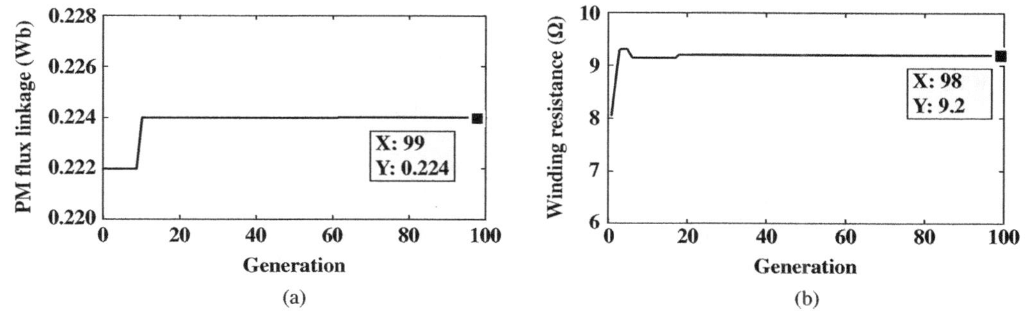

Figure 6.4 Estimated parameters under variable speed control without compensation of VSI non-linearity at low-speed range (S1) [1], IEEE: (a) estimated PM flux linkage and (b) estimated winding resistance.

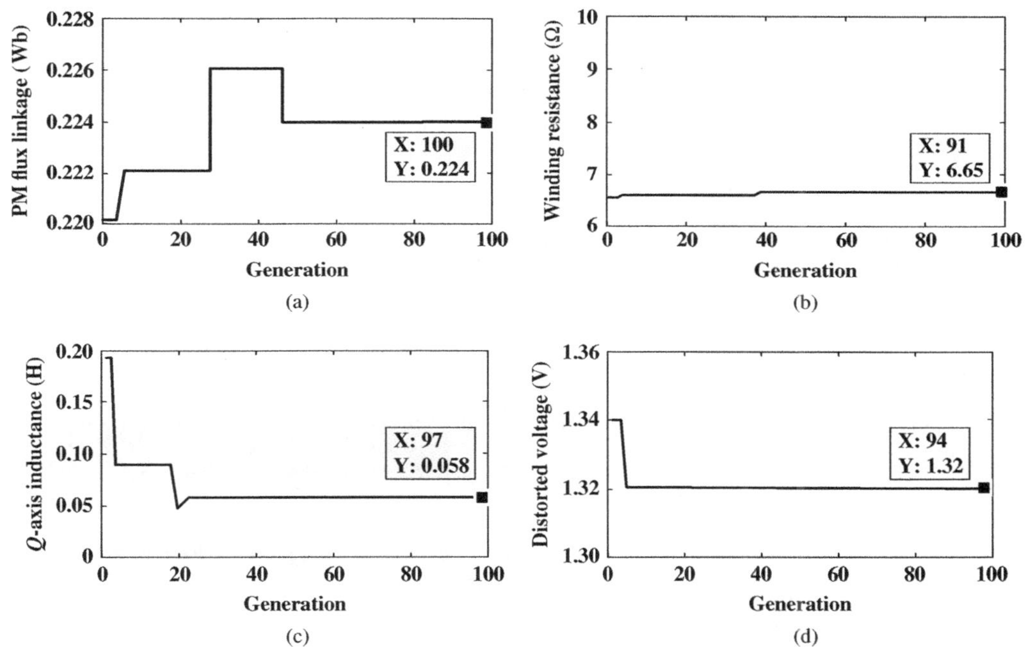

Figure 6.5 Estimated parameters under variable speed control with compensation of VSI non-linearity at low-speed range (S1): (a) estimated PM flux linkage, (b) estimated winding resistance, (c) estimated q-axis inductance, and (d) estimated voltage distortion.

Nevertheless, it is worth mentioning that when the VSI non-linearity is not compensated, the estimated stator resistance (9.20 Ω) is much larger than that with consideration (6.65 Ω), and the nominal stator resistance is 6.20 Ω. The slight difference may be attributed to the influences of temperature rise, on-state resistance of power device [9], and core-loss resistance [10].

At the medium-/high-speed region (S2), the estimated parameters are shown in Figure 6.6. As can be observed, the estimated values are close to the nominal values listed in Table 6.1, which underpins that the described estimation method is not sensitive to low or high speed.

Besides, for the estimation of the d-axis inductance, the load torque will be fixed during the data measurement, and a dc offset $\Delta i_d = 0.2$ A is added into the d-axis current, as shown in Figure 6.7.

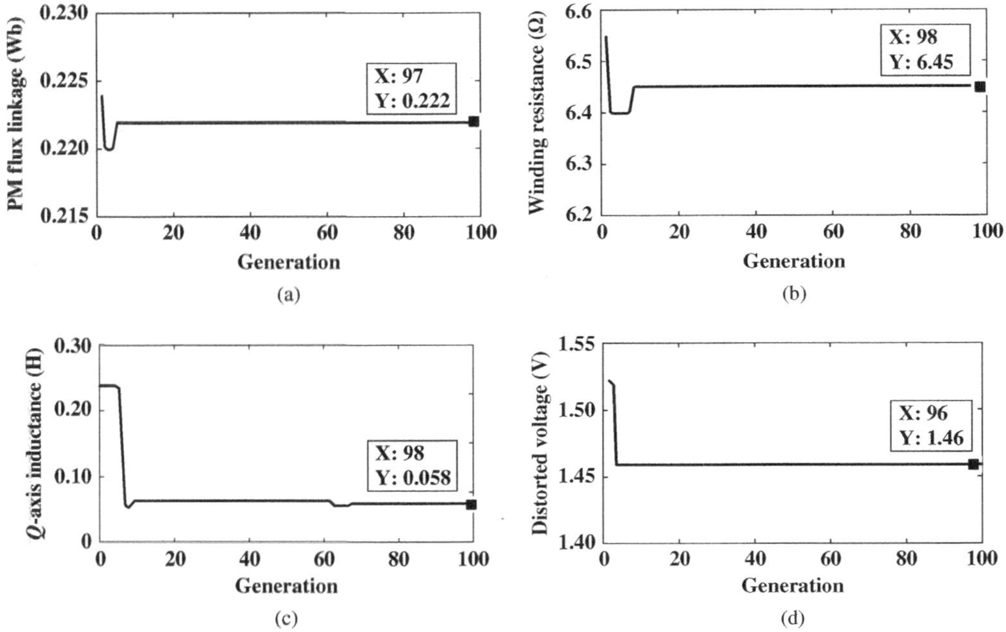

Figure 6.6 Estimated parameters under variable speed control with compensation of VSI non-linearity at the medium-/high-speed range (S2) [1], IEEE: (a) estimated PM flux linkage, (b) estimated winding resistance, (c) estimated q-axis inductance, and (d) estimated voltage distortion.

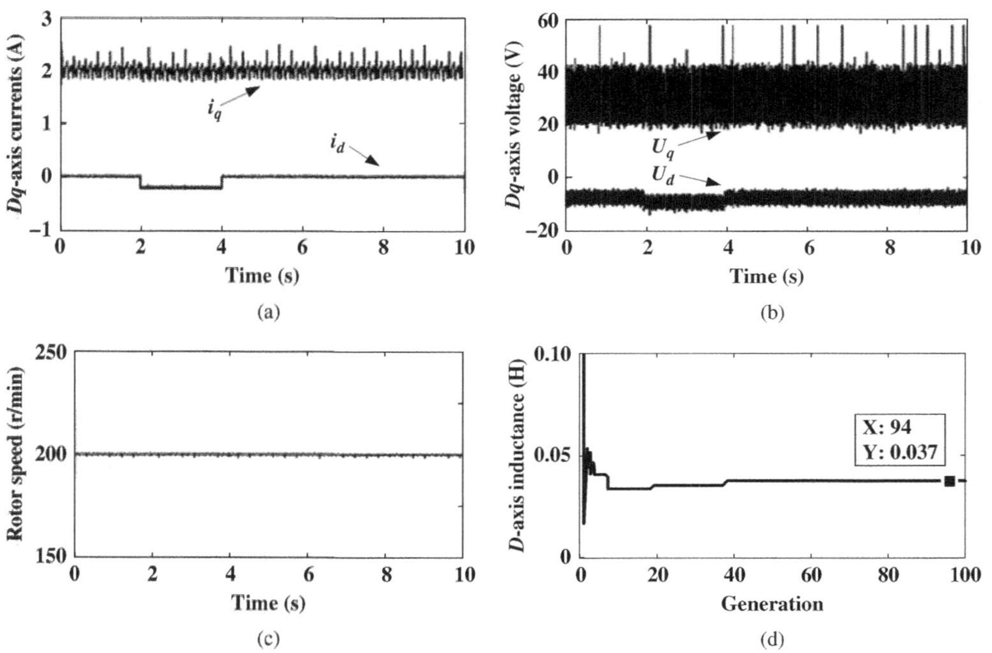

Figure 6.7 Estimation of d-axis inductance under constant load torque [1], IEEE: (a) measured dq-axis currents, (b) measured dq-axis voltages, (c) measured rotor speed, and (d) estimated d-axis inductance.

The estimated rotor PM flux linkage and q-axis inductance are introduced in the estimation, as illustrated in Section 6.2.2. As can be seen from Figure 6.7d, the estimated inductance is 37.0 mH, which agrees well with the nominal value of 38.1 mH.

In addition, the amplitude of Δi_d should not be too large. Otherwise, it will result in significant changes in the rotor PM flux linkage and dq-axis inductances. On the other hand, a too-small amplitude of Δi_d will result in a low ratio of signal versus noise, and an ill-conditioned problem may be introduced into the estimation [11].

6.3 Estimation of *dq*-axis Flux Linkage Maps with Uncertainties of Circuit Resistance and Inverter Non-linearity

The dq-axis flux linkages can be lumped in the PMSM model in Eq. (6.14) to represent the contributions of PMs and current excitations on the dq-axes. The accurate flux linkages can be used to determine the optimal current trajectories of MTPA for improvement in output torque production.

$$u_d = Ri_d + \frac{d\psi_d}{dt} - \psi_q \omega \tag{6.14a}$$

$$u_q = Ri_q + \frac{d\psi_q}{dt} + \psi_d \omega \tag{6.14b}$$

The dq-axis flux linkages can be represented as follows:

$$\psi_d = L_d i_d + L_{dq} i_q + \psi_{pm} \cos(\Delta\theta) \tag{6.15a}$$

$$\psi_q = L_q i_q + L_{qd} i_d + \psi_{pm} \sin(\Delta\theta) \tag{6.15b}$$

where $\Delta\theta$ is the inherent error due to non-ideal position measurement and is usually close to zero.

By using Euler's method, the steady-state PMSM model with discrete sampling instant k can be simplified as

$$u_d(k) = Ri_d(k) - \psi_q(k)\,\omega(k) \tag{6.16a}$$

$$u_q(k) = Ri_q(k) + \psi_d(k)\,\omega(k) \tag{6.16b}$$

Normally, the terminal voltages cannot be directly measured due to high-frequency PWM. Thus, the reference voltages are usually used, whereas the influence of VSI non-linearity must be considered:

$$u_d^*(k) + V_{\text{dead}} Dd(k) = u_d(k) = Ri_d - \psi_q(k)\,\omega(k) \tag{6.17a}$$

$$u_q^*(k) + V_{\text{dead}} Dq(k) = u_q(k) = Ri_q + \psi_d(k)\,\omega(k) \tag{6.17b}$$

where V_{dead} is the distorted voltage. Dd and Dq are the gains of VSI non-linearity.

6.3.1 Modelling of Cost Functions for Identification of Flux Linkages

Under variable speed control, two sets of data, Data 1 and Data 2, corresponding to two sets of rotor speeds, are recorded for estimation (see Figure 6.1). If the ith and jth indices are the sampling indices of Data 1 and Data 2, two time-discrete d-axis voltage equations can be obtained as follows:

$$u_d^*(i) + V_{\text{dead}}(i)\,Dd(i) = Ri_d(i) - \psi_q(i)\,\omega(i) \tag{6.18a}$$

$$u_d^*(j) + V_{\text{dead}}(j)\,Dd(j) = Ri_d(j) - \psi_q(j)\,\omega(j) \tag{6.18b}$$

Therefore, the q-axis flux linkage can be obtained as follows:

$$\begin{aligned}\psi_q(j)\omega(j) - \psi_q(i)\omega(i) = {} & R[\,i_d(j) - i_d(i)] + [\,V_{\text{dead}}(i)Dd(i) - V_{\text{dead}}(j)Dd(j)] \\ & + [\,u_d^*(i) - u_d^*(j)]\end{aligned} \tag{6.19}$$

Accounting for the total sampling number N for each data set, Eq. (6.19) becomes:

$$\begin{aligned}\sum_{i=j=1}^{N} \left[\psi_q(j)\omega(j) - \psi_q(i)\omega(i)\right] = {} & \sum_{i=j=1}^{N} \{R[\,i_d(j) - i_d(i)] \\ & + [\,V_{\text{dead}}(i)Dd(i) - V_{\text{dead}}(j)Dd(j)] \\ & + [\,u_d^*(i) - u_d^*(j)]\}\end{aligned} \tag{6.20}$$

Since the dq-axis currents are fixed by the PI regulators, V_{dead} is regarded as a constant. Thus, Eq. (6.20) can be simplified as follows:

$$N\psi_q \sum_{i=j=1}^{N} [\omega(j) - \omega(i)] = \sum_{i=j=1}^{N} [\,u_d^*(i) - u_d^*(j)] \tag{6.21}$$

Ultimately, the q-axis flux linkage ψ_q can be identified by minimizing the cost function P_{fq}, which is expressed as

$$P_{fq}\left(i_d, i_q\right) = \sum_{i=j=1}^{N} \left[\widehat{\psi}_q(i_d, i_q) - \frac{u_d^*(i) - u_d^*(j)}{\omega(j) - \omega(i)}\right] \tag{6.22}$$

where $\widehat{\psi}_q(i_d, i_q)$ is the identified q-axis flux linkage.

Similarly, two discrete-time q-axis voltage equations can be obtained from Data 1 and Data 2:

$$u_q^*(i) + V_{\text{dead}}(i)Dq(i) = Ri_q(i) + \psi_d(i)\,\omega(i) \tag{6.23a}$$

$$u_q^*(j) + V_{\text{dead}}(j)Dq(j) = Ri_q(j) + \psi_d(j)\,\omega(j) \tag{6.23b}$$

The d-axis flux linkage can be identified by minimizing cost function P_{fd}:

$$P_{fd}\left(i_d, i_q\right) = \sum_{i=j=1}^{N} \left[\widehat{\psi}_d(i_d, i_q) \;-\; \frac{u_q^*(i) - u_q^*(j)}{\omega(j) - \omega(i)}\right] \tag{6.24}$$

where $\widehat{\psi}_d(i_d, i_q)$ is the identified d-axis flux linkage.

With the determined flux linkages at different load points, the whole family of flux linkages can be approximately obtained by using the surface-fitting/curve-fitting techniques [12], which can also mitigate the ill-conditioned issue [11]. Here, a two-variable quadratic polynomial function is employed to represent the non-linear feature of dq-axis flux linkage maps for surface fitting:

$$f(x, y) = ax + by + cx^2 + dy^2 + exy + g \tag{6.25}$$

where x and y represent the dq-axis currents with a, b, c, d, e, and g are the to-be-identified fitting coefficients, which can be determined by optimization algorithm.

For globally searching the six unknown constants, two cost functions, P_d and P_q, are designed as follows:

$$P_d = \sum_{m=1}^{M}\sum_{h=1}^{H}\begin{bmatrix} a_q i_d(m) + b_q i_q(h) + c_q i_d^2(m) + d_q i_q^2(h) \\ +e_q i_d(m) i_q(h) + g_q - \hat{\psi}_d(i_d(m), i_q(h)) \end{bmatrix} \quad (6.26)$$

$$P_q = \sum_{m=1}^{M}\sum_{h=1}^{H}\begin{bmatrix} a_d i_d(m) + b_d i_q(h) + c_d i_d^2(m) + d_d i_q^2(h) \\ +e_d i_d(m) i_q(h) + g_d - \hat{\psi}_d(i_d(m), i_q(h)) \end{bmatrix} \quad (6.27)$$

where a_i, b_i, c_i, d_i, and e_i with $i = d$ and q are the optimal constants. M and H are the lengths of the sampling indices. As can be observed, there is no uncertain circuit resistance and inverter non-linearity terms in the above cost functions.

6.3.2 Minimization of Cost Functions

In the parameter identification of electrical machines, EAs, e.g., genetic algorithm, particle swarm optimization, and QGA, are widely employed [1–4, 13]. It has been reported that the hybrid of immune clonal algorithm (ICA) and QGA has superior performance in jumping out of local optimum [14], where the hypermutation operator, the clonal selection operator, and the diversity operator are added to the conventional QGA to improve the diversity of each generation of the population. Thus, the immune clonal-based quantum genetic algorithm (ICQGA) is used in this section to minimize the designed cost functions P_{fd}, P_{fq}, P_d, and P_q. In addition, since the designed cost function is generic, other EAs can also be employed.

The gene length of $\hat{\psi}_i$, a_i, b_i, c_i, d_i, e_i, and g_i with $i = d$ and q are set to 16 bits, respectively, and their values can be represented in the form of Q-bits and expressed as follows:

$$Q = \begin{bmatrix} \alpha_1 & \alpha_2 & \dots & \alpha_i & \dots & \alpha_{16} \\ \beta_1 & \beta_1 & \dots & \beta_i & \dots & \beta_{16} \end{bmatrix} \quad (6.28)$$

where $\alpha_i^2 + \beta_i^2 = 1$. α_i^2 and β_i^2 represent the probability binary "0" and "1" states in the Q-bit, respectively.

On the one hand, for the minimization of P_{fd} and P_{fq}, the values of the encoded $\hat{\psi}_d$ and $\hat{\psi}_q$ are in the range of $[-2\psi_{pm}, 2\psi_{pm}]$, where ψ_{pm} is the nominal value of the rotor PM flux linkage.

On the other hand, for the surface fitting of identified flux linkages, the principle of encoding of all unknown coefficients, i.e., a_i, b_i, c_i, d_i, e_i, and g_i with $i = d$ and q, is introduced below.

1. The dq-axis flux linkages are usually approximately proportional to dq-axis currents when the machine is not deeply saturated. Thus, the first-order polynomial terms, $b_i i_i + g_i$ with $i = d$ and q, are dominant, and they are in the range of $[-2\psi_{pm}, 2\psi_{pm}]$.
2. The terms $a_q i_d$, $c_q i_d^2$, $b_d i_q$, and $d_d i_q^2$ represent the cross-coupling effect and have minor contributions to the total flux linkages. Therefore, their values are limited to a small range, e.g., $[-0.025\psi_{pm}, 0.025\psi_{pm}]$.
3. The terms $d_q i_q^2$, $e_q i_d i_q$, $c_d i_d^2$, and $e_d i_d i_q$ represent the influence of saturation with moderate contributions to the total flux linkage. Thus, their values are set in the range of $[-0.05\psi_{pm}, 0.05\psi_{pm}]$.

Based on the above encoding scheme, the surface fitting of identified dq-axis flux linkages at all load points can be achieved by minimizing Eqs. (6.26) and (6.27).

6.3.3 Experimental Results

The designed estimation scheme is applied to the same IPMSM Prototype Machine 2 (see Section 6.2) based on a vector control system consisting of a dSPACE DS1006 process board and a SEMIX 71GD12E4s inverter. The measurement sampling frequency is 10 kHz.

6.3.3.1 Data Recording and FE-predicted Flux Linkage Maps

In the data measurement, the IPMSM operates under variable speed control, and the waveforms of the measured signals are shown in Figure 6.8. The *dq*-axis currents are fixed by the PI regulators at each load point. The complete data recording process of all load points is depicted in Figure 6.9, where there are 24 sets of data (1 s for each set) are recorded for estimation. The FE-predicted *dq*-axis flux linkages are shown in Figure 6.10, which is used as the benchmark to validate the following estimated results. It is worth mentioning that the influence of the end effect is considered [15].

6.3.3.2 Identification of *dq*-axis Flux Linkage Maps

By utilizing the surface fitting function (Eq. 6.25), the fitness values of the *dq*-axis flux linkages at 12 different load points in the penalty functions (Eqs. 6.26 and 6.27) are shown in Figure 6.11. As can be observed, the fitness value will not change after 70 s, i.e., the identified fitting coefficients converge to the final values. Furthermore, based on the above 12 load points, the identified *dq*-axis flux linkage maps and their estimation errors are presented in Figure 6.12. It is obvious that the results agree well with the FEA results (see Figure 6.10), and the maximum errors are kept within 4% and 2% in the *dq*-axes. Therefore, although the identified surface fitting results are only one of

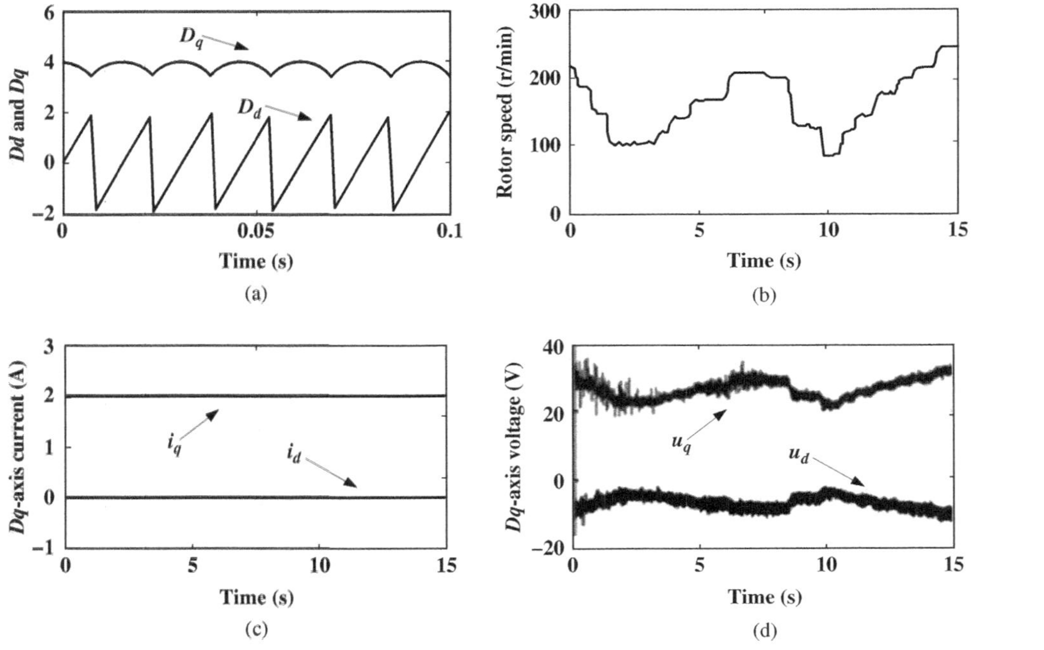

Figure 6.8 Measured data during variable speed control at load point $i_d = 0$, $i_q = 2$ A [2], IEEE: (a) *Dd* and *Dq*, (b) rotor speed, (c) *Dq*-axis currents, and (d) *Dq*-axis voltages.

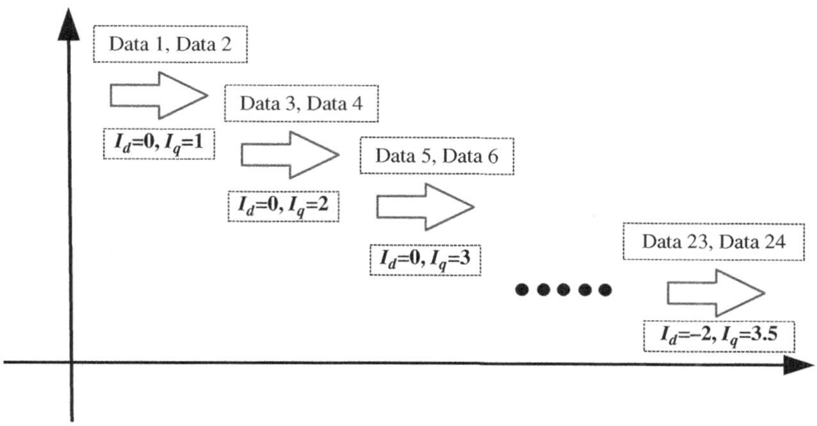

Figure 6.9 Complete data recording process at different load points for estimation of *dq*-axis flux linkage maps [2], IEEE.

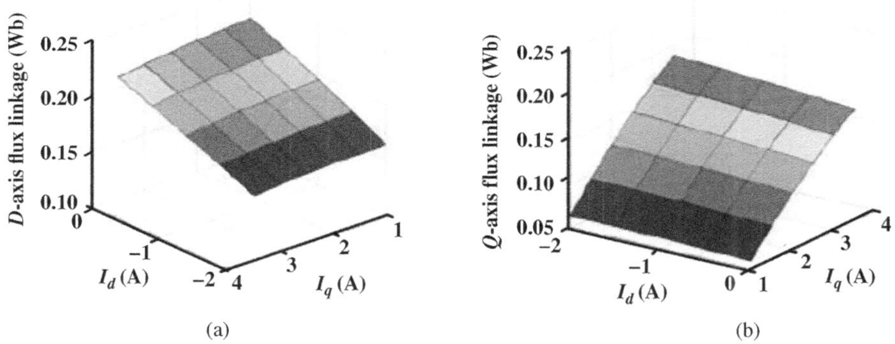

Figure 6.10 FE-predicted *dq*-axis flux linkage maps (end effect included): (a) *D*-axis flux linkage map and (b) *Q*-axis flux linkage map.

the best sub-optimal solutions due to the rank-deficient cost functions, i.e., one equation has six unknown fitting coefficients (a, b, c, d, e, g), it can still be effective in improving the performance of MTPA control because those sub-optimum are close to the FEA results. Meanwhile, the identification method requires much less computational time compared to the FEA method.

In the application of the drive systems with quite large mechanical inertia, e.g., fast-speed train, almost all possible combinations of *dq*-axis currents can be reached, since the range of load torque is wide. By way of example, the braking time for a fast-speed train running at 350 km/h is usually approximately 135 s. When the data measurement for each load point needs 2 s, the described method can record approximately 67 load points, which can cover the major load condition.

6.3.3.3 Influence of Uncertain Circuit Resistance

For an inverter-fed drive system, the circuit resistance consists of winding resistance, on-state resistance of power devices, and equivalent core-loss resistance. The on-state resistance of power devices usually does not change at a fixed current, while the core loss resistance usually varies with the excitation frequency non-linearly [10].

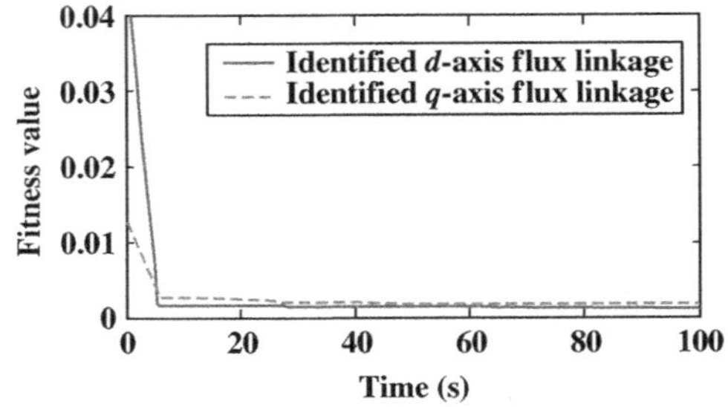

Figure 6.11 Fitness of ICQGA-based identification of *dq*-axis flux linkage maps [2], IEEE.

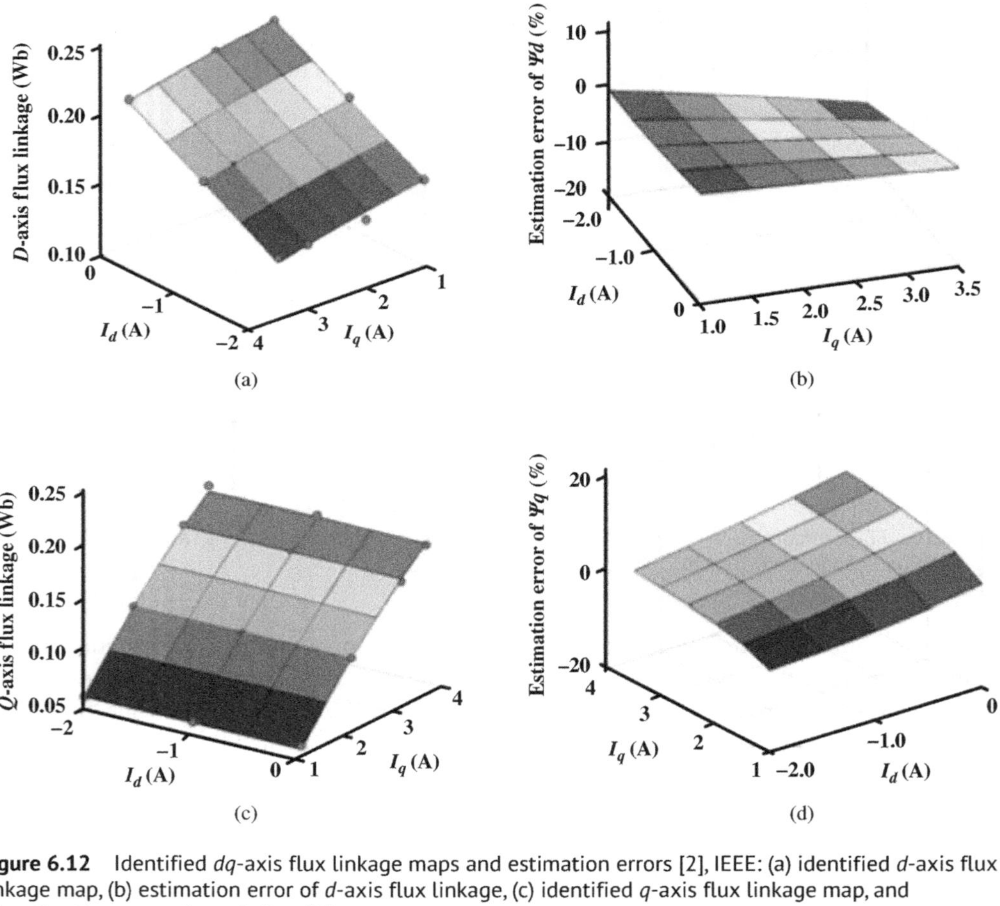

Figure 6.12 Identified *dq*-axis flux linkage maps and estimation errors [2], IEEE: (a) identified *d*-axis flux linkage map, (b) estimation error of *d*-axis flux linkage, (c) identified *q*-axis flux linkage map, and (d) estimation error of *q*-axis flux linkage.

As can be seen from Eqs. (6.22) and (6.24), the terms of stator resistance have been cancelled, and the *dq*-axis flux linkages depend on the difference of two transient rotor speeds, i.e., $\omega(j) - \omega(i)$. Thus, the speed difference should be limited to minimize the influence of variation of core loss. In practice, it is suggested that the rotor speed difference should be limited to a small value, e.g., 5 Hz.

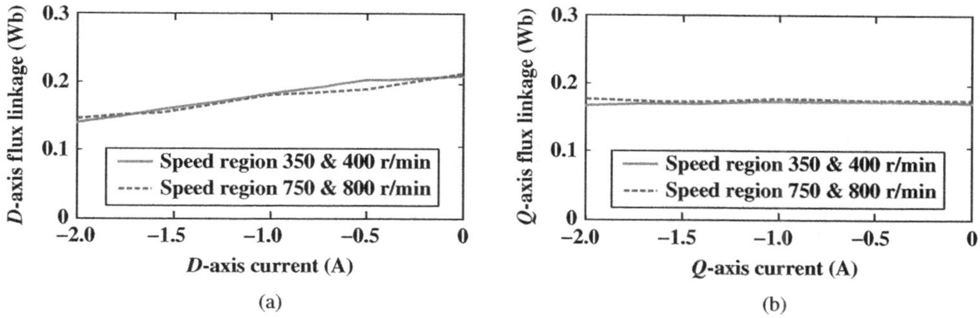

Figure 6.13 Estimated dq-axis flux linkages at different speed regions ($i_q = 3$ A) [2], IEEE: (a) estimated d-axis flux linkage and (b) estimated q-axis flux linkage.

Figure 6.13 shows the variations of the identified dq-axis flux linkages with d-axis currents at different rotor speed ranges of 350 and 400 r/min and 750 and 800 r/min, and the rotor speed difference is set to 50 r/min, i.e., 2.5 Hz of the excitation frequency. It is assumed that the winding resistance during the data measurement (2 s) remains constant and can be self-cancelled due to the large thermal time constant. It is easily found that the estimated dq-axis flux linkages at different rotor speeds are almost the same.

Moreover, Figure 6.14 presents the identified dq-axis flux linkages accounting for the uncertain circuit resistance due to temperature dependency, where the temperature of the stator winding is increased by setting the dq-axis currents to $i_q = 0$ and $i_d = 4$ A, respectively. At time points $t = 0, 200$, 400, 600, and 800 s, the PMSM is excited by the dq-axis currents of $i_q = 3$ A and $i_d = -1$ A temporarily, while the machine states will be recorded for following identification. As can be observed, the measured winding resistance increases from 6.2 to 7.7 Ω, i.e., a ~5 V change in the stator voltage, but there is almost no change in the identified dq-axis flux linkages. Therefore, it can be concluded that the identification method is not affected by the uncertain circuit resistance.

6.3.3.4 Derivation of *dq*-axis Inductances

Based on the identified flux linkages, the dq-axis apparent inductances can be calculated [16] as follows:

$$L_q(i_d, i_q) \approx \frac{\psi_q(i_d, i_q)}{i_q} \tag{6.29a}$$

$$L_d(i_d, i_q) \approx \frac{\psi_d(i_d, i_q) - \psi_{pm}(i_d, i_q)}{i_d} \tag{6.29b}$$

Figure 6.15 shows the calculated maps of the q-axis inductance predicted by the FE and the ICQGA-based method, where the difference between both results is relatively small, i.e., the mean average error is kept within 4.46%. The maximum error occurs when i_q approaches zero, and it is attributed to the non-ideal measurement of the rotor position.

Besides, since ψ_{pm} cannot be directly derived when $i_d \neq 0$ (Eq. 6.29b) and the d-axis flux linkage is linear to the d-axis current, i.e., the machine is unsaturated, and the d-axis incremental inductance L_{dh} is calculated as follows (Eq. 6.30):

$$L_{dh}(i_d, i_q) \approx \frac{\psi_d(i_d + \Delta i_d, i_q) - \psi_d(i_d, i_q)}{\Delta i_d} \tag{6.30}$$

where the incremental d-axis current Δi_d is set to 0.1 A.

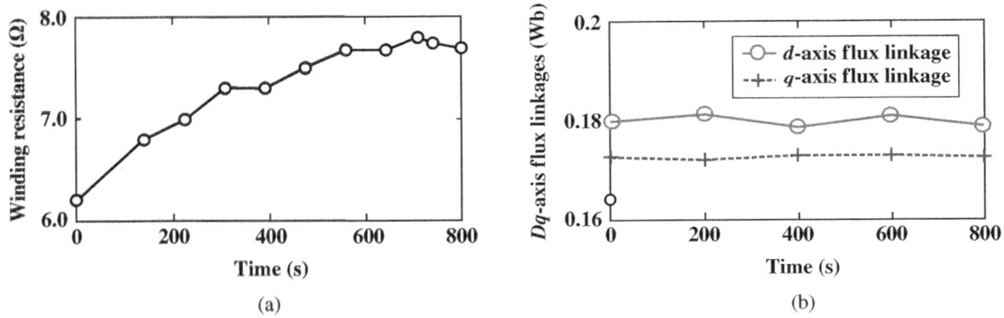

Figure 6.14 Estimated *dq*-axis flux linkages under uncertain circuit resistance [2], IEEE: (a) variation of winding resistance due to temperature rise and (b) estimated *dq*-axis flux linkages.

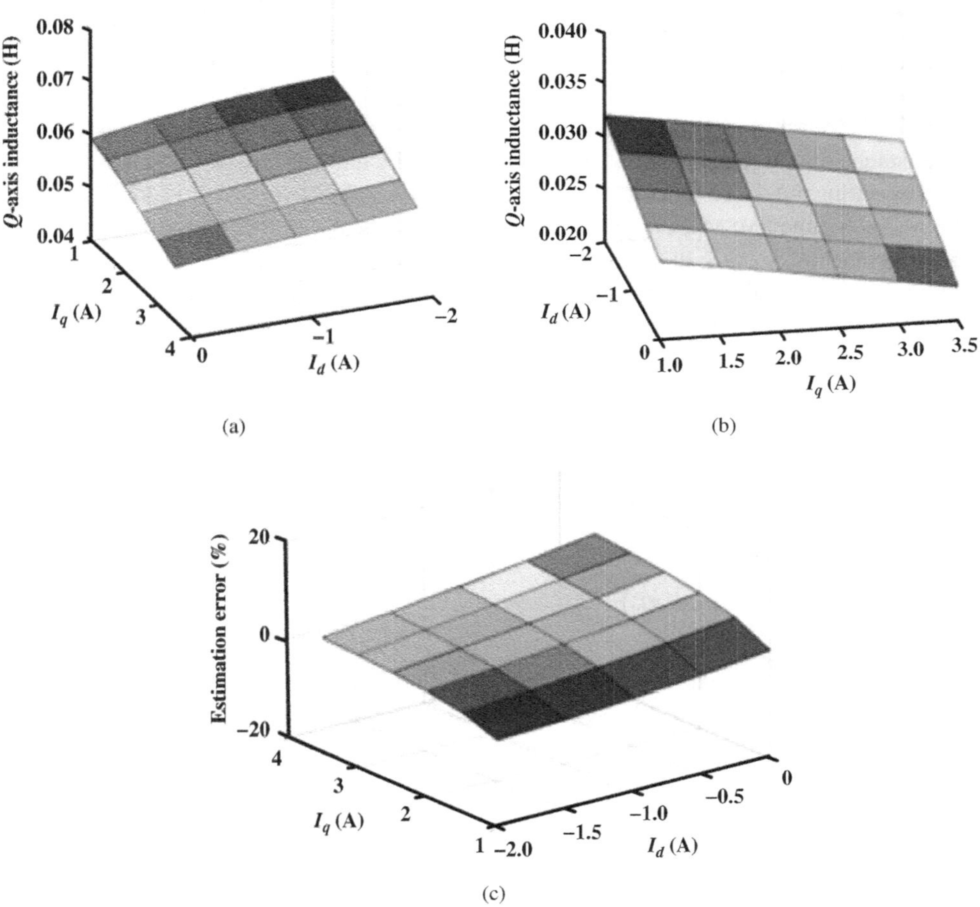

Figure 6.15 *Q*-axis inductance maps predicted by FE and ICQGA-based method [2], IEEE: (a) FE predicted, (b) ICQGA estimated, and (c) estimation error of *q*-axis inductance.

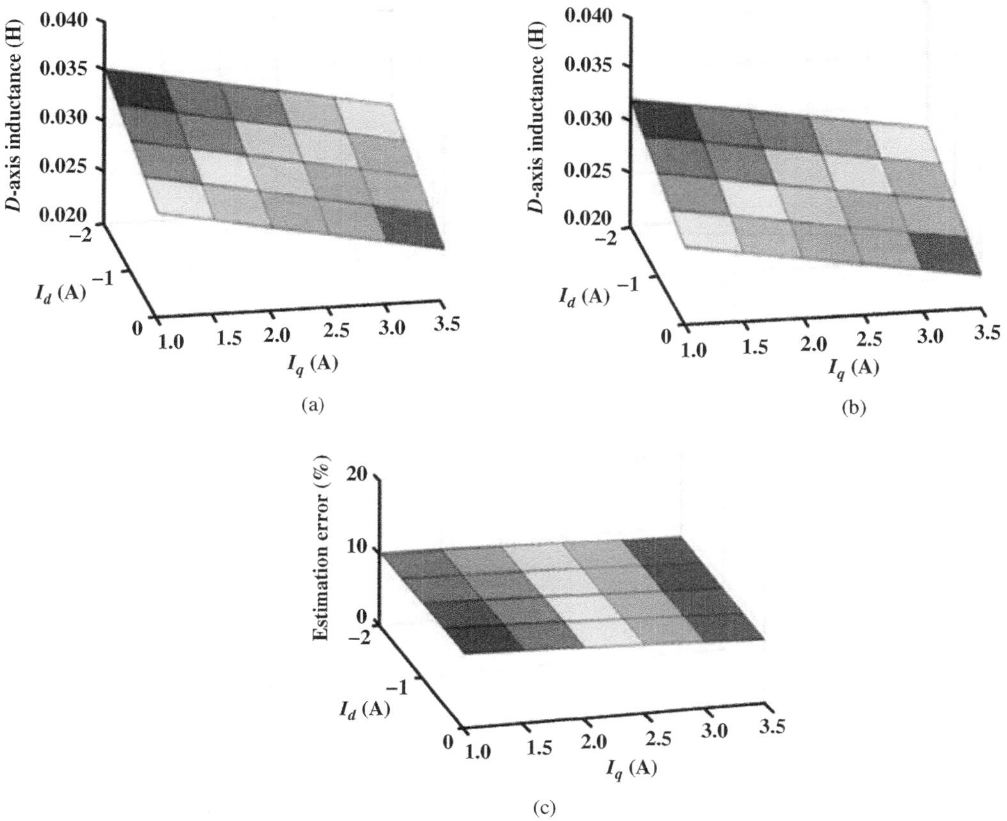

Figure 6.16 *D*-axis inductance maps predicted by FE and ICQGA-based method [2], IEEE: (a) FE predicted, (b) ICQGA estimated, and (c) estimation error of *d*-axis inductance.

The FE predicted and ICQGA estimated L_{dh} are shown in Figure 6.16, where the error is approximately 9% and does not change too much at different load points.

6.3.3.5 MTPA Current Trajectory Based on *dq*-axis Flux Linkage Maps

In order to derive the MTPA current trajectory, the expression of flux linkage maps (Eq. 6.31) and (Eq. 6.32) should be substituted into the partial differential of torque T_e in Eq. (6.33):

$$\psi_d(i_d, i_q) = a_d i_d + b_d i_q + c_d i_d^2 + d_d i_q^2 + e_d i_d i_q + g_d \tag{6.31}$$

$$\psi_q(i_d, i_q) = a_q i_d + b_q i_q + c_q i_d^2 + d_q i_q^2 + e_q i_d i_q + g_q \tag{6.32}$$

$$\frac{\partial T_e}{\partial i_d} = \frac{1.5p(\psi_d i_q - \psi_q i_d)}{\partial i_d} = 0 \tag{6.33}$$

where a_i, b_i, c_i, d_i, e_i, and g_i with $i = d$ and q are the fitting coefficients in the dq-axis flux linkage maps.

By solving Eqs. (6.31)–(6.33), the optimal dq-axis currents are expressed in Eq. (6.34) and shown in Figure 6.17. It can be easily found from Figure 6.17 that the MTPA trajectory derived by the

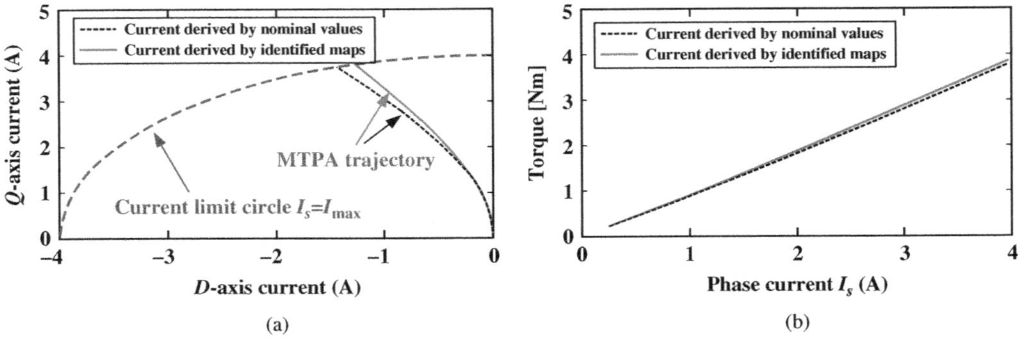

Figure 6.17 MTPA current trajectory and output torque [2], IEEE: (a) MTPA current trajectory and (b) measured output torque.

surface-fitting results can achieve a 3.2% higher maximum output torque than that derived by using the nominal dq-axis inductances and rotor PM flux linkage:

$$\begin{cases} i_d = \dfrac{-B + \sqrt{B^2 - 4AC}}{2A} \\ i_q = \sqrt{I_s^2 - i_d^2} \end{cases} \tag{6.34a}$$

with

$$A = b_q - a_d + 2(d_q - e_d)i_q - 3c_q i_q \tag{6.35b}$$

$$B = (2c_d - 3d_d - 2e_q)i_q^2 - 2(a_q + b_d)i_q - g_d + (e_q - c_d)(I_s^2 - i_q^2) \tag{6.35c}$$

$$C = (a_d - b_q)i_q^2 + (e_d - d_q)i_q^3 - g_q i_q \tag{6.35d}$$

where I_s is the amplitude of phase current.

6.4 Summary

This chapter has described two estimation methods for PMSMs under variable speed control. The PMSM states at different rotor speeds have been recorded and used for identification, and thus, additional disturbance signals are not required.

In terms of estimating the multi-parameter in Section 6.2, electrical parameters, i.e., stator resistance, dq-axis inductances, and rotor PM flux linkage, can be identified progressively, where the influence of VSI non-linearity has also been considered. Meanwhile, the estimated stator resistance rotor PM flux linkage can be used for the thermal condition monitoring of the most vulnerable stator winding and rotor PMs.

In terms of estimating the dq-axis flux linkage maps in Section 6.3, it can quickly identify the flux linkages maps of PMSMs for determination of optimal current trajectory. Since the influences of uncertainties of circuit resistance and VSI non-linearity have been cancelled in the designed estimation scheme, the estimated flux linkages have exhibited good agreement with the FE results. Meanwhile, the estimation method does not require detailed information on geometric parameters and material magnetic properties of PMSMs.

These two methods have been applied to an IPMSM-based drive system, and experimental results have been presented for illustrative purposes. In addition, the described methods are generic estimation schemes and can be performed by all EAs.

References

1 K. Liu and Z. Q. Zhu, "Quantum genetic algorithm-based parameter estimation of PMSM under variable speed control accounting for system identifiability and VSI nonlinearity," *IEEE Trans. Ind. Electron.*, vol. 62, no. 4, pp. 2363–2371, April 2015.

2 K. Liu, J. Feng, S. Guo, L. Xiao, and Z. Q. Zhu, "Identification of flux linkage map of permanent magnet synchronous machines under uncertain circuit resistance and inverter nonlinearity," *IEEE Trans. Ind. Inform.*, vol. 14, no. 2, pp. 556–568, Feb. 2018.

3 F. Alonge, F. D'lppolito, G. Ferrante, and F. M. Raimondi, "Parameter identification of induction motor model using genetic algorithms," *IEE Proc. Control Theory Appl.*, vol. 145, no. 6, pp. 587–593, Nov. 1998.

4 X. Xiao, Q. S. Xu, J. Tao, P. G. Tian, and G. H. Wang, "An improved method of interior permanent magnet synchronous motor parameter identification based on particle swarm optimization," In *Proc. 15th Eur. Conf. Power Electron. Appl.*, France, 2013, pp. 1–9.

5 K. Liu and Z. Q. Zhu, "Online estimation of rotor flux linkage and voltage source inverter nonlinearity in permanent magnet synchronous machine drives," *IEEE Trans. Power Electron.*, vol. 29, no. 1, pp. 418–427, Jan. 2014.

6 H. W. Kim, M. J. Youn, and K. Y. Cho, "New voltage distortion observer of PWM VSI for PMSM," *IEEE Trans. Ind. Electron.*, vol. 52, no. 4, pp. 1188–1192, Aug. 2005.

7 S. Y. Kim, W. Lee, M. S. Rho, and S. Y. Park, "Effective dead-time compensation using a simple vectorial disturbance estimator in PMSM drives," *IEEE Trans. Ind. Electron.*, vol. 57, no. 5, pp. 1609–1614, May 2010.

8 K. H. Han and J. H. Kim, "Quantum-inspired evolutionary algorithm for a class of combinatorial optimization," *IEEE Trans. Evol. Comput.*, vol. 6, no. 6, pp. 580–593, Dec. 2002.

9 K. Liu, Z. Q. Zhu, Q. Zhang, and J. Zhang, "Influence of nonideal voltage measurement on parameter estimation in permanent magnet synchronous machines," *IEEE Trans. Ind. Electron.*, vol. 59, no. 6, pp. 2438–2447, Jun. 2012.

10 I. Lar and M. M. Radulescu, "Equivalent core-loss resistance identification for interior permanent-magnet synchronous machines," In *Proc. IEEE Int. Conf. Elect. Mach.*, 2012, vol. 2, pp. 1665–1669.

11 M. A. Jabbar, J. Dong, and Z. Liu, "Determination of machine parameters for internal permanent magnet synchronous motors," In *Proc. Power Electron., Mach. Drives*, 2004, vol. 2, pp. 805–810.

12 B. Stumberger, G. Stumerger, D. Dolinar, A. Hamler, and M. Trlep, "Evaluation of saturation and cross-magnetization effects in interior permanent magnet synchronous motor," *IEEE Trans. Ind. Appl.*, vol. 39, no. 5, pp. 1264–1271, Sep./Oct. 2003.

13 J. W. Kim and S. W. Kim, "Parameter identification of induction motors using dynamic encoding algorithm for searches (DEAS)," *IEEE Trans. Energy Convers.*, vol. 20, no. 1, pp. 16–24, Mar. 2005.

14 L. N. de Castro and F. J. Von Zuben, "Learning and optimization using the clonal selection principle," *IEEE Trans. Evol. Comput.*, vol. 6, no. 3, pp. 239–251, Jun. 2002.

15 M. Liwschitz-Garik and C. Whipple, *Alternating-Current Machines*. Cambridge MA, USA: Van Nostrand Reinhold Inc., 1961.

16 G. Qi, J. T. Chen, Z. Q. Zhu, D. Howe, L. B. Zhou, and C. L. Gu, "Influence of skew and cross-coupling on flux-weakening performance of permanent magnet brushless AC machines," *IEEE Trans. Magn.*, vol. 45, no. 5, pp. 2110–2117, May 2009.

7

Estimation of Magnetic Saturation and Cross-coupling Based on High-frequency Signal Injection

7.1 Introduction

Magnetic saturation occurs in almost all permanent magnet synchronous machines (PMSMs) and may be divided into self-magnetic saturation and mutual-magnetic saturation (cross-coupling). Under magnetic saturation, the permeability along the *d*- or *q*-axis alone is not constant during PMSM operation (self-magnetic saturation), while the flux linkages produced by the magnetomotive force (MMF) that lies along the *d*- or *q*-axis may also have components in both the *d*- and *q*-axes (mutual-magnetic saturation) [1]. Indeed, both the self- and mutual-magnetic saturations have severely detrimental effects on electrical machines, which may decrease the output torque and deteriorate sensored and sensorless control performance [2–6].

From the perspective of parameter estimation, the self- and mutual-saturation effects have a great impact on the identification of *dq*-axis self- and mutual inductances. Generally, inductances, classified as apparent L^{ap} and incremental L^{inc}, can be illustrated in Figure 7.1.

As can be seen, the apparent inductance is defined as the increase of flux linkage (from zero current) divided by the current. In comparison, the incremental inductance is calculated by the slope at the operating point, e.g., Point B in Figure 7.1. Briefly, the incremental inductances represent the dynamic behaviour of the machine, while the apparent inductances represent the amount of the *dq*-axis flux linkages at specific *dq*-axis currents. When the magnetic circuit is unsaturated, the incremental inductance is the same as the apparent inductances, whereas once the magnetic circuit is saturated, the incremental inductance becomes smaller.

Therefore, when the PMSM is saturated, the neglect of magnetic saturation and cross-coupling will cause the misestimation of *dq*-axis inductances, which could further deteriorate the estimation accuracy of other electrical parameters [7–9]. Although the magnetic field distributions of electrical machines can be obtained by the finite element (FE) method [10–12], detailed machine specifications and material magnetic properties are not always available. Meanwhile, sometimes, the calculated values are also different from the measured values due to manufacturing tolerances and uncertain material properties.

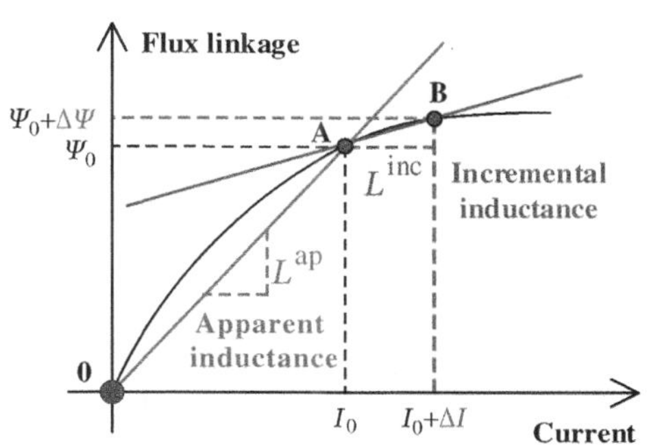

Figure 7.1 Illustration of incremental and apparent inductances.

Parameter Estimation of Permanent Magnet Synchronous Machines, First Edition. Zi Qiang Zhu, Kan Liu, and Dawei Liang.

To date, several parameter estimation methods have been developed accounting for self- and/or mutual-saturation effects.

In an offline manner, the *dq*-axis flux linkages can be estimated by injecting voltage pulses in *d*- and *q*-axes in turn [5] or by the measured *dq*-axis currents at constant rotor speed [13]. Afterwards, the *dq*-axis incremental inductances are calculated by the partial derivatives of *dq*-axis flux linkages to currents. Moreover, by using alternating current (ac) signal excitation, both the incremental and apparent inductances can be estimated, where the direct current (dc) bias point sets the magnetic operating point [14] and the induced voltages and internal load angles are measured in Ref. [15], where both the self- and mutual-saturation can also be considered. A detailed description of offline parameter estimation methods can be found in Chapter 8.

In an online manner, most online methods are concerned with the apparent inductance estimation [7], as described in Chapters 4 and 5, whereas only the self-saturation can be considered. For the cross-coupling effect, the mutual-inductances can be derived based on the estimated *dq*-axis flux linkage maps described in Chapter 6 or can also be modelled as polynomial functions and then identified based on perturbation signal injections [16].

Besides, the HF signal injection-based method is another mainstream to estimate HF self- and mutual-incremental inductances, which have been widely used in rotor position sensorless control, torque estimation, and PM temperature estimation [17–29]. Moreover, once the HF incremental inductances are estimated, the apparent inductances can be calculated subsequently by simple integration of current increments. The knowledge of apparent inductances plays a critical role in the high-performing machine drive performance, which is detailed in Chapter 11.

In this chapter, four HF voltage injection methods will be described to estimate HF self- and mutual-incremental inductances, i.e., (a) rotating voltage injection, (b–c) pulsating voltage injection with and without using a position estimator, and (d) combined pulsating and rotating voltage injection methods, of which injection sequences of the HF voltage are shown in Figure 7.2.

More specifically, Section 7.2 introduces the fundamental mathematical model of PMSMs accounting for the cross-coupling magnetic saturation and the time delay effect in the HF signal injection method. Section 7.3 describes the HF sinusoidal rotating voltage injection method in the stationary reference frame, and the values of HF inductances are estimated based on a position estimator. Likewise, a position estimator is also utilized in Section 7.4.1, where the HF pulsating

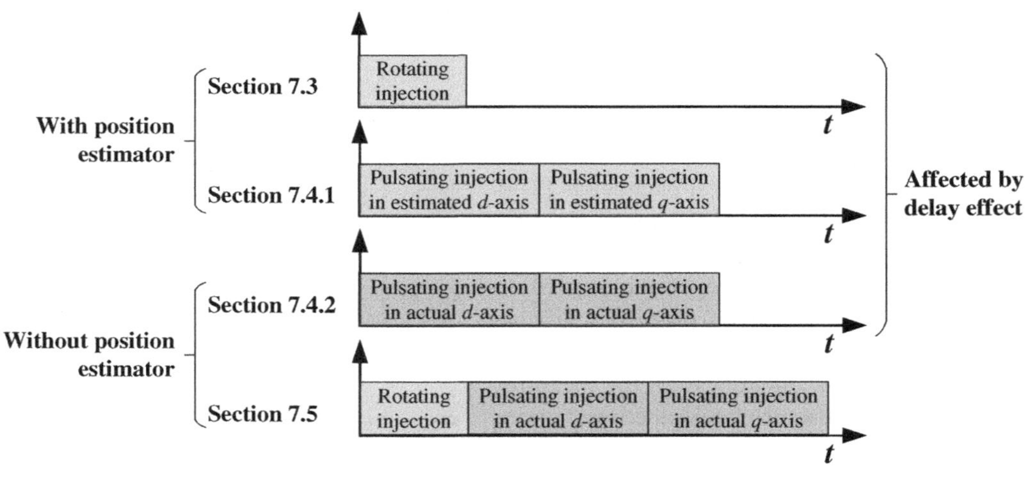

Figure 7.2 Comparison of described injection sequences of HF voltage methods.

voltage is injected in the estimated synchronous rotating dq-axes. Furthermore, to get rid of the utilization of the position estimator, Section 7.4.2 describes an open-loop estimation method by injecting HF pulsating voltages in the actual dq-axes, of which the position is obtained from an encoder. Nevertheless, the above three methods may suffer from the time delay effect caused by the non-linear implementation. This issue will be addressed by a combined HF pulsating and rotating injection method described in Section 7.5. In addition, all the above four methods will be evaluated experimentally based on a PMSM drive system.

7.2 Magnetic Saturation Modelling and Time Delay Effect in HF Signal Injection

7.2.1 Fundamental Mathematical Model

Although the fundamental mathematical model has been presented in Chapter 1, it is briefly duplicated here for convenience.

When both the self- and mutual-saturation effects are considered, the dq-axis voltage equations can be written as:

$$u_d = R_s i_d + L_{dd}^{\text{inc}} \frac{di_d}{dt} + L_{dq}^{\text{inc}} \frac{di_q}{dt} - \omega_r \psi_q \tag{7.1a}$$

$$u_q = R_s i_q + L_{qq}^{\text{inc}} \frac{di_q}{dt} + L_{qd}^{\text{inc}} \frac{di_d}{dt} + \omega_r \psi_d \tag{7.1b}$$

where

$$\psi_d = L_{dd}^{\text{ap}}(i_d, i_d)\, i_d + L_{dq}^{\text{ap}}(i_d, i_d)\, i_q + \psi_{pm}(i_d, i_d) \tag{7.2a}$$

$$\psi_q = L_{qq}^{\text{ap}}(i_d, i_d)\, i_q + L_{qd}^{\text{ap}}(i_d, i_d)\, i_d \tag{7.2b}$$

where u_d, u_q, i_d, and i_q are the dq-axis voltages and currents. R_s is the stator resistance. ψ_{pm} is the rotor PM flux linkage. ω_r is the electrical rotor speed. L_{dd}^{i}, L_{qq}^{i}, L_{dq}^{i}, and L_{qd}^{i} with $\{i = \text{"ap" and "inc"}\}$ are the dq-axis apparent/incremental self- and mutual-inductances, respectively. Meanwhile, the relationship between the dq-axis incremental and apparent inductances can be expressed as [19]:

$$\boldsymbol{L}_{\boldsymbol{dq}}^{\textbf{inc}} = \begin{bmatrix} L_{dd}^{\text{inc}} & L_{dq}^{\text{inc}} \\ L_{qd}^{\text{inc}} & L_{qq}^{\text{inc}} \end{bmatrix} \tag{7.3}$$

Afterwards, Eq. (7.3) can be transferred to the diagonal matrix $\boldsymbol{L}_{\boldsymbol{dq},\textbf{dia}}^{\textbf{inc}}$.

$$\boldsymbol{L}_{\boldsymbol{dq},\textbf{dia}}^{\textbf{inc}} = \begin{bmatrix} \dfrac{L_{dd}^{\text{inc}} \Delta i_d + L_{dq}^{\text{inc}} \Delta i_q}{\Delta i_d} & 0 \\ 0 & \dfrac{L_{qq}^{\text{inc}} \Delta i_q + L_{qd}^{\text{inc}} \Delta i_d}{\Delta i_q} \end{bmatrix} = \begin{bmatrix} L_{dd,\text{dia}}^{\text{inc}} & 0 \\ 0 & L_{qq,\text{dia}}^{\text{inc}} \end{bmatrix} \tag{7.4}$$

Thus, the equivalent apparent inductance, i.e., $\boldsymbol{L}_{\boldsymbol{dd}}^{\textbf{ap}}$ and $\boldsymbol{L}_{\boldsymbol{qq}}^{\textbf{ap}}$, can be calculated by:

$$L_{dd}^{\text{ap}} = \frac{\int L_{dd,\text{dia}}^{\text{inc}}(i_d)\, di_d}{i_d} \tag{7.5a}$$

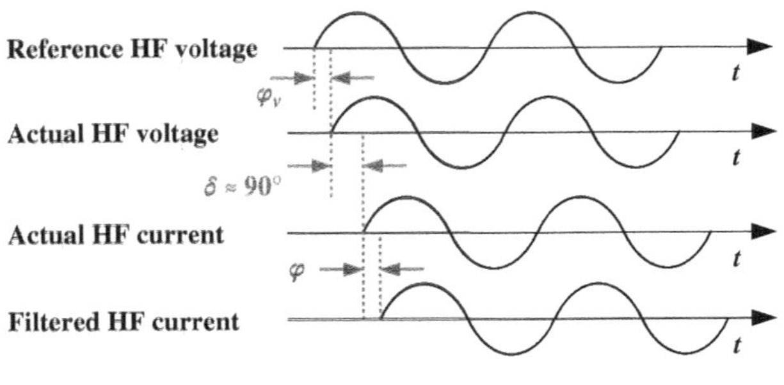

Figure 7.3 Time delay effects in HF voltage injection methods [26], IEEE.

$$L_{qq}^{\text{ap}} = \frac{\int L_{qq,\text{dia}}^{\text{inc}}(i_q)\, di_d}{i_q} \tag{7.5b}$$

For estimation of incremental inductances at standstill and low-speed range by HF voltage injection technique, the speed-dependent terms and the resistance voltage drop in Eq. (7.1) can be neglected compared to the HF reactance [19].

Therefore, the HF voltage equation in the dq-axes can be written as:

$$\begin{bmatrix} u_{dh}^r \\ u_{qh}^r \end{bmatrix} = \begin{bmatrix} L_{dh} & L_{dqh} \\ L_{qdh} & L_{qh} \end{bmatrix} \frac{d}{dt} \begin{bmatrix} i_{dh}^r \\ i_{qh}^r \end{bmatrix} \tag{7.6}$$

where u_{dh}^r, u_{qh}^r, i_{dh}^r, and i_{qh}^r are the HF voltages and currents. The superscript "r" denotes synchronous rotating reference frame.

7.2.2 Time Delay Effect in HF Signal Injection Methods

The time delay effect, as illustrated in Figure 7.3, indicates the angle difference between the actual HF voltages/currents with their reference waveforms. In the HF voltage injection process, the actual applied HF voltage usually lags behind the reference HF voltage, as denoted by φ_v in Figure 7.3, due to voltage synthesis and pulse-width-modulation (PWM) update delay [21]. Moreover, due to sampling and digital filters, the filtered HF currents also lag the actual HF currents, as denoted by φ_i in Figure 7.3. However, the accurate value of the delay angle is hard to determine since it is affected by the current levels, sampling hardware, and the injection frequency [25]. Consequently, either inappropriate compensation or neglect of the delay angle will deteriorate the estimation accuracy of the estimated HF inductances.

7.3 HF Rotating Voltage Injection Method

In the stationary $\alpha\beta$-reference frame, the HF voltage Eq. (7.6) can be obtained by [19]:

$$\begin{bmatrix} i_{\alpha h}^s \\ i_{\beta h}^s \end{bmatrix} = R^{-1}(\theta_r) \begin{bmatrix} L_{dh} & L_{dqh} \\ L_{qdh} & L_{qh} \end{bmatrix}^{-1} R(\theta_r) \int \begin{bmatrix} u_{\alpha h}^s \\ u_{\beta h}^s \end{bmatrix} dt \tag{7.7a}$$

$$R(\theta_r) = \begin{bmatrix} \cos(\theta_r) & \sin(\theta_r) \\ -\sin(\theta_r) & \cos(\theta_r) \end{bmatrix} \tag{7.7b}$$

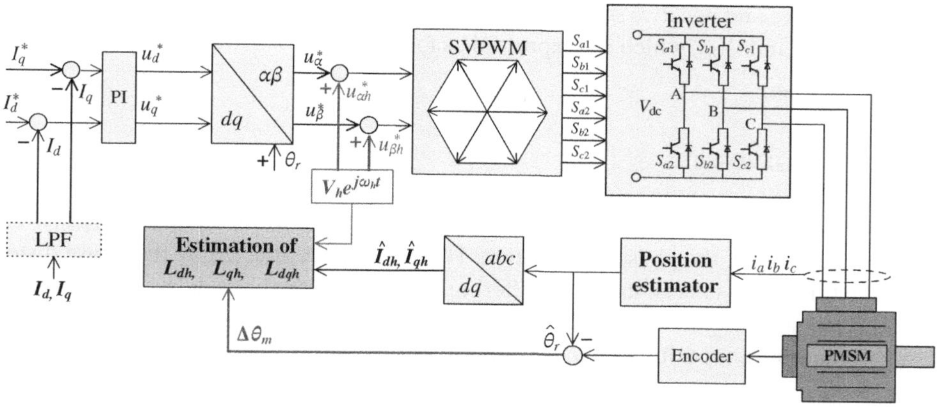

Figure 7.4 Control diagram of HF inductance estimation based on HF rotating voltage injection.

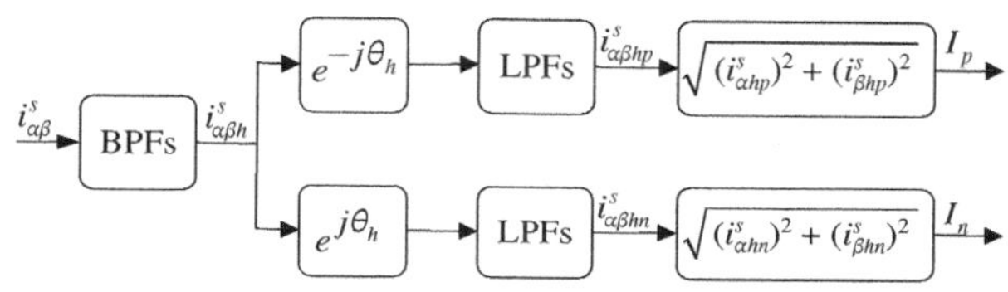

Figure 7.5 Demodulation of amplitudes of positive and negative sequence HF currents excited by HF rotating voltage injection [26], IEEE.

When a HF rotating voltage (Eq. 7.8) is injected into the α- and β-axes subsequently, the control diagram can be presented in Figure 7.4:

$$u^s_{\alpha h} + j u^s_{\beta h} = V_h e^{j\omega_h t} \tag{7.8}$$

where V_h and ω_h are the amplitude and angular frequency of the injected HF voltage. $\theta_h = \omega_h t$ is the injection angle.

The resulting HF currents can be obtained from Eqs. (7.7) and (7.8), and expressed as:

$$i^s_{\alpha h} + j u^s_{\beta h} = I_p e^{j(\theta_h - \pi/2)} + I_n e^{-j[\theta_h - 2(\theta_r - \Delta\theta_m) - \pi/2]} \tag{7.9a}$$

$$\begin{cases} I_p = \dfrac{V_h L_\Sigma}{\omega_h (L_{dh} L_{qh} - L^2_{dqh})} \\ I_n = \dfrac{V_h \sqrt{L^2_\Delta - L^2_{dqh}}}{\omega_h (L_{dh} L_{qh} - L^2_{dqh})} \end{cases} \tag{7.9b}$$

$$\begin{cases} L_\Sigma = \dfrac{L_{dh} + L_{qh}}{2} \\ L_\Delta = \dfrac{L_{qh} - L_{dh}}{2} \end{cases} \tag{7.9c}$$

$$\Delta\theta_m = 0.5 \tan^{-1}\left(\frac{2L_{dqh}}{L_{qh} - L_{dh}}\right) \tag{7.9d}$$

where $\Delta\theta_m$ is the angle offset caused by the cross-coupling effect. I_p and I_n are the amplitudes of the positive and negative sequence currents, respectively, which can be extracted by the method in Figure 7.5.

Firstly, the positive and negative sequence currents are transferred to low-frequency domain and filtered out as $i^s_{\alpha\beta hp}$ and $i^s_{\alpha\beta hn}$, which are expressed as follows:

$$i^s_{\alpha\beta hp} = I_p e^{j(-\pi/2)} \tag{7.10a}$$

$$i^s_{\alpha\beta hn} = I_n e^{j[\theta_r - 2\Delta\theta_m + \pi/2]} \tag{7.10b}$$

According to Eq. (7.9), the values of I_p and I_n can be obtained from the moduli of $i^s_{\alpha\beta hp}$ and $i^s_{\alpha\beta hn}$, which is independent of the phase information. Hence, it is not affected by the phase distortion caused by the time delay effects, including filtering and voltage synthesis delays.

Secondly, the angle $\Delta\theta_m$ can be obtained by measuring the difference between the actual rotor position obtained from a position encoder and the estimated rotor position obtained by sensorless position estimation technique incorporating heterodyning demodulation, as shown in Figure 7.6a, where $\hat{\theta}_r$ is the estimated rotor position. k_p and k_i are the estimator parameters.

The HF component after heterodyning operation is filtered by a low-pass filter (LPF), and the error ε is obtained as follows:

$$\varepsilon = I_n \sin\left[2(\theta_r - \Delta\theta_m) - 2\hat{\theta}_r\right] \tag{7.11}$$

Afterwards, as can be seen from Figure 7.6b, the angle $\Delta\theta_m$ can be obtained from the difference between the rotor position θ_r measured by the encoder and the estimated position $\hat{\theta}_r$, which is expressed as follows:

$$\Delta\theta_m = \theta_r - \hat{\theta}_r \tag{7.12}$$

Ultimately, the HF inductances, i.e., L_{dh}, L_{qh}, and L_{dqh}, can be obtained based on the information of I_p, I_n, and $\Delta\theta_m$ from Eq. (7.13). Obviously, this method depends on the estimated position, which is affected by the time delay effects [22]:

$$L_{dh} = \frac{V_h\left(I_p - I_n/\sqrt{1+\tan^2(2\Delta\theta_m)}\right)}{\omega_h\left(I_p^2 - I_n^2\right)} \tag{7.13a}$$

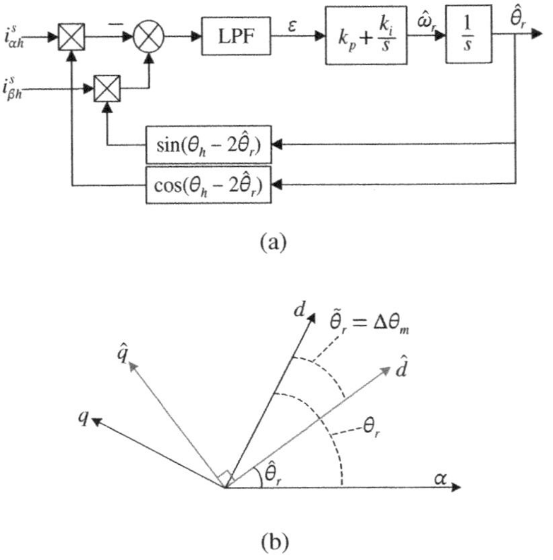

Figure 7.6 Sensorless position estimation based on HF rotating voltage injection [26], IEEE: (a) position estimator and (b) estimated position and error.

$$L_{qh} = \frac{V_h\left(I_p + I_n/\sqrt{1+\tan^2(2\Delta\theta_m)}\right)}{\omega_h\left(I_p^2 - I_n^2\right)} \tag{7.13b}$$

$$L_{dqh} = \frac{V_h I_n \tan(2\Delta\theta_m)}{\omega_h\left(I_p^2 - I_n^2\right)\sqrt{1+\tan^2(2\Delta\theta_m)}} \tag{7.13c}$$

In terms of the influence of time delay effect, the angle θ_h used for position demodulation (see Figure 7.6a) has a delay to the desired demodulation angle by $(\varphi_v + \varphi_i)$, which will result in an error of $0.5(\varphi_v + \varphi_i)$ in the estimated rotor position if it is not compensated [22]. However, the accurate value of the delay angle, i.e., $(\varphi_v + \varphi_i)$, is hard to determine since it is affected by the current levels, sampling hardware, and the injection frequency [25].

7.4 HF Pulsating Voltage Injection Method

The HF pulsating voltage injection technique can also be used for measuring the HF inductances [23, 24]. In this section, two HF inductances measuring methods will be introduced, which are all based on pulsating voltage injection with and without requiring a position estimator, as presented below.

7.4.1 With the Aid of Position Estimator

As aforementioned, the dq-axis inductance matrix (Eq. 7.6) can be transferred to the following diagonal form:

$$\begin{bmatrix} L_{dh} & L_{dqh} \\ L_{qdh} & L_{qh} \end{bmatrix} = R(\Delta\theta_m)\begin{bmatrix} X_1 & 0 \\ 0 & X_2 \end{bmatrix} R^{-1}(\Delta\theta_m) \tag{7.14a}$$

$$X_1 = L_\Sigma - \sqrt{L_\Delta^2 + L_{dqh}^2} \tag{7.14b}$$

$$X_2 = L_\Sigma + \sqrt{L_\Delta^2 + L_{dqh}^2} \tag{7.14c}$$

where X_1 and X_2 are the characteristic roots.

Therefore, the HF inductances can be estimated by injecting HF pulsating voltages subsequently, i.e., $V_h\cos(\omega_h t)$, in the dq-axes based on Eq. (7.14) [23], and the corresponding control diagram is shown in Figure 7.7.

Therefore, the current responses in the estimated dq-axes can be expressed as:

$$i_{dh}^{\hat{r}} = I_{dh}^{\hat{r}}\sin(\omega_h t) \tag{7.15a}$$

$$i_{dh}^{\hat{r}} = I_{qh}^{\hat{r}}\sin(\omega_h t) \tag{7.15b}$$

where "$\hat{r}$" denotes the estimated synchronous rotating reference frame, $I_{dh}^{\hat{r}}$ and $I_{qh}^{\hat{r}}$ are the amplitudes of the HF currents, which can be extracted by the synchronous demodulation [23], as calculated in Eq. (7.16) and shown in Figure 7.8.

$$I_{dh}^{\hat{r}} = \frac{V_h(X_1 + X_2)}{2\omega_h X_1 X_2} + \frac{V_h(X_2 - X_1)}{2\omega_h X_1 X_2}\cos\left[2(\theta_r - \Delta\theta_m) - 2\hat{\theta}_r\right] \tag{7.16a}$$

$$I_{qh}^{\hat{r}} = \frac{V_h(X_2 - X_1)}{2\omega_h X_1 X_2}\sin\left[2(\theta_r - \Delta\theta_m) - 2\hat{\theta}_r\right] \tag{7.16b}$$

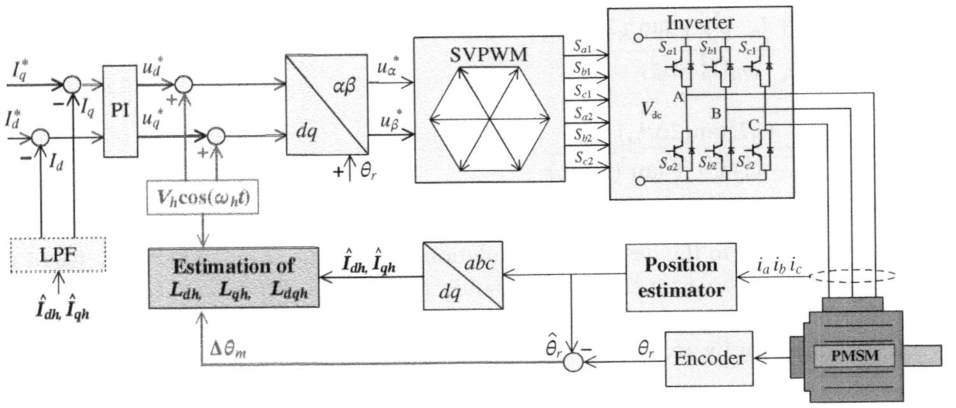

Figure 7.7 Estimation of HF inductances based on HF pulsating voltage injection with position estimator.

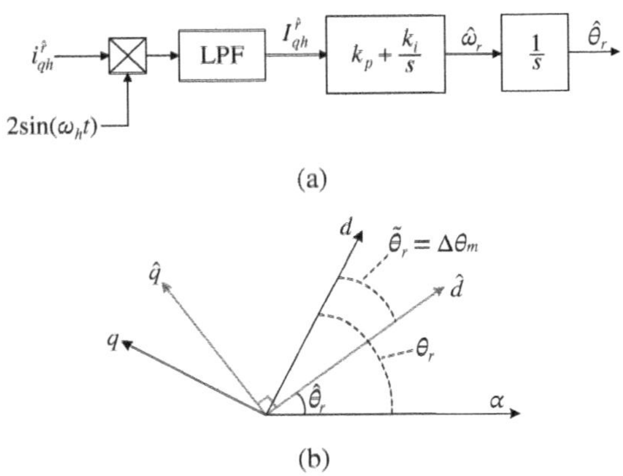

Figure 7.8 Sensorless position estimation based on HF pulsating voltage injection [26], IEEE: (a) position estimator and (b) estimated position and error.

The estimated position $\hat{\theta}_r$ can be obtained by controlling $I_{qh}^{\hat{r}} = 0$, as shown in Figure 7.8b, and thus, $\Delta\theta_m$ can be obtained from the difference between actual and estimated positions, which is the same as Eq. (7.12).

Furthermore, the value of X_1 can be calculated from the measured steady-state d-axis current amplitude:

$$X_1 = V_h / \left(\omega_h I_{dh}^{\hat{r}}\right) \tag{7.17}$$

On the other hand, the value of X_2 is calculated by taking the estimated q-axis, and the dq-axis HF current components are calculated.

$$I_{dh}^{\hat{r}} = \frac{V_h(X_2 - X_1)}{2\omega_h X_1 X_2} \cos\left[2(\theta_r - \Delta\theta_m) - 2\hat{\theta}_r\right] \tag{7.18a}$$

$$I_{qh}^{\hat{r}} = \frac{V_h(X_1 + X_2)}{2\omega_h X_1 X_2} - \frac{V_h(X_2 - X_1)}{2\omega_h X_1 X_2} \cos\left[2(\theta_r - \Delta\theta_m) - 2\hat{\theta}_r\right] \tag{7.18b}$$

Afterwards, by controlling $I_{dh}^{\hat{r}} = 0$, X_2 can be obtained from $I_{qh}^{\hat{r}}$ at steady state:

$$X_2 = \frac{V_h}{\omega_h I_{qh}^{\hat{r}}} \tag{7.19}$$

Clearly, two separate injections are necessary to calculate X_1 and X_2 to estimate the HF inductance. Similar to the method described in Section 7.3, a position estimator is required.

In addition, it is worth noting that although the time delay effects will not affect the estimated rotor position $\hat{\theta}_r$ and the observed $\Delta\theta_m$, it may cause error in the measured HF current amplitude $I_{dh}^{\hat{r}}$, i.e.,

$$\text{LPF}[2\sin(\omega_h t) I_{dh}^{\hat{r}} \sin(\omega_h t + \varphi_{\text{delay}})] = I_{dh}^{\hat{r}} \cos(\varphi_{\text{delay}}) \tag{7.20}$$

where "LPF" is the abbreviation of low-pass filter. φ_{delay} is the actual phase delay between the filtered HF current and the demodulation signal $\sin(\omega_h t)$. As can be seen from Eq. (7.20), a scaling factor, i.e., $\cos^{-1}(\varphi_{\text{delay}})$, needs to be introduced in the measured HF inductances with pulsating voltage injections to compensate the delay. However, the actual value of φ_{delay} is hard to be determined [25].

7.4.2 Without the Aid of Position Estimator

To avoid the utilization of the position estimator, which requires a complicated parameter tuning process, a simple measuring method proposed in Ref. [24] utilizes the injection pulsating voltage $V_h \cos(\omega_h t)$ in the actual d-axis and q-axis, respectively. The control diagram is shown in Figure 7.9.

In this case, the HF inductances can be calculated from the measured HF current amplitudes and expressed as:

$$\begin{bmatrix} L_{dh} & L_{dqh} \\ L_{qdh} & L_{dh} \end{bmatrix} = \frac{V_h}{\omega_h} \begin{bmatrix} I_{dh}^{I} & I_{dh}^{\text{II}} \\ I_{qh}^{I} & I_{qh}^{\text{II}} \end{bmatrix}^{-1} \tag{7.21}$$

where I_{dh}^{k} and I_{qh}^{k} with $\{k = \text{I and II}\}$ are the HF current amplitudes on the dq-axes due to HF voltage injection in the d-axis (I) and q-axis (II), respectively, which can be extracted by synchronous demodulation and expressed as:

$$I_{dh}^{I} = \text{LPF}\left[2\sin(\omega_h t) i_{dh}^{I}\right], I_{qh}^{I} = \text{LPF}\left[2\sin(\omega_h t) i_{qh}^{I}\right] \tag{7.22a}$$

$$I_{dh}^{\text{II}} = \text{LPF}\left[2\sin(\omega_h t) i_{dh}^{\text{II}}\right], I_{qh}^{\text{II}} = \text{LPF}\left[2\sin(\omega_h t) i_{qh}^{\text{II}}\right] \tag{7.22b}$$

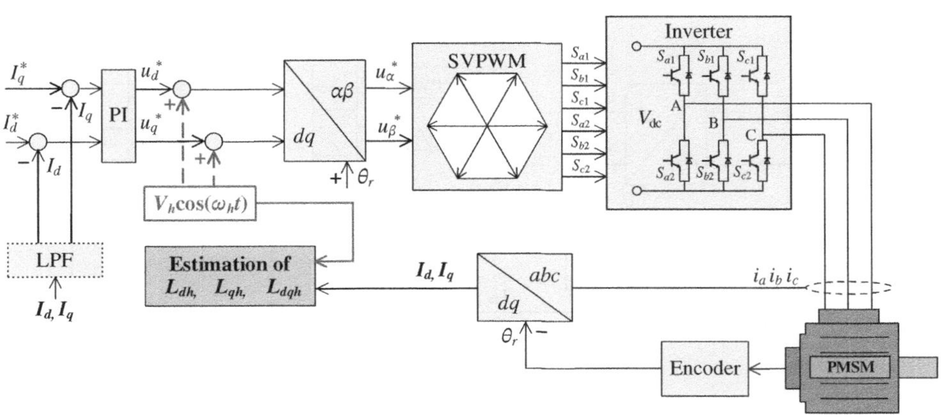

Figure 7.9 Estimation of HF inductances based on HF pulsating voltage injection without position estimator.

Therefore, the above method features an open-loop type implementation, and thus, the sensorless position estimator is not required. Nevertheless, the measured results are also affected by the delay effects described in Section 7.2.2, which should be compensated by a scaling factor:

$$\begin{bmatrix} L_{dh} & L_{dqh} \\ L_{qdh} & L_{dh} \end{bmatrix} = \frac{1}{\cos(\varphi_{\text{delay}})} \frac{V_h}{\omega_h} \begin{bmatrix} I^{I}_{dh} & I^{II}_{dh} \\ I^{I}_{qh} & I^{II}_{qh} \end{bmatrix}^{-1} \tag{7.23}$$

7.5 Combined HF Rotating and Pulsating Voltage Injection Method

This section introduces a combined HF rotating and pulsating voltage injection method [26] to estimate HF inductance, and its control diagram is shown in Figure 7.10.

The complete estimation scheme consists of three stages. Firstly, a HF rotating voltage is injected in the $\alpha\beta$ axes at first, and then HF pulsating voltages are injected in the d- and q-axes subsequently, as illustrated in Figure 7.11, where V_h and ω_h are the amplitude and angular frequency of the injected HF voltages. I_p and I_n are the measured amplitudes of the positive and negative sequence HF currents due to rotating voltage injection. I_{dh1} and I_{qh1} are the measured dq-axis HF current amplitudes due to HF pulsating voltage injection in the d-axis. I_{dh2} and I_{qh2} are the measured dq-axis HF current amplitudes due to HF pulsating voltage injection in the q-axis. For each current point (i_d, i_q), the above three injection stages are repeated.

Therefore, the HF self- and mutual-inductances can be calculated as follows:

$$L_{dh} = \frac{L_{dh} + L_{qh}}{2}\left(1 - \frac{L_{qh} - L_{dh}}{L_{qh} + L_{dh}}\right) \tag{7.24}$$

$$L_{qh} = \frac{L_{dh} + L_{qh}}{2}\left(1 + \frac{L_{qh} - L_{dh}}{L_{qh} + L_{dh}}\right) \tag{7.25}$$

$$L_{dqh} = -2V_h I_p \frac{I_{qh1}(I_{dh2})}{\omega_h\left(I_p^2 - I_n^2\right)\left(I_{dh1} + I_{qh2}\right)} \tag{7.26}$$

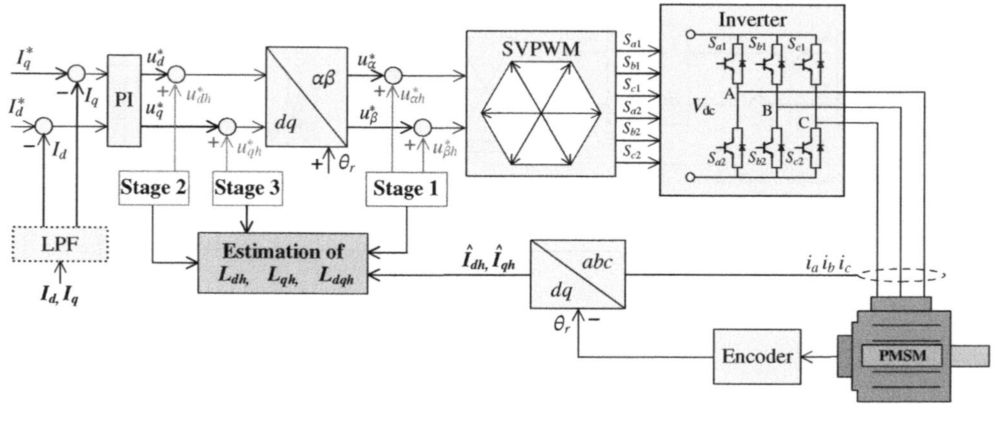

Figure 7.10 Estimation of HF inductances based on combined HF rotating and pulsating voltage injection.

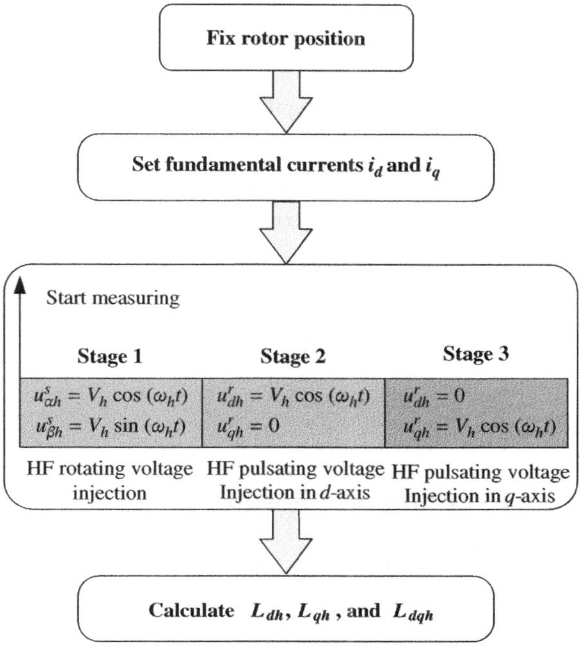

Figure 7.11 Estimation of HF inductances based on combined HF rotating and pulsating voltage injection.

As can be seen from Eqs. (7.24) and (7.25), the values of L_{dh}, L_{qh} can be obtained by measuring the values of $(L_{dh} + L_{qh})$ and $(L_{qh} - L_{dh})/(L_{qh} + L_{dh})$, respectively.

According to Eq. (7.9b), the value of $(L_{dh} + L_{qh})$ can be obtained from the measured HF current amplitudes I_p and I_n in the HF rotating voltage injection stage as:

$$L_{dh} + L_{qh} = \frac{2V_h I_p}{\omega_h\left(I_p^2 - I_n^2\right)} \tag{7.27}$$

The value of $(L_{qh} - L_{dh})/(L_{qh} + L_{dh})$ can be obtained from I_{dh1} and I_{qh2}, which are the d-axis HF current amplitude in Stage 2, and the q-axis HF current amplitude in Stage 3, respectively, and can be calculated by:

$$I_{dh1} = \frac{V_h L_{qh}}{\omega_h\left(L_{dh}L_{qh} - L_{dqh}^2\right)} \tag{7.28}$$

$$I_{qh2} = \frac{V_h L_{dh}}{\omega_h\left(L_{dh}L_{qh} - L_{dqh}^2\right)} \tag{7.29}$$

Therefore, $(L_{qh} - L_{dh})/(L_{qh} + L_{dh})$ can be expressed as:

$$\frac{L_{qh} - L_{dh}}{L_{qh} + L_{dh}} = \frac{I_{dh1} - I_{qh2}}{I_{dh1} + I_{qh2}} \tag{7.30}$$

By substituting $(L_{dh} + L_{qh})$ and $(L_{qh} - L_{dh})/(L_{qh} + L_{dh})$ in Eqs. (7.24) and (7.25) with Eqs. (7.27) and (7.30), L_{dh} and L_{qh} can be calcualted as follows:

$$L_{dh} = \frac{2V_h I_p I_{qh2}}{\omega_h\left(I_p^2 - I_n^2\right)\left(I_{dh1} + I_{qh2}\right)} \tag{7.31}$$

$$L_{qh} = \frac{2V_h I_p I_{dh1}}{\omega_h \left(I_p^2 - I_n^2\right)\left(I_{dh1} + I_{qh2}\right)} \tag{7.32}$$

It is also worth emphasizing that the delay φ_{delay} can be cancelled in Eq. (7.30), as illustrated by the following:

$$\frac{I_{dh1}\cos(\varphi_{\text{delay}}) - I_{qh2}\cos(\varphi_{\text{delay}})}{I_{dh1}\cos(\varphi_{\text{delay}}) + I_{qh2}\cos(\varphi_{\text{delay}})} = \frac{I_{dh1} - I_{qh2}}{I_{dh1} + I_{qh2}} \tag{7.33}$$

Therefore, the measured results are immune to the delay effects. After determining L_{dh} and L_{qh}, the mutual-inductance L_{dqh} can be obtained from $-I_{qh1}/I_{dh1}$ or $-I_{dh2}/I_{qh2}$, which is the ratio between L_{dqh} and L_{qh}, as demonstrated in Ref. [27].

$$\frac{L_{dqh}}{L_{qh}} = -\frac{I_{qh1}}{I_{dh1}} = -\frac{I_{dh2}}{I_{qh2}} \tag{7.34}$$

7.6 Experimental Results

The above four HF signal injection methods are implemented based on a dSPACE platform DS1006, as shown in Figure 7.12, where the parameters of the Prototype Machine 2 are shown in Appendix B. The resolution of the analogue-to-digital converter for sampling current is 16 bits. An incremental encoder of 1024 pulses per revolution is used to obtain the rotor position. The PWM and the sampling frequency are all set as 10 kHz.

7.6.1 Evaluation of Estimation Performance

In this section, the estimation performance of all four HF signal injection methods is compared experimentally under the same current point $(i_d, i_q) = (0\text{A}, 2\text{A})$, where the rotor is locked at $\theta_r = 10°$ during the estimation process. The amplitudes and frequencies of the rotating sinusoidal voltage and/or pulsating sinusoidal voltage are all chosen as 20 V/500 Hz considering sampling accuracy and saliency level [28], and the selection criteria of the injected voltage is discussed in Section 7.6.3.

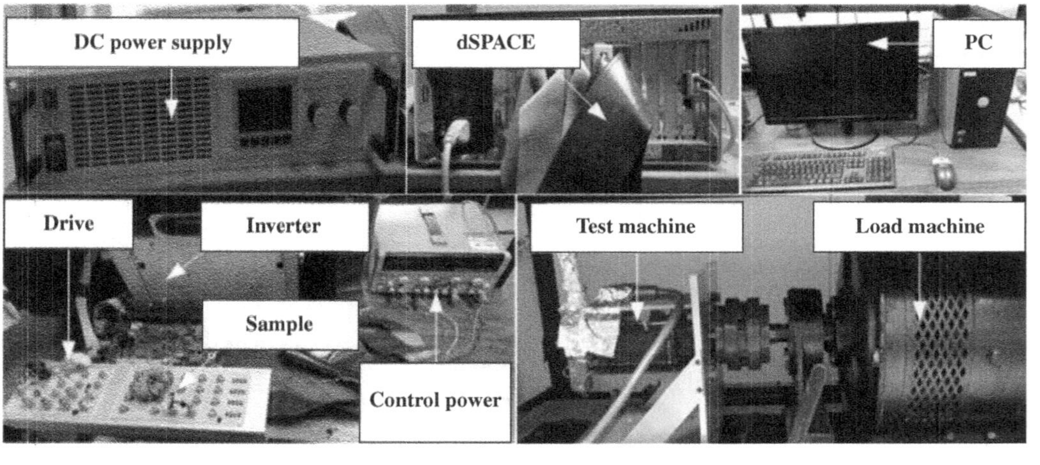

Figure 7.12 Experimental setup.

To evaluate the delay effect caused by filtering and voltage synthesis delays, a delay compensation angle φ_{comp} is subtracted from the HF angle θ_h in the demodulation of the HF current amplitudes. The applied delay will also affect the measured HF current amplitudes of $I_{dh1}, I_{qh1}, I_{dh2}$, and I_{qh2}:

$$\text{LPF}\left[2\,\sin\left(\omega_h t + \varphi_{\text{comp}}\right) I\,\sin\left(\omega_h t + \varphi_{\text{delay}}\right)\right] = I\,\cos\left(\varphi_{\text{delay}} - \varphi_{\text{comp}}\right) \tag{7.35}$$

where φ_{delay} is the actual delay in the system.

In the experiment, a delay compensation angle φ_{comp} is added to the injection angle θ_h for compensating the delay effects, which is typically approximated as the delay over 1.5 times of the PWM periods, i.e., $1.5 \times 2\pi \times 500 \times 10^{-4}$ rad $= 27°$. However, since the actual delay is hard to determine accurately, the applied delay compensation angle varies from 0° to 40° in this test to evaluate the influence of the delay effects, as presented in Figure 7.13.

When the applied delay is set to zero, i.e., the delay effects are not considered, the estimated results by all HF injection methods are compared with the finite element analysis (FEA) results, as listed in Table 7.1. The detailed FE calculation can be found in Appendix A. As can be observed, the measured results of the former three methods vary with the increase of the applied delay angle. In comparison, the HF inductances estimated by the combined HF rotating and pulsating injection method are independent of the delay effect and do not require a position estimator, but the injection process is more complicated.

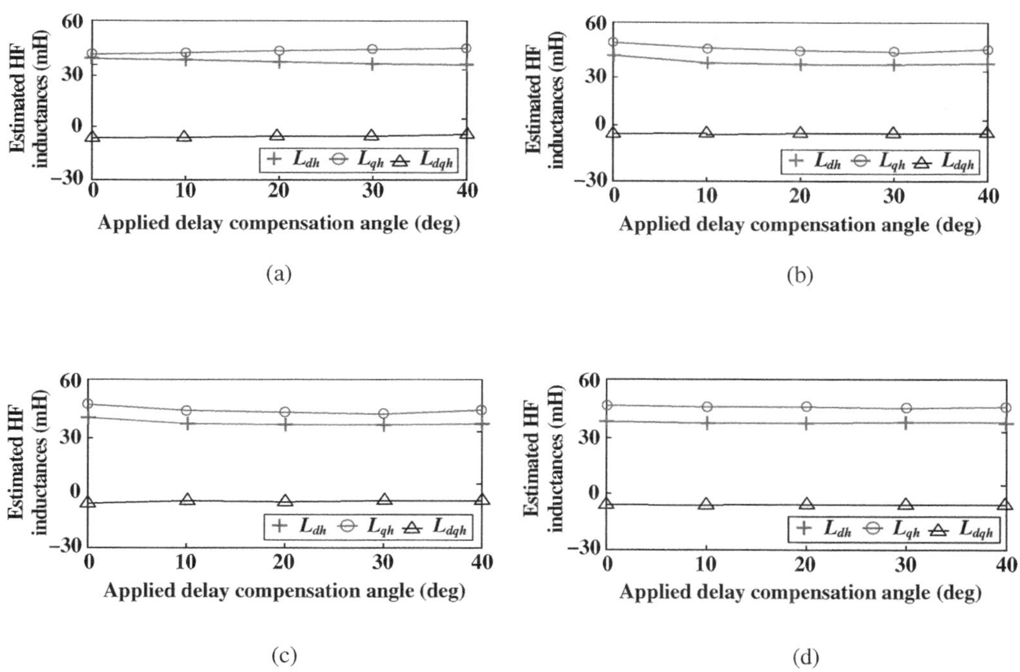

Figure 7.13 Variation of estimated HF inductances by different HF signal injection methods at $(i_d, i_q) = (0\text{ A}, 2\text{ A})$ with different delay compensation angles [26], IEEE: (a) HF rotating voltage injection with position estimation, (b) HF pulsating voltage injection with position estimator, (c) HF pulsating voltage injection without position estimator, and (d) combined HF rotating and pulsating injection without position estimator.

Table 7.1 Calculated HF inductances by various methods at $i_d = 0$ A and $i_q = 2$ A.

Methods	L_{dh}	L_{qh}	L_{dqh}
HF rotating voltage injection	42.0 mH	44.0 mH	−6.7 mH
HF pulsating voltage injection with position estimator	43.8 mH	52.1 mH	−6.8 mH
HF pulsating voltage injection without position estimator	44.1 mH	51.9 mH	−6.6 mH
Combined HF rotating and pulsating injection	38.7 mH	47.9 mH	−6.9 mH
FEA	38.7 mH	47.9 mH	−7.4 mH

7.6.2 Comparison with FE Results

Since the combined HF pulsating and rotating injection method is independent of the time delay effect, it is further used to compare with the FE results. First, the dq-axis currents during the voltage injection at $i_d = 0$ A and $i_q = 2$ A are shown in Figure 7.14.

As can be seen, the variation of the q-axis HF current during d-axis voltage injection is due to the cross-coupling effect, and vice versa. When HF rotating voltage is injected into the stationary reference frame, both dq-axis currents contain HF currents. Meanwhile, the duration for each injection stage should be long enough for the HF currents to be stable and for the measuring strategy to extract the HF current amplitudes.

Figure 7.15 shows the measured HF inductances at different rotor positions at the same working point $i_d = 0$ A and $i_q = 2$ A. It is found that the measured HF inductances vary with the rotor position, and contain sixth harmonics, which is caused by the inverter non-linearities and spatial harmonics [29]. The ripples in measured inductances can be reduced by increasing the injection voltage amplitude at the cost of reduced voltage margins. The spatial inductance harmonics also contribute to the ripples in the measured inductances. In this section, the average HF inductances

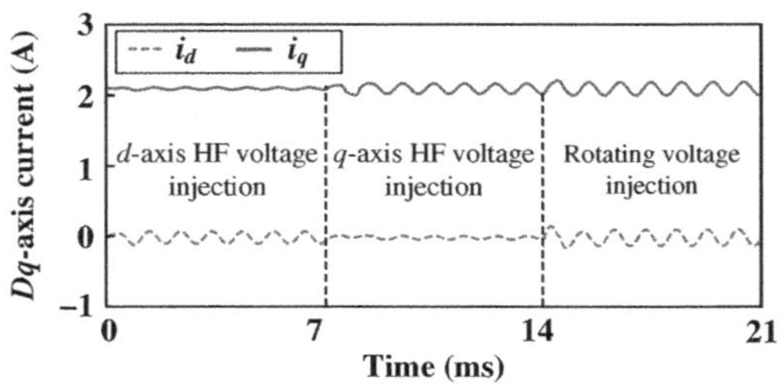

Figure 7.14 Dq-axis current sequences under combined HF pulsating and rotating injection at $i_d = 0$ A and $i_q = 2$ A [26], IEEE.

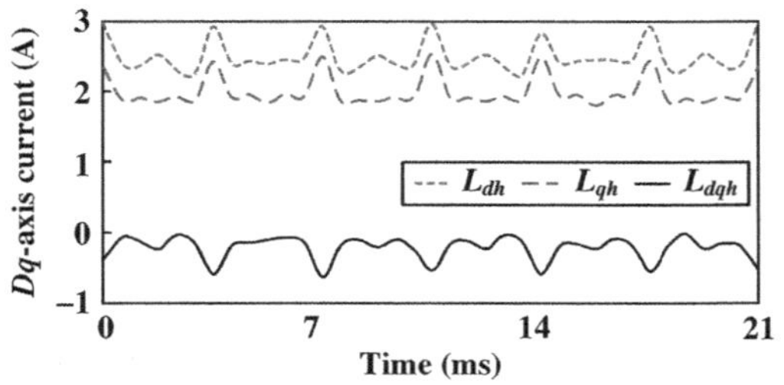

Figure 7.15 Variation of measured HF inductances with different rotor positions at $i_d = 0$ A and $i_q = 2$ A [26], IEEE.

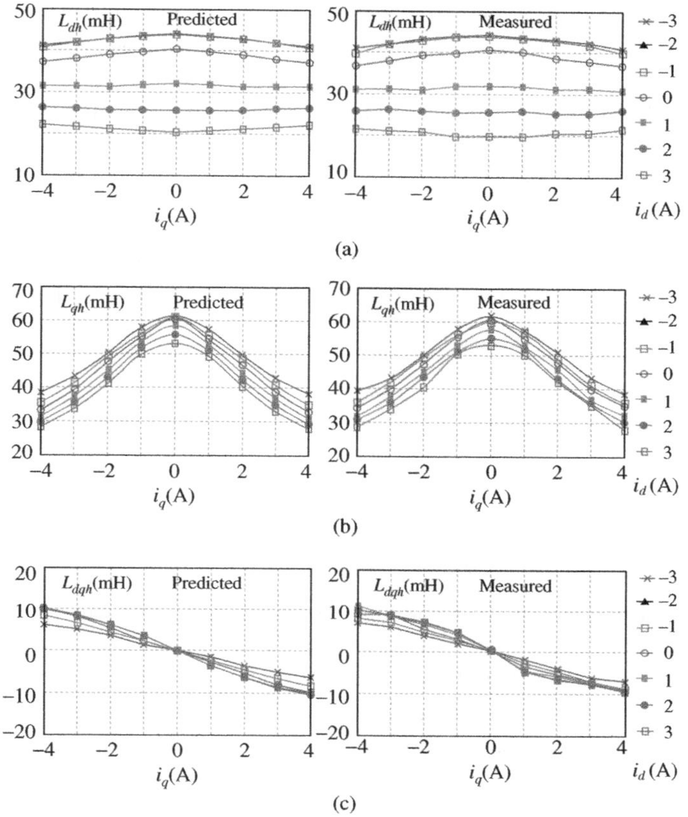

Figure 7.16 Comparison of HF inductances predicted by FEA and measured based on combined HF pulsating and rotating method [26], IEEE: (a) L_{dh}, (b) L_{qh}, and (c) L_{dqh}.

is calculated by the average values of the measured HF currents amplitudes of the three injection stages during one electrical period.

Figure 7.16 compares the HF inductances predicted by the FE method, detailed in Appendix A, with those measured based on the combined HF pulsating and rotating voltage method described in Section 7.5. Clearly, the measured results show a good agreement with the FEA predicted values. The d-axis HF inductance changes slightly with the variation of q-axis current, and the d-axis and q-axis HF inductances are almost symmetrical along the current $i_q = 0$ A.

7.6.3 Selection of Injected Voltage

The selection of the frequency and amplitude of injected voltage should consider the estimation time and voltage utilization. The estimation time can be reduced with higher injection frequency as it usually takes several injection periods for extracting the HF current amplitudes. However, with higher injection frequency, higher voltage amplitude is also required to maintain the amplitudes of the HF currents, which decreases the available voltage for fundamental currents. Moreover, the optimum voltage amplitude is dependent on the resolution of the sampling circuit and the machine characteristics.

A feasible way to determine a suitable voltage amplitude is to gradually increase the injection voltage amplitude under the same injection frequency until the estimated results are almost unchanged. According to the above guidelines, 20V/500Hz are chosen in this chapter for both HF rotating and pulsating voltages under a dc-link voltage of 150V.

Besides, the resolution of the current sampling circuit is important for improving the accuracy of the estimation. For the maximum current sampling range $[-I_{max}, I_{max}]$ with sampling resolution N bits, the sampled current resolution can be calculated as follows:

$$I_{\mathrm{LSB}} = \frac{2I_{max}}{2^N} \tag{7.36}$$

According to Ref. [28], the measurement error of the current due to quantization is $0.5I_{\mathrm{LSB}}$. Thus, the estimation errors of the estimated HF inductances can be approximated as follows:

$$\frac{\Delta L_{dh}}{L_{dh}} = \frac{0.5I_{\mathrm{LSB}}}{I_{qh2}} * 100\% \tag{7.37a}$$

$$\frac{\Delta L_{qh}}{L_{qh}} = \frac{0.5I_{\mathrm{LSB}}}{I_{dh1}} * 100\% \tag{7.37b}$$

$$\frac{\Delta L_{dqh}}{L_{dqh}} = \frac{0.5I_{\mathrm{LSB}}}{I_{dh2}(I_{qh1})} * 100\% \tag{7.37c}$$

By using the above equations, the estimation error can be easily quantified. For instance, for the estimation at $i_d = 0$ A and $i_q = 2$ A with 16 bits sampling resolution in the experiment ($I_{\mathrm{LSB}} = 0.224$ mA and $I_{qh2} = 55$ mA), the estimation error of L_{dh} is calculated as $0.5 * 0.224/55 \times 100\% = 0.20\%$. In practice, the 12-bit sampling circuit resolution is commonly used in commercial inverters, and thus, the estimation error caused by sampling quantization of L_{dh} increases to 3.2%.

7.7 Summary

This chapter has described four HF voltage injection methods for the estimation of HF self- and mutual-incremental inductances accounting for magnetic saturation and cross-coupling, including the rotating voltage injection in the stationary $\alpha\beta$-axes, the pulsating voltage injections in the estimated and actual dq-axes, as well as the combined rotating and pulsating voltage injections in the $\alpha\beta$- and actual dq-axes. The features of all estimation methods are summarized in Table 7.2. According to the experimental results, it has been shown that the latter two methods have simple

Table 7.2 Comparison of HF inductance estimation methods.

	Rotating	Pulsating		Combined
	[19]	[23]	[24]	[26]
Injection axis	α- and β-	d- and q-	d- and q-	α- and β- d- and q-
Non-requirement of position estimator	✗	✗	✔	✔
Immune to delay effect	✗	✗	✗	✔

estimation procedures since the position estimator is required. Meanwhile, since the delay effect caused by the non-ideal hardware implementation can be eliminated by the combined HF injection method, it has exhibited similar estimation performance against FE results.

References

1 P. Vas, K. E. Hallenius, and J. E. Brown, "Cross-saturation in smooth-air-gap electrical machines," *IEEE Trans. Energy Convers.*, vol. EC-1, no. 1, pp. 103–112, March 1986.

2 A. M. El-Serafi and A. S. Abdallah, "Effect of saturation on the steady state stability of a synchronous machine connected to an infinite bus system," *IEEE Trans. Energy Convers.*, vol. 6, pp. 514–521, Sept. 1991.

3 D. Wu and Z. Q. Zhu, "Influence of slot and pole number combinations on voltage distortion in surface-mounted permanent magnet machines with local magnetic saturation," *IEEE Trans. Energy Convers.*, vol. 30, no. 4, pp. 1460–1471, Dec. 2015.

4 G. Qi, J. T. Chen, Z. Q. Zhu, D. Howe, L. B. Zhou, and C. L. Gu, "Influence of skew and cross-coupling on flux-weakening performance of permanent-magnet brushless AC machines," *IEEE Trans. Magn.*, vol. 45, no. 5, pp. 2110–2117, May 2009.

5 B. Stumberger, G. Stumberger, D. Dolinar, A. Hamler, and M. Trlep, "Evaluation of saturation and cross-magnetization effects in interior permanent-magnet synchronous motor," *IEEE Trans. Ind. Appl.*, vol. 39, no. 5, pp. 1264–1271, Sept.-Oct. 2003.

6 Z. Azar, Z. Q. Zhu, and G. Ombach, "Influence of electric loading and magnetic saturation on cogging torque, back-EMF and torque ripple of PM machines," *IEEE Trans. Magn.*, vol. 48, no. 10, pp. 2650–2658, Oct. 2012.

7 K. Liu, "Research on the online multi-parameter identification of permanent magnet synchronous machines," Ph.D dissertation, Hunan Univ., Changsha, China, 2010.

8 K. Liu, Q. Zhang, J. Chen, Z.Q. Zhu, and J. Zhang, "Online multiparameter estimation of nonsalient-pole PM synchronous machines with temperature variation tracking," *IEEE Trans. Ind. Electron.*, vol. 58, no. 5, pp. 1776–1788, 2011.

9 M. Rashed, P. Macconnell, A. Stronach, and P. Acarnley, "Sensorless indirect-rotor-field-orientation speed control of a permanent-magnet synchronous motor with stator-resistance estimation," *IEEE Trans. Ind. Electron.*, vol. 54, no. 3, pp. 1664–1675, Jun. 2007.

10 X. Liang, A. M. El-Serafi, and S. O. Faried, "Application of the finite-element method for the determination of the parameters representing the cross-magnetizing in saturated synchronous machines," *IEEE Trans. Energy Convers.*, vol. 25, no. 1, pp. 70–79, March 2010.

11 Y. Mollet, G. Barbierato, and J. Gyselinck, "Finite-element and magnetic-equivalent-circuit modelling of brushed wound-field DC machines with cross-saturation," *2016 XXII Int. Conf. Electr. Mach. (ICEM)*, Lausanne, Switzerland, 2016.

12 G. H. Bayazıt, E. I. Caarls, and E. A. Lomonova, "Inductance map regression of doubly excited electrical machines considering cross-saturation," *IEEE Trans. Magn.*, vol. 59, no. 11, pp. 1–5, Nov. 2023.

13 K. M. Rahman and S. Hiti, "Identification of machine parameters of a synchronous motor," *IEEE Trans. Ind. Appl.*, vol. 41, no. 2, pp. 557– 565, Mar./Apr. 2005.

14 S. A. Odhano, R. Bojoi, S. G. Rosu, and A. Tenconi, "Identification of the magnetic model of permanent-magnet synchronous machines using DC-biased low-frequency AC signal injection," *IEEE Trans. Ind. Appl.*, vol. 51, no. 4, pp. 3208–3215, Jul./Aug. 2015.

15 B. Stumberger, B. Kreca, and B. Hribernik, "Determination of parameters of synchronous motor with permanent magnets from measurement of load conditions," *IEEE Trans. Energy Convers.*, vol. 14, no. 4, pp. 1413–1416, Dec. 1999.

16 G. Feng, C. Lai, and N. C. Kar, "A novel current injection-based online parameter estimation method for PMSMs considering magnetic saturation," *IEEE Trans. Magn.*, vol. 52, no. 7, pp. 1–4, Jul. 2016.

17 K. Liu, Z.Q. Zhu, and D. Stone, "Parameter estimation for condition monitoring of PMSM stator winding and rotor permanent magnets," *IEEE Trans. Ind. Electron.*, vol. 60, no. 12, pp. 5902–5913, 2013.

18 G. Lin, J. Zhang, and Z. Liu, "Parameter identification of PMSM using immune clonal selection differential evolution algorithm," *Math. Probl. Eng.*, vol. 2014, 2014.

19 W. Xu and R. D. Lorenz, "High-frequency injection-based stator flux linkage and torque estimation for DB-DTFC implementation on IPMSMs considering cross-saturation effects," *IEEE Trans. Ind. Appl.*, vol. 50, no. 6, pp. 3805–3815, Nov./Dec. 2014.

20 J. Liu and Z. Q. Zhu, "Improved sensorless control of permanent-magnet synchronous machine based on third-harmonic back EMF," *IEEE Trans. Ind. Appl.*, vol. 50, no. 3, pp. 1861–1870, May./Jun. 2014.

21 B. Bae and S. Sul, "A compensation method for time delay of full-digital synchronous frame current regulator of PWM AC drives," *IEEE Trans. Ind. Appl.*, vol. 39, no. 3, pp. 802–810, May./Jun. 2003.

22 M. Moghadam and F. Tahami, "Sensorless control of PMSMs with tolerance for delays and stator resistance uncertainties," *IEEE Trans. Power Electron.*, vol. 28, no. 3, pp. 1391–1399, Mar. 2013.

23 S. Kuehl and R. M. Kennel, "Measuring magnetic characteristics of synchronous machines by applying position estimation techniques," *IEEE Trans. Ind. Appl.*, vol. 50, no. 6, pp. 3816–3824, Nov./Dec. 2014.

24 Y. Li, Z. Q. Zhu, D. Howe, and C. M. Bingham, "Modelling of cross-coupling magnetic saturation in signal injection-based sensorless control of permanent-magnet brushless AC motors," *IEEE Trans. Magnet.*, vol. 43, no. 6, pp. 2552–2554, Jun. 2007.

25 P. Xu and Z. Q. Zhu, "Carrier signal injection-based sensorless control for permanent magnet synchronous machine drives with tolerance of signal processing delays," *IET Electr. Power Appl.*, vol. 11, no. 6, pp. 1140–1149, 2017.

26 B. Shuang and Z. Q. Zhu, "A novel method for estimating the high frequency incremental *dq*-axis and cross-coupling inductances in interior permanent magnet synchronous machines," *IEEE Trans. Ind. Appl.*, vol. 57, no. 5, pp. 4913–4923, Sept.-Oct. 2021.

27 Y. Li, Z. Q. Zhu, D. Howe, C. M. Bingham, and D. A. Stone, "Improved rotor-position estimation by signal injection in brushless AC motors, accounting for cross-coupling magnetic saturation," *IEEE Trans. Ind. Appl.*, vol. 45, no. 5, pp. 1843–1850, Sept. 2009.

28 Z. Q. Zhu and L. M. Gong, "Investigation of effectiveness of sensorless operation in carrier signal-injection based sensorless-control methods," *IEEE Trans. Ind. Electron.*, vol. 58, no. 8, pp. 3431–3439, Aug. 2011.

29 Y. Kwon, J. Lee, and S. -K. Sul, "Recent advances in sensorless drive of interior permanent-magnet motor based on pulsating signal injection," *IEEE J. Emerg. Sel. Top. Power Electron.*, vol. 9, no. 6, pp. 6577–6588, Dec. 2021.

8

Offline and Multi-step Parameter Estimation Methods

8.1 Introduction

As described in the previous chapters, online parameter estimation for permanent magnet synchronous machines (PMSMs) is affected by voltage source inverter (VSI) non-linearity, magnetic saturation, temperature effect, etc., which may even interact with each other [1]. Therefore, the offline parameter estimation methods [2–9] have also been widely investigated to simplify the estimation scheme and to eliminate the uncertainties during the machine operation.

In the last few decades in the 20th century, the research of offline parameter estimation methods has already gained great interest. As early as the 1980s, the estimation of the reactance and excitation electromotive force (EMF) in the three-phase and *dq*-axes was developed in Refs. [2–5] according to the analyses of the phasor diagram with the aid of the measured power, voltages and currents, where the self- and mutual-magnetic saturations are also considered, and the estimated inductances belong to apparent type. Besides, based on the transient characteristics of the damper windings, the step direct current (dc) voltage excitations are employed for PMSMs at standstill in Refs. [6] and [7] to estimate *dq*-axis incremental inductances and stator resistance.

More recently, when the PMSM is at a standstill, the *dq*-axis incremental inductances and the stator winding resistance can be estimated by injecting *dq*-axis square voltages [8], the *dq*-axis high-frequency (HF) currents [9], as well as dc biased low-frequency ac signals [10]. Therefore, the *dq*-axis flux linkage maps excited by currents can be further calculated by integral incremental inductance with respect to the current increments. In this case, the estimation accuracy is strongly affected by the VSI non-linearity although it can be easily determined by calculating the voltage drops in the *dq*-axes in an offline manner [9].

In terms of estimation of rotor PM flux linkage, which can only be activated during the operation, the two-step [11] and three-step [9] estimation methods have been developed, which combine offline measurement and online parameter update. More specifically, in the two-step estimation method, the stator winding resistance and the rotor PM flux linkage are fixed to the nominal/measured values alternatively to estimate the other one. However, as will be presented later in this chapter, the two-step estimation easily suffers from the parameter-mismatch issue due to the disturbance of magnetic and thermal effects. In contrast, in the three-step estimation method, with the aid of the estimated stator winding resistance and the distorted voltage due to VSI non-linearity, the rotor PM flux linkage and *q*-axis inductance are estimated under operation in the second step. Afterwards, the stator winding resistance is updated in the third step to track its variation due to temperature changes.

Parameter Estimation of Permanent Magnet Synchronous Machines, First Edition. Zi Qiang Zhu, Kan Liu, and Dawei Liang.

Nowadays, the boundaries between offline and online estimation methods are becoming blurred due to the rapid development of high-performing microprocessors. The estimation of dq-flux linkage maps based on the comprehensive experiments [12–15] can be implemented in an online manner. Therefore, both incremental and apparent inductances can be calculated subsequently, where the influences of magnetic saturation can be considered. A detailed description of the identification of dq-axis flux linkage maps can be found in Chapters 5 and 6.

This chapter describes several classical offline and multi-step estimation methods. In Sections 8.2 and 8.3, the principles of square voltage- and HF current-based offline estimation methods are described and then performed experimentally based on a prototype interior permanent magnet synchronous machine (IPMSM), where the importance of compensation for VSI non-linearity is emphasized. In Section 8.4, both two-step and three-step estimation methods are also implemented experimentally, and the corresponding estimation performance is evaluated by accounting for the influence of parameter variations due to the temperature effect.

8.2 Parameter Estimation at Standstill by Square Voltage Injection

The dq-axis flux linkages are calculated from the recorded dq-axis reference voltages and currents at standstill (i.e. the rotor is locked) while square wave voltages are injected as perturbation signals.

Accounting for self- and mutual-magnetic saturations and the VSI non-linearity, the dq-axis PMSM model can be expressed as follows:

$$u_d^* + DdV_{\text{dead}} = R_s i_d + L_{dd}^{\text{inc}} \frac{di_d}{dt} + L_{dq}^{\text{inc}} \frac{di_q}{dt} - \frac{d\theta_r}{dt} \psi_q \tag{8.1a}$$

$$u_q^* + DqV_{\text{dead}} = R_s i_q + L_{qq}^{\text{inc}} \frac{di_q}{dt} + L_{qd}^{\text{inc}} \frac{di_d}{dt} + \frac{d\theta_r}{dt} (\psi_d^* + \psi_{pm}) \tag{8.1b}$$

where u_d^*, u_q^*, Dd, Dq, i_d, and i_q are the reference voltages, the gain functions of VSI non-linearity, and the currents in the dq-axes. V_{dead} is the distorted voltage caused by the VSI non-linearity. R_s is the stator resistance. ψ_{pm} is the rotor PM flux linkage. θ_r is the rotor position. The total d-axis flux linkage is divided into two parts, i.e., ψ_d^* and ψ_{pm}, which are the flux linkages caused by the d-axis current excitation and the PM field, respectively. ψ_q is the q-axis flux linkage. L_{dd}^{inc}, L_{qq}^{inc}, L_{dq}^{inc}, and L_{qd}^{inc} are the dq-axis apparent/incremental self- and mutual-inductances.

The offline estimation method based on square-wave voltage injection [8] is illustrated in Figure 8.1 and implemented as follows. It is worth noting that during the estimation the rotor is locked to avoid the speed pulsation due to square-wave voltage injection.

Step 1: When the PMSM is supplied by a VSI and performed at locked-rotor position, the rotor-speed related terms $\frac{d\theta_r}{dt}\psi_q$ and $\frac{d\theta_r}{dt}\psi_d$ can be eliminated from Eq. (8.1), which is simplified as follows:

$$u_d^* + DdV_{\text{dead}} = R_s i_d + \frac{d\psi_d^*(i_d, i_q)}{dt} \tag{8.2a}$$

$$u_q^* + DqV_{\text{dead}} = R_s i_q + \frac{d\psi_q(i_d, i_q)}{dt} \tag{8.2b}$$

Step 2: To obtain the q-axis (d-axis) flux linkage in the entire operation range, a square wave is injected in the q-axis (d-axis) to the control system, while the d-axis (q-axis) current is fixed by the current regulator, as illustrated in Figure 8.2.

Step 3: The dq-axis flux linkages are calculated from Eq. (8.2) and expressed as follows:

$$\psi_d(t) = \int_0^t [u_d(\tau) + Dd(\tau)V_{\text{dead}} - R_s i_d(\tau)]d\tau \tag{8.3a}$$

$$\psi_q(t) = \int_0^t \left[u_q(\tau) + Dq(\tau)V_{\text{dead}} - R_s i_d(\tau)\right]d\tau \tag{8.3b}$$

Step 4: The incremental dq-axis inductances can be derived from Eq. (8.4).

$$L_{dd}^{\text{inc}} = \frac{\partial \psi_d^*(i_d,i_q)}{\partial i_d} \quad L_{qq}^{\text{inc}} = \frac{\partial \psi_q(i_d,i_q)}{\partial i_q} \tag{8.4a}$$

$$L_{dq}^{\text{inc}} = \frac{\partial \psi_d^*(i_d,i_q)}{\partial i_q} \quad L_{qd}^{\text{inc}} = \frac{\partial \psi_q(i_d,i_q)}{\partial i_d} \tag{8.4b}$$

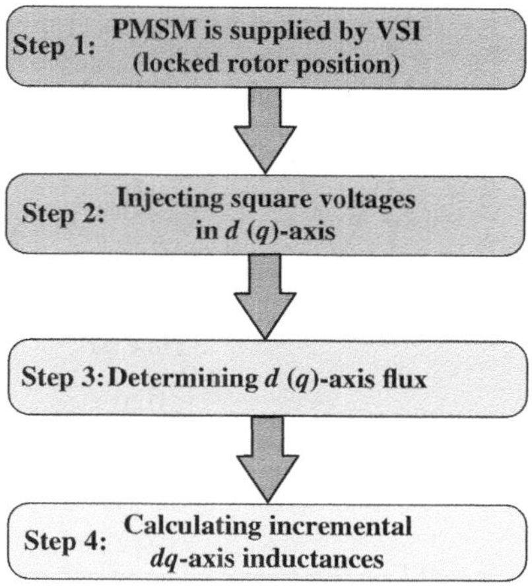

Figure 8.1 Flowchart of square-wave voltage-based estimation method.

As can be seen from Eq. (8.2), the influence of VSI non-linearity should be considered in the estimation. Otherwise, the estimation accuracy will be deteriorated. By way of example, Figure 8.3 compares the q-axis flux linkages ψ_q calculated by Eq. (8.3) with and without considering the VSI non-linearity.

First, as can be seen from Figure 8.3 (I.a), the calculated q-axis flux linkage is not zero mean, since the dc component in q-axis voltage will be integrated during the calculation. This dc component mainly origins from the difference of VSI nonlinearity at positive and negative directions of q-axis currents. Thus, it can be observed that the calculated q-axis flux linkage will increase with time. Moreover, the calculated q-axis flux linkage with respect to the q-axis current in one magnetization cycle is shown in Figure 8.3 (I.b). A simple solution to rectify this dc offset is to use the middle point of two $\psi_q(i_q = 0)$ at $i_q = 0$ to compensate it, by which the q-axis flux linkage around the origin of coordinate ($i_q = 0$, $\psi_q = 0$) will be approximated by a simple polynomial equation $y = ax + b$. After compensation of VSI nonlinearity at both positive and negative directions of q-axis currents and the rectification of dc offset of $\psi_q(i_q = 0)$ at $i_q = 0$, there is almost no dc offset in the calculated q-axis flux linkage, and the curve fitted ψ_q is symmetrical with respect to the origin, as shown in Figure 8.3 (II).

Second, when the VSI non-linearity is not compensated, as shown in Figure 8.3I, the estimated

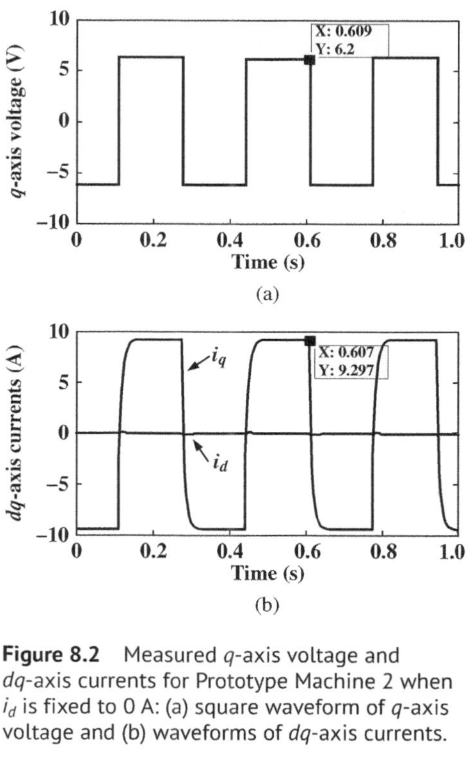

Figure 8.2 Measured q-axis voltage and dq-axis currents for Prototype Machine 2 when i_d is fixed to 0 A: (a) square waveform of q-axis voltage and (b) waveforms of dq-axis currents.

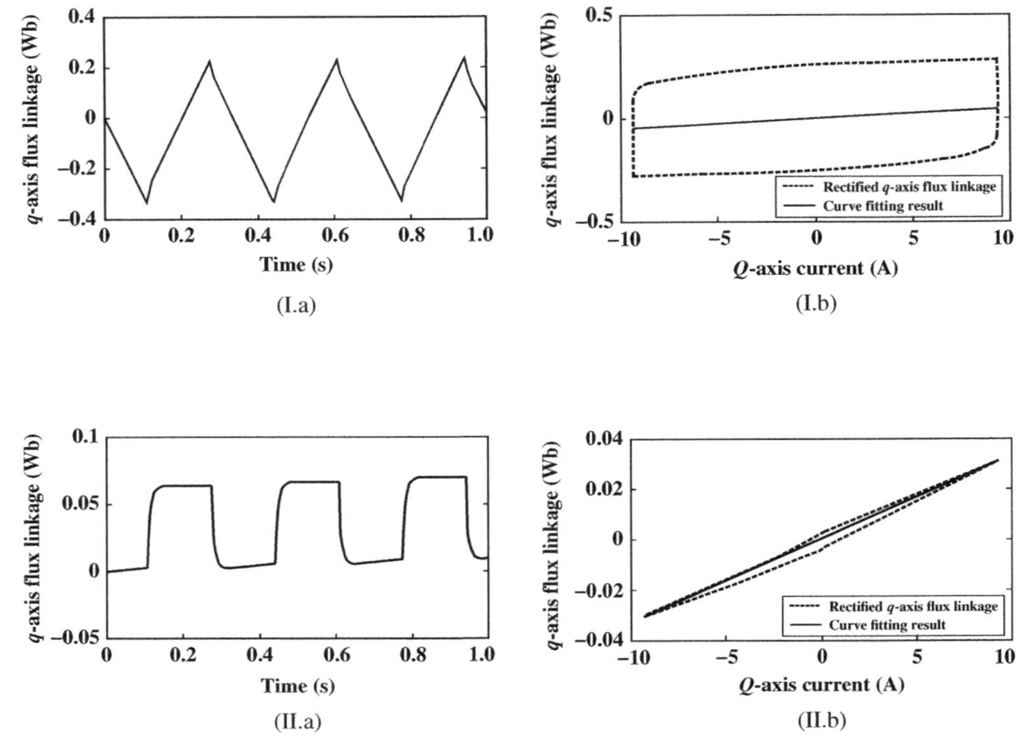

Figure 8.3 Comparison of estimated q-axis flux linkages without (I) and with (II) compensation for VSI non-linearity: (I.a) estimated ψ_q with time and (I.b) rectified and curve-fitted ψ_q; (II.a) estimated ψ_q with time and (II.b) rectified and curve-fitted ψ_q.

q-axis flux linkage has a large deviation from the nominal values. In addition, it is worth noting that the compensation for VSI non-linearity is much simpler than that under load conditions. In this case, the voltage drop due to VSI non-linearity is approximately calculated by using the steady state q-axis current (or d-axis current) and the reference q-axis voltage (or d-axis voltage). For instance, as can be seen from Figure 8.2, the amplitude of the resulting q-axis current due to positive and negative q-axis voltages will be 9.2 and −9.1 A, respectively. Since the winding resistance of Prototype Machine 2 is measured as 0.33 Ω, the voltage drop due to VSI non-linearity at the injection of positive and negative voltages (±35V) are (35V − 6.5Ω × 3.9A) and (−35V − 6.5Ω × 4.1A), respectively.

Figure 8.4 compares the estimated q-axis inductance at standstill with the value calculated by finite element analysis (FEA). Obviously, the estimated L_q decreases with the amplitude of the

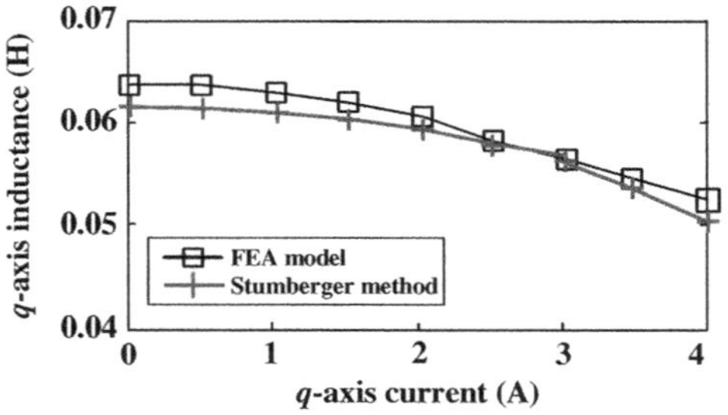

Figure 8.4 Comparison of estimated q-axis inductance against FEA reuslts.

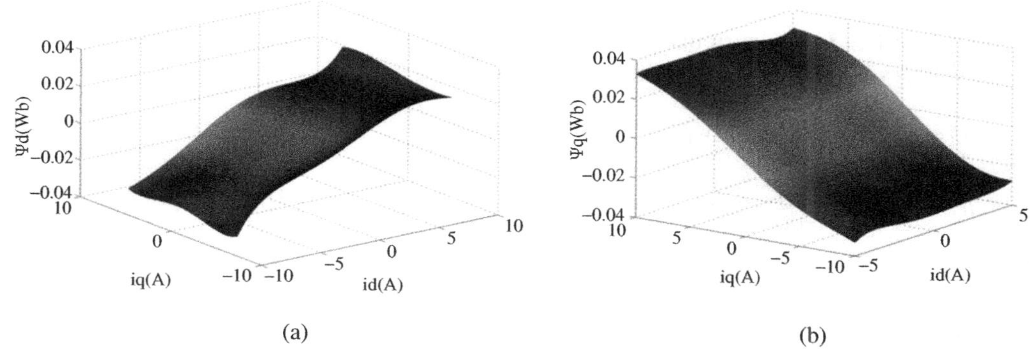

Figure 8.5 Estimated dq-axis flux linkage maps with i_d and i_q: (a) estimated d-axis flux linkage and (b) estimated q-axis flux linkage.

q-axis current due to magnetic saturation, and the maximum estimation error is smaller than 5%, which may be attributed to the measurement errors and uncertainties of the materials' properties. A detailed description of the FEA model can be found in Appendix A.

Furthermore, by setting different dc offsets, the dq-axis flux linkages in the entire operation range can be obtained. In addition, the continuous flux linkages can be easily implemented by using a surface fitting algorithm [16], as shown in Figure 8.5.

8.3 Parameter Estimation at Standstill by HF Current Injection

This section describes the principle of the offline estimation method based on HF current injection at standstill. The conventional estimation method is introduced in Section 8.3.1, where the complete dq-axis current variations and a complex estimation algorithm are required to estimate dq-axis inductances. In Section 8.3.2, the conventional estimation method is simplified by detecting the zero-crossing points of the injected dq-axis HF currents, where the dq-axis inductances can be obtained by simple calculation.

8.3.1 Estimation Based on Current Variations

When the PMSM is at standstill, the winding resistance and dq-axis incremental inductances can also be estimated by using HF current injections in the dq-axes [9]. It consists of two steps and is summarized as follows:

Step 1: Estimating the winding resistance R_s and the distorted voltage V_{dead} by using the steady state dq-axis current and voltage at standstill.

Step 2: Estimating the dq-axis incremental inductances by injecting HF currents.

At the first step, the dq-axis voltages (Eq. 8.1) can be simplified as follows:

$$u_d^* = R_s i_d - DdV_{\text{dead}} \tag{8.5a}$$

$$u_q^* = R_s i_q - DqV_{\text{dead}} \tag{8.5b}$$

Therefore, the unknown parameters, i.e., R_s and V_{dead}, can be estimated from (Eq. 8.5) by using simple estimation algorithms, e.g., Adaline neural networks.

Based on the estimated R_s and V_{dead} in Step 1, the incremental dq-axis inductances are estimated by injection of HF dq-axis currents at standstill. When the cross-coupling effect is neglected, the dq-axis voltages (Eq. 8.1) can be rewritten as follows:

$$u_d^* = R_s i_d + L_d^{\text{inc}} \frac{di_d}{dt} - DdV_{\text{dead}} \tag{8.6a}$$

$$u_q^* = R_s i_q + L_q^{\text{inc}} \frac{di_q}{dt} - DqV_{\text{dead}} \tag{8.6b}$$

When the HF sinusoidal currents, i.e., $i_d = i_d^* \sin(\omega_{h1}t)$ and $i_q = i_q^* \sin(\omega_{h2}t)$, are injected in the dq-axes to estimate the incremental dq-axis inductance L_d^{inc} and L_q^{inc}. In this case, Eq. (8.6) can be rewritten as follows:

$$u_d^* = R_s i_d^* \sin(\omega_{h1}t) + L_d^{\text{inc}} i_d^* \cos(\omega_{h1}t) - DdV_{\text{dead}} \tag{8.7a}$$

$$u_q^* = R_s i_q^* \sin(\omega_{h2}t) + L_q^{\text{inc}} i_q^* \cos(\omega_{h2}t) - DqV_{\text{dead}} \tag{8.7b}$$

Furthermore, the described estimation method is performed based on the Prototype Machine 2, of which the nominal stator winding R_s is 6.2 Ω. The nominal dq-axis inductances L_d and L_q are 32.1 and 55.0 mH, respectively.

First step:

First, the measured quantities and the estimated R_s and V_{dead} are depicted in Figures 8.6 and 8.7, respectively, where the rotor of the PMSM is locked at a specific position (Position I). Furthermore, the above estimation processes are repeated at two different rotor positions, i.e., Positions II and III, and the estimated results are shown in Figure 8.8.

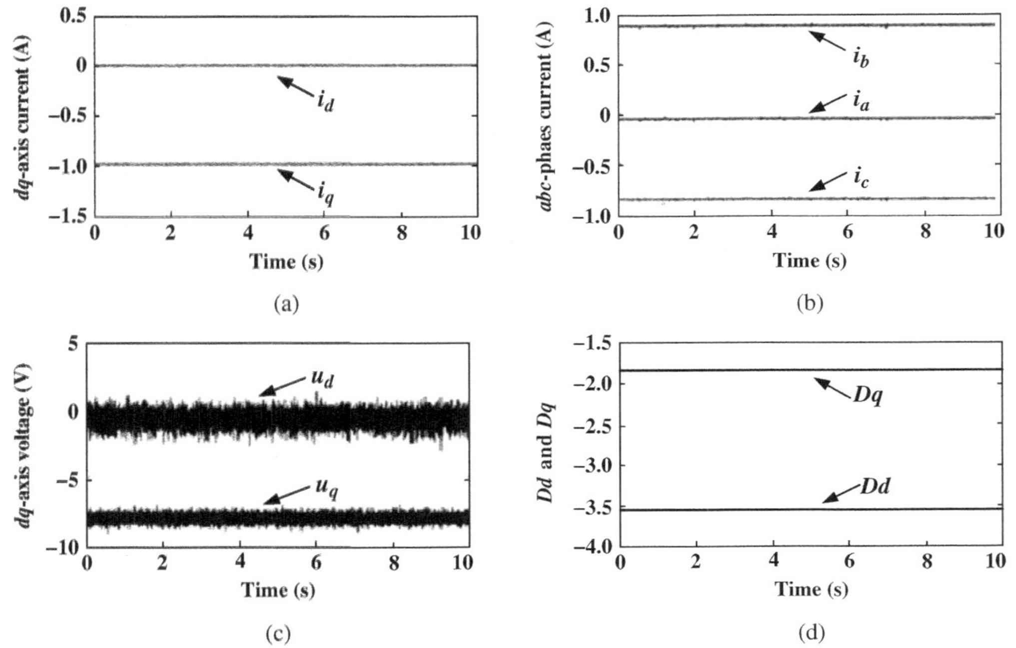

Figure 8.6 Measured electrical quantities for estimation of R_s and V_{dead} at Position I: (a) dq-axis currents, (b) abc three-phase currents, (c) dq-axis voltages, and (d) Dd and Dq.

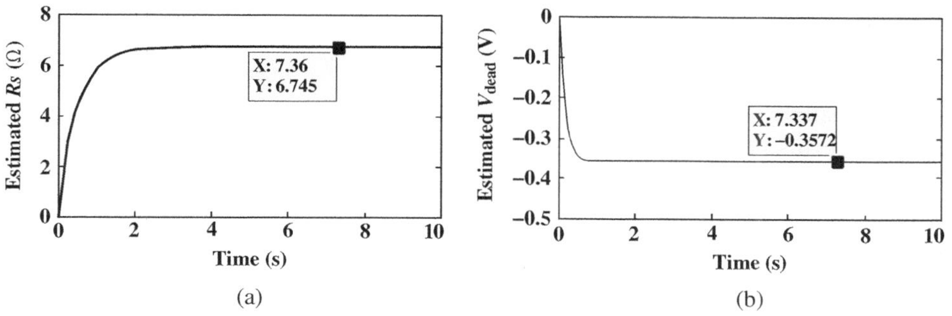

Figure 8.7 Estimation results at Position I: (a) estimated stator resistance R_s and (b) estimated distorted voltage V_{dead}.

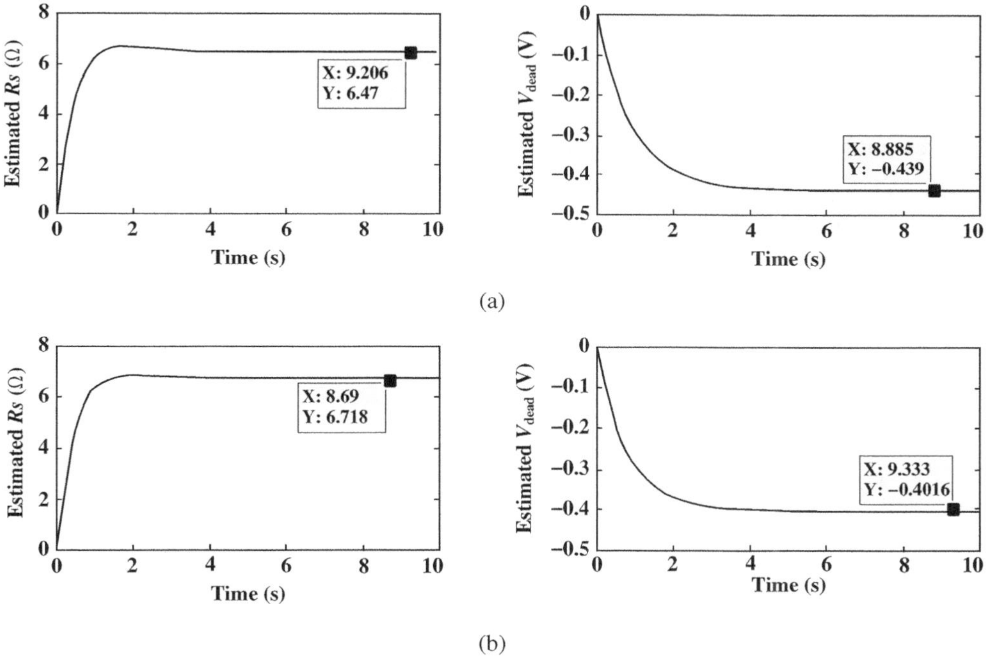

Figure 8.8 Estimation results at Positions II and III: (a) estimated stator resistance R_s and distorted voltage V_{dead} at Position II and (b) estimated stator resistance R_s and distorted voltage V_{dead} at Position III.

As can be seen the estimated stator resistances at three rotor positions are 6.7, 6.5, and 6.7 Ω, which are close to the nominal value. The slight difference in the estimated R at different rotor positions may be attributed to the variation of VSI non-linearity, which is a function of rotor position and current vector angle, as detailed in Chapter 2. In addition, it is worth mentioning that the influence of on-state resistance and core-loss resistance on the estimation of winding resistance may be more significant if the machine has a relatively small winding resistance.

To demonstrate the importance of the VSI non-linearity on the estimation accuracy, Figure 8.9 shows the estimated stator winding resistance without compensating for the VSI non-linearity. In this case, the estimated resistance is 8.0 Ω, which is much larger than the nominal value of 6.2 Ω.

Second step:

During the estimation of the dq-axis inductances for Prototype Machine 2, the dq-axis currents are set as $i_d = \sin(80\pi t)$ and $i_q = \cos(160\pi t)$. It is worth noting that the frequency of i_q should be higher than that of i_d to avoid unexpected rotating since i_q is dominant in the generation of electromagnetic torque. The measured quantities are shown in Figure 8.10, and the estimated dq-axis inductances are shown in Figure 8.11, which is solved according to Eq. (8.6).

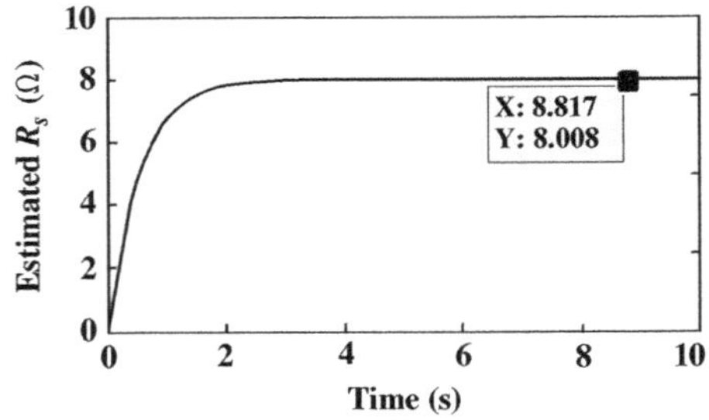

Figure 8.9 Estimated winding resistance R_s without compensating for VSI non-linearity.

Clearly, although the inductances of PMSM are small, the HF current-based offline estimation method still has relatively good performance. The estimated L_d and L_q are 35.12 and 61.4 mH, which are close to the respective nominal values of 32.1 and 55.0 mH, respectively. Meanwhile, it is worth noting that a larger amplitude of current injection will result in a more accurate estimation result due to high signal-to-noise ratio (SNR).

8.3.2 Estimation Based on Zero-crossing Detection

As aforementioned, the method described in Section 8.3.1 requires the utilization of complex identification algorithms to estimate the dq-axis incremental inductances in Eq. (8.7). It will be simplified in this sub-section by detecting zero-crossing points of injected HF currents.

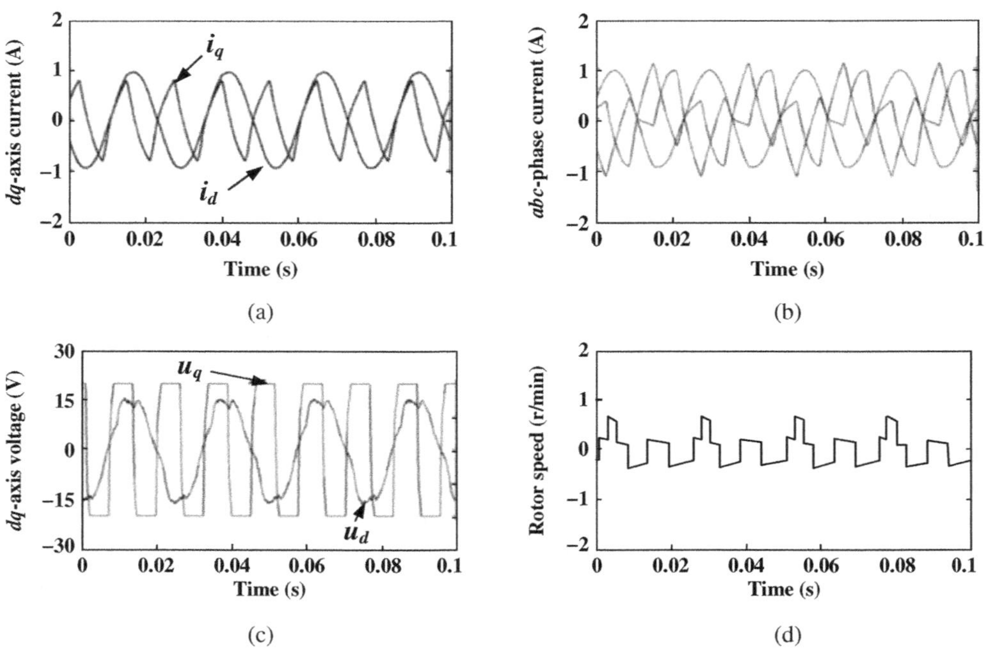

Figure 8.10 Measured quantities for estimation of dq-axis inductances: (a) dq-axis currents, (b) abc three-phase currents, (c) dq-axis voltages, and (d) rotor speed.

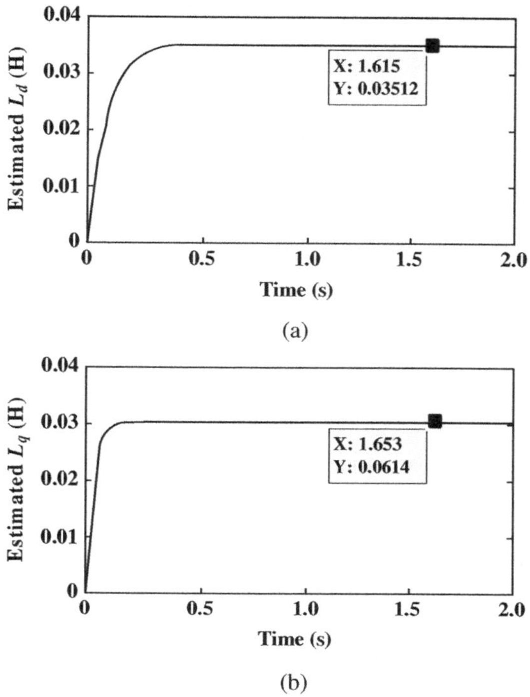

Figure 8.11 Estimated L_d and L_q at standstill: (a) estimated L_d and (b) estimated L_q.

When the HF dq-axis currents, i.e., $i_d = i_d^* \, sin(\omega_{h1}t)$ and $i_q = i_d^* \, sin(\omega_{h2}t)$, are injected, the dq-axis voltage equations at standstill is written as follows:

$$u_d^* = R_s i_d^* \sin(\omega_{h1}t) + L_d^{\text{inc}} i_d^* \cos(\omega_{h1}t) - DdV_{\text{dead}} \tag{8.8a}$$

$$u_q^* = R_s i_q^* \sin(\omega_{h2}t) + L_q^{\text{inc}} i_q^* \cos(\omega_{h2}t) - DqV_{\text{dead}} \tag{8.8b}$$

At the time points of $\omega_h t = 2k\pi$, $k = 0,1,2\ldots$, Eq. (8.8) can be simplified as follows:

$$u_d^* = L_d^{\text{inc}} i_d^* - DdV_{\text{dead}} \tag{8.9a}$$

$$u_q^* = L_q^{\text{inc}} i_q^* - DqV_{\text{dead}} \tag{8.9b}$$

Therefore, the dq-axis incremental inductances can be obtained as follows:

$$L_d^{\text{inc}} = \frac{u_d^*(\omega_{h1}t = 2k\pi) + DdV_{\text{dead}}}{\omega_h i_d^*} \tag{8.10a}$$

$$L_q^{\text{inc}} = \frac{u_q^*(\omega_{h2}t = 2k\pi) + DqV_{\text{dead}}}{\omega_h i_q^*} \tag{8.10b}$$

Obviously, only three points are needed for the calculation of each of the dq-axis inductances, i.e., the real peak value of i_d^* (or i_q^*), the zero-crossing time point ($t = t_0$) of i_d (or i_q), and the reference voltages u_d^* ($t = t_0$) (or u_q^* [$t = t_0$]).

The pseudocode for zero-crossing detection is shown below, where the variables, i.e., "Data," "Polarity," and "ZeroCrossing," are measured data of i_d (i_q), the polarity of measured data, and the sequence number of zero-crossing point, respectively.

Pseudocode of zero-crossing detection:

```
if Data(i) > 0;
  Polarity(i) = 1;
else
  Polarity(i) = −1;
end
if abs(Polarity(i + 1) − Polarity (i)) > 0
  ZeroCrossing(i) = 1;
else
  ZeroCrossing(i) = 0;
end
```

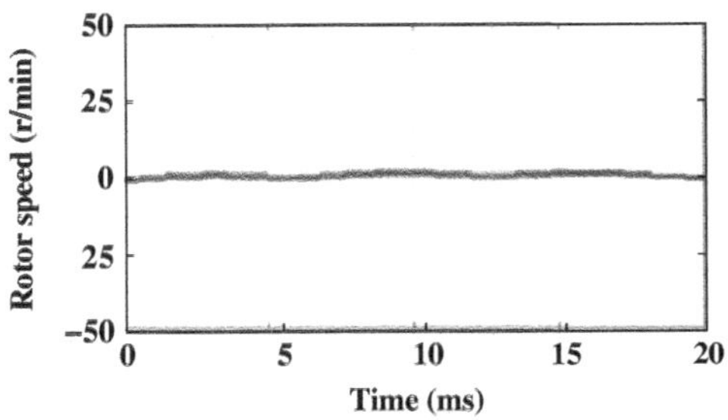

Figure 8.12 Measured rotor speed during HF current injections.

The performance of the simplified estimation method is then performed based on Prototype Machine 2. The measured rotor speed is presented in Figure 8.12, where the rotor speed slightly fluctuates around 0 r/min due to HF current injections. The measured HF zero mean dq-axis currents and voltages are shown in Figure 8.13, where $\omega_{h1} = 120$ Hz and $\omega_{h2} = 150$ Hz, and the dc-link voltage V_{dc} is set to 40 V.

Afterwards, the dq-axis incremental inductances can be obtained at the zero-crossing point, i.e., the 54th sampling instant in the d-axis and the 46th sampling instant in the q-axis. Therefore, the dq-axis incremental inductances can be obtained as follows:

$$L_d^{\text{inc}} = \frac{u_d^*(\omega_{h1}t = 2k\pi) + DdV_{\text{dead}}}{\omega_{h1} i_d^*} = \frac{-12.16 + DdV_{\text{dead}}}{-120 \times 2 \times \pi \times 0.4671} \approx 32.4 \text{ mH} \tag{8.11}$$

$$L_q^{\text{inc}} = \frac{u_q^*(\omega_{h2}t = 2k\pi) + DqV_{\text{dead}}}{\omega_{h2} i_q^*} = \frac{30.89 + DqV_{\text{dead}}}{-150 \times 2 \times \pi \times 0.4826} \approx 66.3 \text{ mH} \tag{8.12}$$

where the distorted voltage due to VSI non-linearity is 0.41 V at $V_{dc} = 40$ V, which can be estimated at standstill, as described in Section 8.3.1.

Overall, by using the measured electrical quantities at the zero-crossing points, the estimated inductances are close to the nominal values of Prototype Machine 2. Meanwhile, the influence of stator winding resistance has also been cancelled during the design process.

8.4 Multi-step Parameter Estimation

The multi-step estimation methods, including two-step [11] and three-step [9] methods, combine offline measurement and online parameter estimation, and the measured parameter is fixed during the operation to address the rank-deficient issue, which will be described in this section together with experimental results.

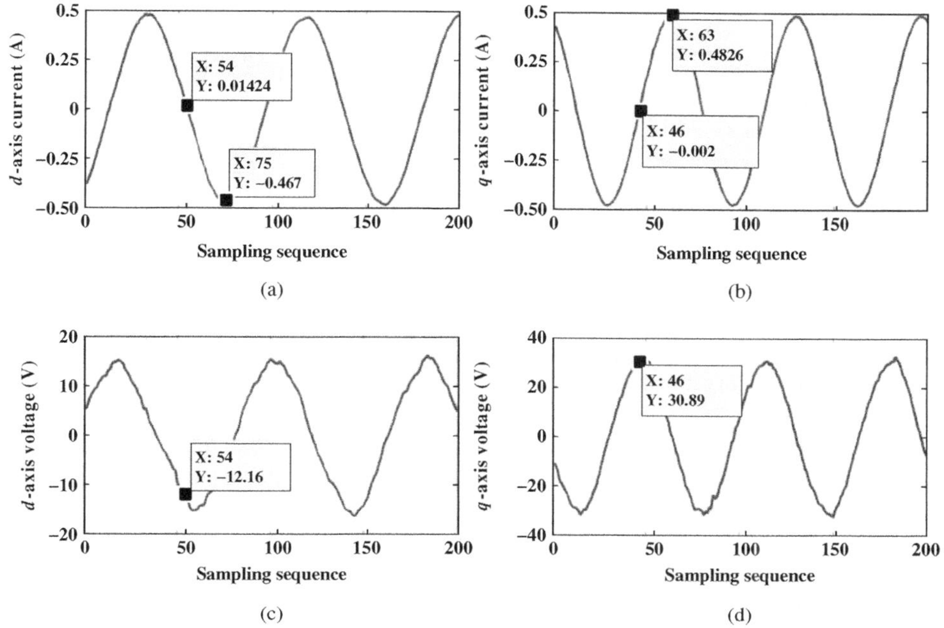

Figure 8.13 *dq*-axis HF currents and voltages: (a) measured *d*-axis current, (b) measured *q*-axis current, (c) *d*-axis reference voltage, and (d) *q*-axis reference voltage.

8.4.1 Two-step Estimation Method

The two-step estimation method is appropriate for $i_d = 0$ control, where only two parameters, i.e., the rotor PM flux linkage and the winding resistance, are in the q-axis equation. It is implemented as follows.

Step 1: The rotor PM flux linkage $\hat{\psi}_{pm}$ is estimated by setting the stator winding resistance to its nominal or measured value R_s^n.

$$\hat{\psi}_{pm} = \frac{u_q^* + DqV_{\text{dead}} - R_s^n i_q}{\omega_r} \tag{8.13}$$

Step 2: The winding resistance $\hat{R}_s$ can be estimated by setting the rotor PM flux linkage to its nominal or measured value ψ_{pm}^n.

$$\hat{R}_s = \frac{u_q^* + DqV_{\text{dead}} - \omega_r \psi_{pm}^n}{i_q} \tag{8.14}$$

Therefore, the estimation in each step needs to fix one parameter as the nominal or measured value to overcome the rank-deficient issue. Meanwhile, the VSI non-linearity needs to be compensated in both steps.

Based on Prototype Machine 1 under $i_d = 0$ control, the estimation results are shown in Figure 8.14, where the nominal values of R_s^n and ψ_{pm}^n are 0.3 Ω and 70.7 mWb, respectively.

It is obvious that when the VSI non-linearity is not compensated in the period of 22–60 s, the estimated rotor PM flux linkage (88.0 mWb) and winding resistance (0.84 Ω) are much larger than their respective nominal values. In contrast, the estimated parameters are 68.7 mWb and 0.28 Ω

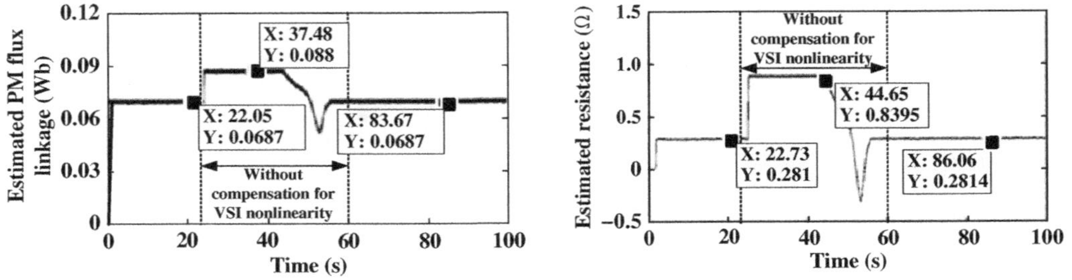

(a) (b)

Figure 8.14 Estimation of winding resistance and rotor PM flux linkage by two-step method: (a) estimated rotor PM flux linkage and (b) estimated winding resistance.

are close to their nominal values. It is worth noting that the estimated stator winding resistance is 12.5% smaller than the nominal value, whereas the actual value should be larger due to the on-state resistance of power devices [17] and the core loss [18]. It may be attributed to the mismatching between nominal and real rotor PM flux linkages.

It should be emphasized that the high accuracy of the two-step estimation can only be ensured by accurate knowledge of nominal values. Figure 8.15 illustrates the misestimation issue in the two-step estimation due to the inaccurate setting of nominal values. As can be seen, when the winding resistance is set to twice the nominal value, i.e., $R_s = 0.32 \times 2\ \Omega$ at the first step estimation, by which the rotor PM flux linkage is converged to 61.0 mWb. Likewise, when the rotor PM flux linkage is set to 85% of the nominal value for the second step, i.e., $\psi_{pm} = 0.85 \times 77.7$ mWb at the second-step estimation, by which the winding resistance is estimated to be 0.66 Ω.

Moreover, accounting for the parameter variation caused by the temperature rise, the two-step estimation method is employed to identify the rotor PM flux linkage and the stator winding resistance after 20 min heating of the PMSM, and the estimation results are shown in Figure 8.16. After heating, the actual winding resistance increases from 0.3 to 0.45 Ω measured by an inductance-capacitor-resistance (LCR) meter, and it is larger than the estimated resistance of 0.42 Ω. Meanwhile, the estimated rotor PM flux linkage is 78.6 mWb, which is even larger than the nominal value of 77.6 mWb and does not conform to the fact, i.e., the value of the rotor PM flux linkage decreases with the increasing temperature.

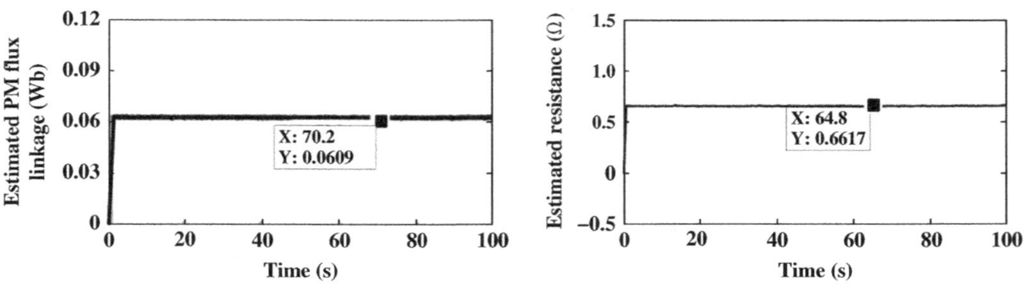

(a) (b)

Figure 8.15 Estimated rotor PM flux linkage and stator winding resistance under inaccurate setting of nominal values (VSI non-linearity is compensated): (a) estimated rotor PM flux linkage and (b) estimated stator winding resistance.

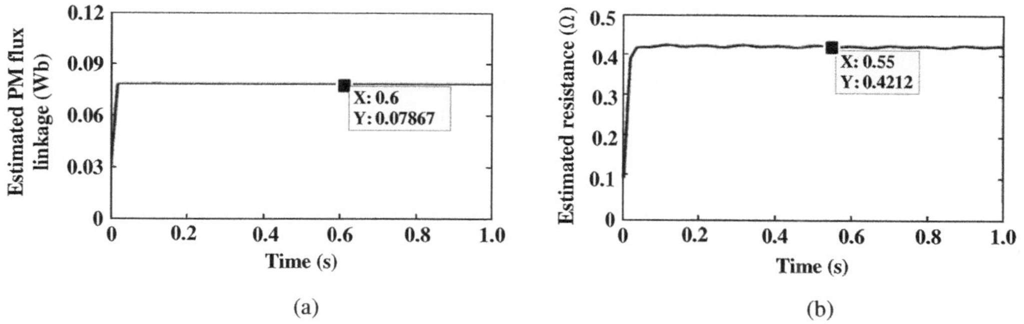

Figure 8.16 Estimated rotor PM flux linkage and winding resistance of PMSM after 20 min heating: (a) estimated rotor PM flux linkage and (b) estimated winding resistance.

Overall, it can be concluded that the two-step method easily suffers from the mismatch issue by using the nominal/measured parameter values. Meanwhile, the values of winding resistance and rotor PM flux linkage are sensitive to the temperature and magnetic saturation effect. Therefore, the estimation accuracy is hard to be guaranteed.

8.4.2 Three-step Estimation Method

The three-step estimation [9] utilizes the machine state at standstill and operation, of which the flowchart is presented in Figure 8.17 and is implemented as follows.

Step 1: When the d-axis currents is injected into PMSM at standstill and $i_q = 0$, the stator winding resistance and the distorted voltage due to VSI non-linearity can be estimated as follows:

$$\begin{cases} u_{d1}^* = R_s i_{d1} - DdV_{\text{dead}} \\ u_{q1}^* = R_s i_{q1} - DqV_{\text{dead}} \end{cases} \tag{8.15}$$

The first-step estimation process is the same as that described in Section 8.3.1.

Step 2: By using the estimated parameters in Step 1, the rotor PM flux linkage is estimated by Eq. (8.16), where the PMSM is operating under $i_d = 0$ control:

$$\psi_{pm} = \frac{1}{\omega_r}\left(u_{q2}^* + Dq\,\widehat{V}_{\text{dead}} - \widehat{R}_s i_{q2}\right) \tag{8.16}$$

where $\widehat{V}_{\text{dead}}$ and $\widehat{R}_s$ are the distorted voltage and the stator winding resistance in the first step, respectively.

Step 3: By using the estimated rotor PM flux linkage in Step 2 and the distorted voltage in Step 1, the q-axis inductances and the stator resistance can be further estimated as follows:

$$\begin{cases} L_q = \dfrac{u_{d3}^* + Dd\,\widehat{V}_{\text{dead}}}{\omega_r i_{q3}} \\ R_s = \dfrac{u_{q3}^* + Dq\,\widehat{V}_{\text{dead}} - \omega_r \widehat{\psi}_{pm}}{i_{q3}} \end{cases} \tag{8.17}$$

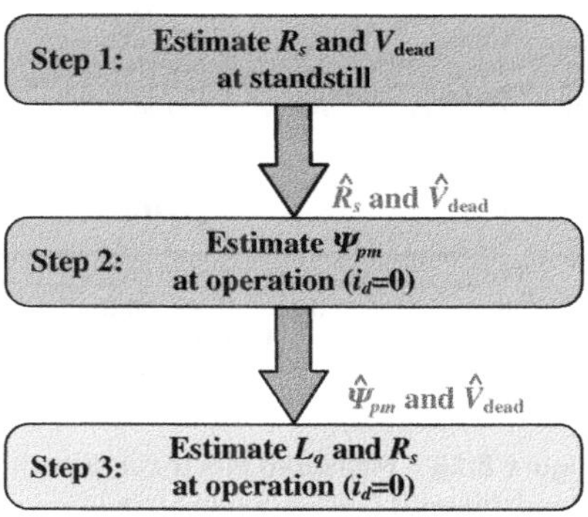

Figure 8.17 Flowchart of three-step estimation method.

Unlike the two-step method, which only involves the estimation of R_s and ψ_{pm} based on the q-axis equation under $i_d = 0$ control in Eqs. (8.13) and (8.14), the three-step method can estimate both L_q and R_s in Step 3 and has taken into account the estimation and compensation of VSI nonlinearity. It is worth noting that the estimated L_q is apparent inductance since it is obtained based on the steady-state Eq. (8.17). Meanwhile, the nominal values, which are required in the two-step method, are no longer needed in the three-step method.

The three-step estimation is performed based on Prototype Machine 1. The measured electrical quantities and estimation results in the first step are shown in Figures 8.18 and 8.19, respectively, and the VSI non-linearity has been compensated in a prior. As can be observed, the estimated winding resistance is converged to 0.396 Ω, which is slightly larger than the nominal value of 0.32 Ω as a result of parasitic resistances due to power devices and core loss.

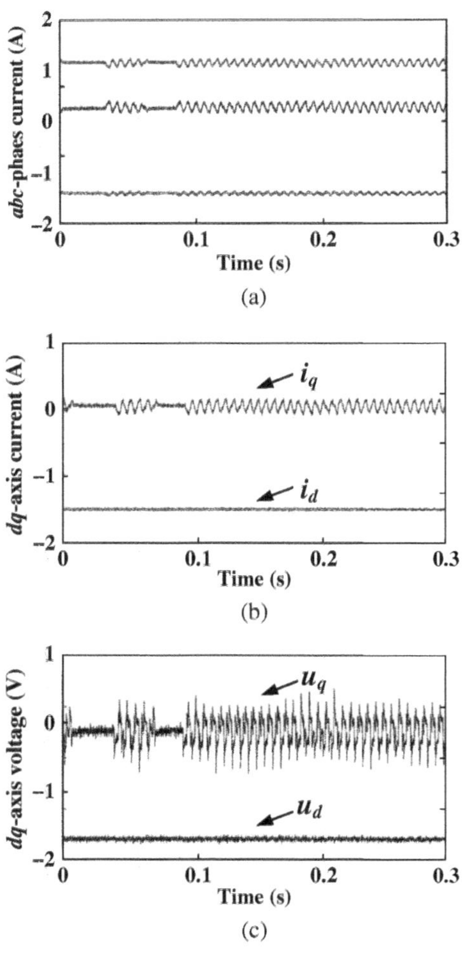

Figure 8.18 Measured electrical quantities at standstill with $i_d = -1.5$ A in Step 1: (a) measured abc three-phase currents, (b) measured dq-axis currents, and (c) measured dq-axis voltages.

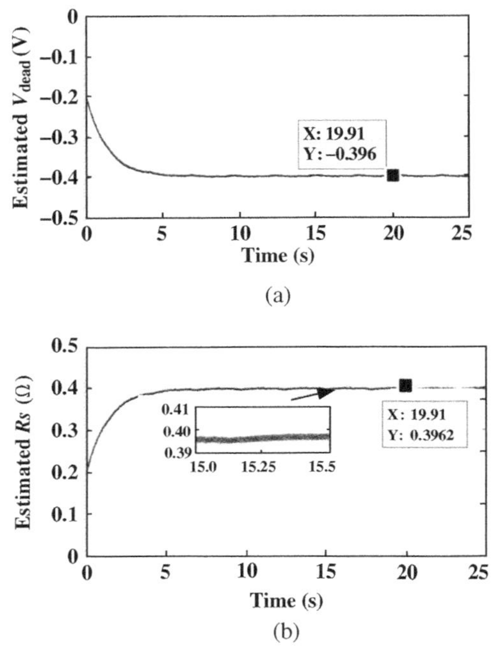

Figure 8.19 Estimation results at standstill in Step 1: (a) estimated distorted voltage and (b) estimated winding resistance.

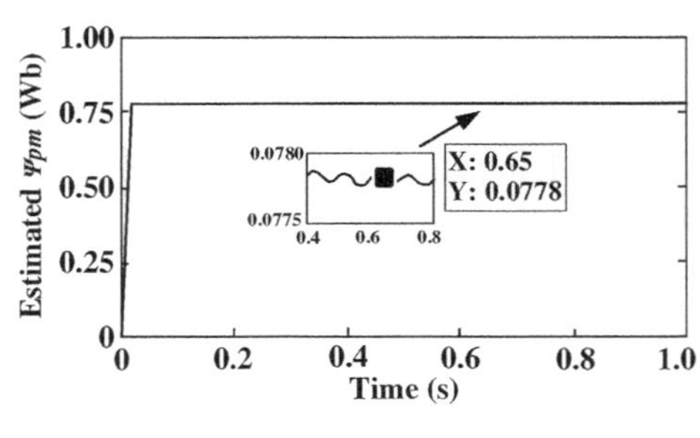

Figure 8.20 Estimated rotor PM flux linkage at operating in Step 2.

Afterwards, by using the estimation results in Figure 8.19, the rotor PM flux linkage is estimated when the PMSM is operating under $i_d = 0$ control, as presented in Figure 8.20. The estimated value is 77.8 mWb, which is almost the same as the nominal value of 77.6 mWb.

Third, the estimated distorted voltage (Step 1) and rotor PM flux linkage (Step 2) are employed to estimate the stator winding resistance. To evaluate the tracking capability of the estimated resistance, a resistor with a resistance of $R_p = 0.414\ \Omega$ is connected in series to each phase winding of the *abc* three-phase current at time $t = 28$ s, as shown in Figure 8.21. As can be seen, the estimated resistance change is 0.39 Ω, which is only slightly lower than the connected series resistance.

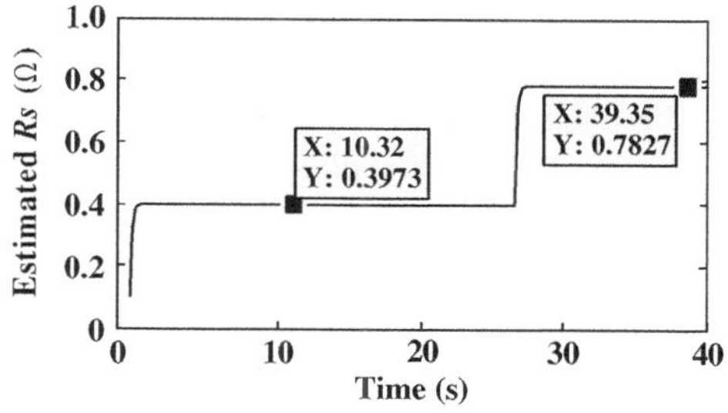

Figure 8.21 Estimated stator winding resistance in Step 3.

8.5 Summary

This chapter has described several classical offline and multi-step estimation methods for estimating stator winding resistance, VSI non-linearity, *dq*-axis inductances, and/or rotor PM flux linkage. It has shown that both the square voltage- and HF current-based offline methods can achieve good estimation performance when the VSI non-linearity is compensated.

By combining offline measured results and the PMSM state during operation, the two-step and three-step estimation methods have been described and implemented experimentally. It has been shown that the two-step estimation method easily suffers from the misestimation issue due to parameter mismatch caused by temperature rises and/or magnetic saturation. In comparison, the three-step method has the better tracking capability for estimating stator winding resistance, since it is updated during operation accounting for the thermal effect.

References

1 Z. Q. Zhu, D. Liang, and K. Liu, "Online parameter estimation for permanent magnet synchronous machines: An overview," *IEEE Access*, vol. 9, pp. 59059–59084, 2021.

2 B. J. Chalmers, S. A. Hamed, and G. D. Baines, "Parameters and performance of a high-field permanent-magnet synchronous motor for variable frequency operation," *IEE Proc. B, Electr. Power Appl.*, vol. 132, no. 3, pp. 117–124, 1985.

3 A. Consoli and A. Abela, "Transient performance of permanent magnet AC motor drives," *IEEE Trans. Ind. Appl.*, vol. IA-22, no. 1, pp. 32–41, Jan. 1986.

4 A. M. El-Serafi and J. Wu, "Determination of the parameters representing the cross-magnetizing effect in saturated synchronous machines," *IEEE Trans. Energy Convers.*, vol. 8, no. 3, pp. 333–342, Sept. 1993.

5 B. Stumberger, B. Kreca, and B. Hribernik, "Determination of parameters of synchronous motor with permanent magnets from measurement of load conditions," *IEEE Trans. Energy Convers.*, vol. 14, no. 4, pp. 1413–1416, Dec. 1999.

6 P. Viarouge, M. Ferfra, I. Kamwa, Y. Lefevre, and H. Le-Huy, "Modelling and identification of permanent magnet synchronous machines from standstill time response tests using a non-linear

method," *Proc. 1993 IEEE Ind. Appl. Conf. Twenty-Eighth IAS Annual Meet.*, Toronto, Canada, 1993, pp. 65–72, vol. 1.

7 S. Weisgerber, A. Proca, and A. Keyhani, "Estimation of permanent magnet motor parameters," *Proc. IEEE Ind. Appl. Soc. Annu. Meeting*, 1997, vol. 1, pp. 29–34.

8 B. Stumberger, G. Stumerger, D. Dolinar, A. Hamler, and M. Trlep, "Evaluation of saturation and cross-magnetization effects in interior permanent-magnet synchronous motor," *IEEE Tran. Ind. Appl.*, vol. 39, no. 5, pp. 1264–1271, Sep./Oct. 2003.

9 S. Morimoto, M. Sanada, and Y. Yakeda, "Mechanical sensorless drives of IPMSM with online parameter identification," *IEEE Trans. Ind. Appl.*, vol. 42, no. 5, pp. 1241–1248, Sept./Oct. 2006.

10 S. A. Odhano, R. Bojoi, Ş. G. Roşu, and A. Tenconi, "Identification of the magnetic model of permanent-magnet synchronous machines using DC-Biased low-frequency AC signal injection," *IEEE Trans. Ind. Appl.*, vol. 51, no. 4, pp. 3208–3215, July/Aug. 2015.

11 M. Rashed, P. MacConnell, A. Stronach, and P. Acarnley, "Sensorless indirect-rotor-field-orientation speed control of a permanent-magnet synchronous motor with stator-resistance estimation," *IEEE Trans. Ind. Electron.*, vol. 54, no. 3, pp. 1664–1675, Jun. 2007.

12 K. M. Rahman and S. Hiti, "Identification of machine parameters of a synchronous motor," *IEEE Trans. Ind. Appl.*, vol. 41, no. 2, pp. 557–565, March-April 2005.

13 E. Armando, R. I. Bojoi, P. Guglielmi, G. Pellegrino, and M. Pastorelli, "Experimental identification of the magnetic model of synchronous machines," *IEEE Trans. Ind. Appl.*, vol. 49, no. 5, pp. 2116–2125, Sept./Oct. 2013.

14 K. Liu and Z. Q. Zhu, "Quantum genetic algorithm-based parameter estimation of PMSM under variable speed control accounting for system identifiability and VSI nonlinearity," *IEEE Trans. Ind. Electron.*, vol. 62, no. 4, pp. 2363–2371, April 2015.

15 K. Liu, J. Feng, S. Guo, L. Xiao, and Z. Q. Zhu, "Identification of flux linkage map of permanent magnet synchronous machines under uncertain circuit resistance and inverter nonlinearity," *IEEE Trans. Ind. Inform.*, vol. 14, no. 2, pp. 556–568, Feb. 2018.

16 *MATLAB*, MathWorks, Accessed: 2022 [Online]. Available: https://uk.mathworks.com/.

17 J. Choi and S. Sul, "Inverter output voltage synthesis using novel dead-time compensation," *IEEE Trans. Power Electron.*, vol. 11, no. 2, pp. 221–227, Mar. 1996.

18 I. Lar and M. Radulescu, "Equivalent core-loss resistance identification for interior permanent-magnet synchronous machines," *Proc. IEEE Int. Conf. Electr. Mach.(ICEM)*, 2012, vol. 2, pp. 1665–1669.

9

Estimation of Mechanical Parameters

9.1 Introduction

In the previous chapters, various online estimation methods of permanent magnet synchronous machine (PMSM) electrical parameters, i.e., resistance, inductance, and rotor PM flux linkage, have been introduced. The mechanical constants, i.e., moment of inertia and viscous friction coefficient, are equally important in the PMSM drive system, particularly for controller design [1–9]. However, since the mechanical parameters are usually related to the load characteristics, i.e., dimensions and the shape of mechanical loads, their information is not always available. Consequently, the mismatched mechanical parameters may lead to inferior dynamic performance of the drive system. Most servo drives have an upper limit to the ratio of load inertia versus rotor inertia. For example, the upper limit of the ratio of load inertia versus rotor inertia of a Panasonic MINAS A4 Series servo drive is 100 [10]. Besides, for the high-frequency injection-based sensorless control of low-speed PMSMs, the correlation between the load inertia and the frequency/amplitude of injected current also needs to be considered during the controller design [11, 12]. Against this background, it is preferred that the machine drives can estimate these parameters either online or offline, which has been investigated for both induction machines and PMSMs over several decades [7–8, 12–24].

For offline estimating mechanical parameters, the constant acceleration or deceleration characteristic of the rotor speed is widely employed [12–14]. In this case, the moment of inertia and/or viscous coefficient can be calculated under no-load conditions according to the slope of the rotor speed [12] and the kinetic energy along the start-up duration [13, 14].

For online estimating mechanical parameters, the electromagnetic torque and rotor speed are usually required [15–20]. The rotor speed can be measured directly by the position sensor, while the electromagnetic torque can be obtained by different methods, e.g., recursive least-squares (RLS) estimator [16], disturbance, and sliding-mode observers [15–20]. It has been widely reported that the above methods exhibit better accuracy and robustness than the constant acceleration/deceleration method, while the design of the observers is usually complex, and the computation time is also considerable. Meanwhile, it is worth emphasizing that the torque constant usually varies with the load condition due to magnetic saturation and temperature variation, which should be considered [21].

In addition, it has also been reported in Refs [21–24] that the moment of inertia can also be estimated based on periodic perturbations of rotor speed, torque, or current, etc., by which the transient torque due to the moment of inertia will not be zero but a forward–reverse variable

Parameter Estimation of Permanent Magnet Synchronous Machines, First Edition. Zi Qiang Zhu, Kan Liu, and Dawei Liang.

with respect to the time. This type of solution can be easily implemented by using different identification algorithms in either online or offline manners, e.g., model reference adaptive system (MRAS) [25], extended Kalman filter, and RLS.

In this chapter, Section 9.2 describes and evaluates three widely used online or offline mechanical parameter estimation techniques, i.e., acceleration/deceleration-based, observer-based, and fundamental motion equation–based methods. The latter two methods are implemented under zero-mean sinusoidal rotor variations excited by q-axis current injection. All the above three methods are implemented based on a PMSM drive system and evaluated experimentally, as presented in Section 9.3. Furthermore, the design of the proportional–integral (PI) speed regulator based on the estimated mechanical parameters is described in Section 9.4. The performance of the speed-loop PI regulators constructed by different mechanical parameter estimation methods is also evaluated experimentally.

9.2 Mechanical Parameter Estimation Methods

The electrical and mechanical equations of PMSM in the dq-axis reference frame are expressed in Eqs. (9.1) and (9.2), respectively.

Electrical equations:

$$u_d = R_s i_d + L_d \frac{di_d}{dt} - \omega_r L_q i_q \tag{9.1a}$$

$$u_q = R_s i_q + L_q \frac{di_q}{dt} + \omega_r \left(L_d i_d + \psi_{pm}\right) \tag{9.1b}$$

where u_d, u_q, i_d, and i_q are the dq-axis voltages and currents, respectively. L_d, L_q, and ψ_{pm} are the dq-axis inductances and the rotor PM flux linkage. ω_r is the electrical angular speed.

Mechanical equations:

$$T_e = 1.5p\left[\psi_{pm} i_q + (L_d - L_q) i_d i_q\right] \tag{9.2a}$$

$$J\frac{d\omega_r}{dt} = T_e - F\omega_r - T_L \tag{9.2b}$$

where p is the pole pair number. J is the moment of inertia. F is the viscous friction coefficient. T_e and T_L are the electromagnetic and shaft torques, respectively. In addition, the torque constant $K_e = \psi_{pm} + (L_d - L_q)i_d$. As can be found from Eq. (9.2), the inertia term $Jd\omega_r/dt$ requires a variation in the rotor speed to be activated. Therefore, to estimate the combined moment of inertia J, the PMSM needs to be accelerated/decelerated or with continuously variable rotor speed by injecting perturbation signals, as introduced in the following.

9.2.1 Acceleration/Deceleration-Based Method

When the influence of $F\omega_r$ is ignored, Eq. (9.2b) can be simplified as follows:

$$J\frac{d\omega_r}{dt} = 1.5K_e i_q - T_m \tag{9.3}$$

First, by utilizing both the constant acceleration and deceleration processes with the same period, the combined moment of inertia can be estimated by integrating both sides in Eq. (9.3).

For a constant acceleration process, from ω_0 to ω_e, the integral equation in the time-discrete system can be calculated as follows:

$$J(\omega_e - \omega_0) + \sum_{n=1}^{N} T_m[n]\,T_s = \sum_{n=1}^{N} 1.5K_e i_q[n]\,T_s \tag{9.4}$$

where T_s is the sampling time. Eq. (9.3) is integrated from n to N.

For another deceleration process, from ω_e to ω_0, the corresponding integral equation is expressed as follows:

$$J(\omega_0 - \omega_e) + \sum_{n=1}^{N} T_m[n]\,T_s = \sum_{n=1}^{N} 1.5K_e i'_q[n]\,T_s \tag{9.5}$$

By subtracting Eq. (9.4) from Eq. (9.5), the integral term of the shaft torque can be eliminated, and thus, J can be estimated as follows:

$$J = \frac{\sum_{n}^{N} 1.5K_e i_q[n]T_s - \sum_{n}^{N} 1.5K_e i'_q[n]T_s}{2(\omega_e - \omega_0)} \tag{9.6}$$

To simplify the estimation procedure, it has been reported in Ref. [26] that the combined moment of inertia can be simply estimated by a single acceleration or deceleration process, where the open-loop transfer function of the mechanical model (Eq. 9.2) of the PMSM drive is regarded as a first-order system with the gain of $1.5pK_e$. Thus, when the influence of $F\omega_r$ is neglected, the transfer function can be simplified as an integral element and the gain of the input signal is $1.5pK_e/J$, as shown in Figure 9.1.

Therefore, for a time-discrete system, the output of an integral element during the unit step response at the nth sampling time point can be expressed as $t = T = nT_s$, where T_s is the sampling time. Thus, when the influence of $F\omega_r$ can be ignored, J can be approximately calculated as follows:

$$J = \frac{1.5pK_e \sum_{1}^{n=N} i_q(t = nT_s)\,nT_s}{\sum_{1}^{n=N} w_r(t = nT_s)} \tag{9.7}$$

9.2.2 MRAS Observer-Based Method

The MRAS observer-based method is widely used to estimate mechanical parameters due to good global convergence and stability. First, under time-varying rotor speed conditions, the variable model can be derived from Eq. (9.2b) and expressed as follows:

$$\frac{d\tilde{\omega}_r}{dt} = \frac{1}{\tilde{J}}(\underbrace{1.5p\hat{\psi}_{pm} i_q}_{T_e} - \tilde{F}\tilde{\omega}_r) \tag{9.8}$$

where $\tilde{\omega}_r$, $\tilde{J}$, and $\tilde{F}$ are the estimated angular velocity of rotor, the combined moment of inertia, and the viscous friction coefficient of rotor and load, respectively.

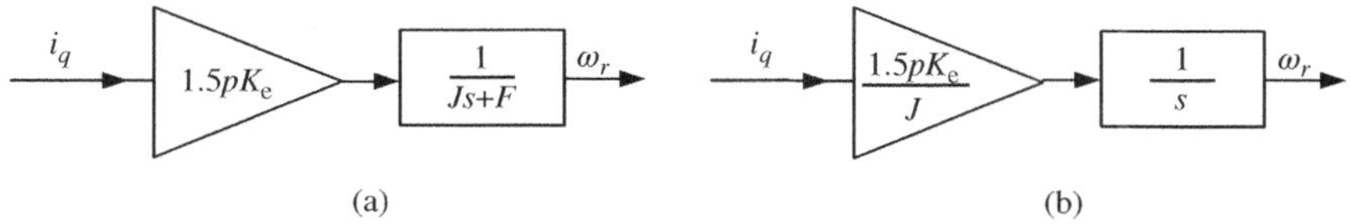

Figure 9.1 Transfer function of mechanical model of PMSM drive system [5], IEEE: (a) considering friction coefficient F and (b) without considering friction coefficient F.

The rotor angular velocity error ε can be derived by Eqs. (9.2b) and (9.8) as follows:

$$\frac{d\varepsilon}{dt} = \frac{d(\omega_r - \widetilde{\omega}_r)}{dt} - \frac{F}{J}\varepsilon + \left(\frac{1}{J} - \frac{1}{\widetilde{J}}\right)T_e + \left(\frac{\widetilde{F}}{\widetilde{J}} - \frac{F}{J}\right)\widetilde{\omega}_r \tag{9.9}$$

Therefore, the MRAS estimators for J and F can be derived and expressed as follows [27]:

$$\widetilde{J} = \frac{1}{1/J_0 + P_J(eT_e) + I_J\int eT_e dt} \tag{9.10a}$$

$$\widetilde{F} = \widetilde{J}\left(\frac{F_0}{J_0} - P_F(e\widetilde{\omega}_r) - I_F\int e\widetilde{\omega}_r dt\right) \tag{9.10b}$$

where $e = \omega_r - \widetilde{\omega}_r$. P_J, I_J, P_F, and I_F are the PI constants of the MRAS observers of J and F. J_0 and F_0 are the initial values.

9.2.3 Fundamental Motion Equation-Based Estimation Method

When the rotor speed varies sinusoidally excited by a sinusoidal perturbation q-axis current, J and F can be directly estimated according to the fundamental motion equation (9.2b). The complete estimation flowchart is presented in Figure 9.2.

Estimation of combined moment of inertia J
First, the dq-axis currents are set as follows:

$$\begin{cases} i_d = 0 \\ i_q = i_{qp}^* \sin(\omega_h t) \end{cases} \tag{9.11}$$

Afterwards, the resulting mechanical angular speed ω_r can be expressed as follows:

$$\omega_r = \Omega_p \sin(\omega_h t + \varphi) \tag{9.12}$$

where Ω_p is the amplitude and φ is the phase shift between i_q and ω_r.

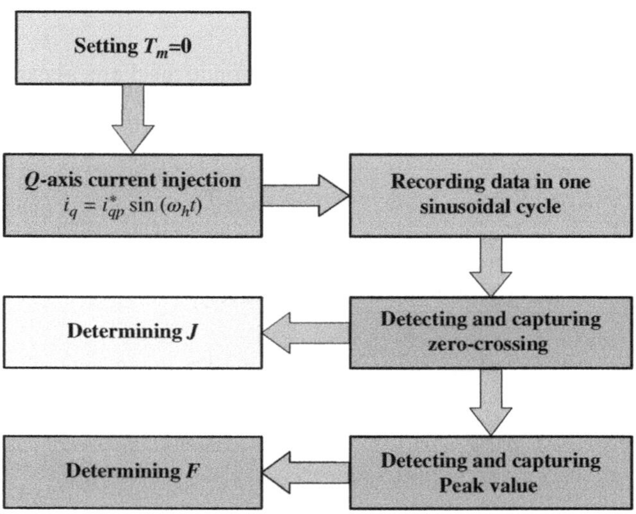

Figure 9.2 Flowchart of fundamental motion equation-based estimation method.

Therefore, when the disturbance torque is zero, $T_L = 0$, Eq. (9.2b) can be rewritten as follows:

$$J\frac{d\left[\Omega_p \sin(\omega_h t + \varphi)\right]}{dt} = J\Omega_p \omega_h \cos(\omega_h t + \varphi) = 1.5p\psi_{pm} i_{qp}^{*} \sin(\omega_h t + \varphi) - F\omega_p \sin(\omega_h t + \varphi) \tag{9.13}$$

At the transient of zero-crossing of rotor speed, i.e., $\omega_r(t = t_0) = \Omega_p \sin(\omega_h t_0 + \varphi) = 0$ and $\cos(\omega_h t_0 + \varphi) = 1$, the torque component caused by $F\omega_r$ will also be zero at this time point.

Finally, Eq. (9.13) can be simplified as follows:

$$J\Omega_p \omega_h = 1.5p\psi_{pm} i_{qp}^{*} \sin(\omega_h t + \varphi) = 1.5p\psi_{pm} i_q(t = t_0) \tag{9.14a}$$

$$J = \frac{1.5p\psi_{pm} i_q(t = t_0)}{\Omega_p \omega_h} \tag{9.14b}$$

As can be seen, this method only requires $i_q(t = t_0)$ and Ω_p to calculate the moment of inertia J. The rotor speed does not need to be differentiated. Therefore, compared to the MRAS observer-based method, the computational burden can be significantly eased.

Estimation of friction coefficient F

The viscous friction coefficient F can be estimated at the peak value of the rotor speed $t = t_p$, where $\omega_r\left(t = t_p\right) = \Omega_p \sin(\omega_h t_0 + \varphi) = \Omega_p$ and $\cos(\omega_h t_0 + \varphi) = 0$. Thus, Eq. (9.13) can be simplified as follows:

$$F\Omega_p = 1.5p\psi_{pm} i_{qp}^{*} \sin(\omega_h t + \varphi) = 1.5p\psi_{pm} i_q(t = t_p) \tag{9.15a}$$

$$F = \frac{1.5p\psi_{pm} i_q(t = t_p)}{\Omega_p} \tag{9.15b}$$

However, the result of Eq. (9.15a) is the transient value of F, which is related to the specific transient rotor speed. To obtain the exact correlation between F and rotor speed, the mean values of F can be further calculated as follows:

$$F \approx \frac{\text{mean}(T_e)}{\text{mean}(\omega_r)} = \frac{\text{mean}(1.5pK_e i_q)}{\text{mean}(\omega_r)} \tag{9.16}$$

where the disturbance torque $T_m = 0$, and thus, $T_e \approx F\omega_r$.

Moreover, it is worth mentioning that the viscous friction coefficient F can also be estimated at steady state of $\omega_r \neq 0$ [21]. In this case, Eq. (9.2b) can be rewritten as follows:

$$T_e = 1.5p\psi_{pm} i_q = F\omega_r \tag{9.17}$$

As can be seen, since $J(d\omega_r/dt) \approx 0$, the influence of J on the rotor speed is minor on the estimation accuracy of F [4].

9.2.4 Critical Issue of Torque Prediction

Amongst the above three methods, the appropriate prediction of electromagnetic torque plays an important role in the accuracy of estimated mechanical parameters [21]. Usually, the electromagnetic torque is calculated based on the measured q-axis current and the value of rotor PM flux linkage.

For the described MRAS observer–based and motion equation–based methods, the rotor speed is set as zero-mean sinusoidal waveform by injecting sinusoidal q-axis current. Therefore, the

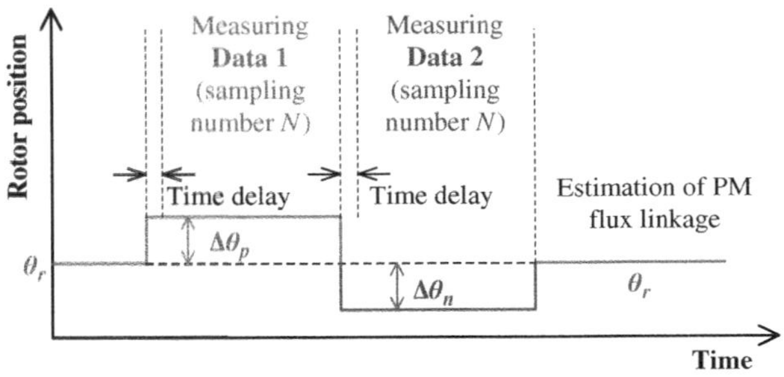

Figure 9.3 Data measurement process of POPE.

amplitude of the injected current should be as large as possible to reach a higher signal-to-noise ratio. However, a large variation in the q-axis current could give rise to a variation of rotor PM flux linkage due to magnetic saturation.

Therefore, the value of the rotor PM flux linkage is preferred to be known in a prior, which can be easily achieved by using the position-offset-based parameter estimation (POPE) method [21]. In the POPE, the positive and negative rotor position offsets are injected subsequently, as shown in Figure 9.3, and the rotor PM flux linkage can be obtained by using the deviation of the d-axis voltage equations with position and negative position offsets in Eq. (9.18), which has been detailed in Chapter 5.

$$\psi_{pm} = \frac{\text{mean}[\, u_{d1}^*(k) - u_{d2}^*(k)\,]}{\text{mean}[\omega(k)\,]2\cos(\Delta\theta)\sin(\Delta\theta)} \tag{9.18}$$

where "mean" is the mean operator. u_{d1}^* and u_{d2}^* are the reference d-axis voltages corresponding to the position and negative position offsets, respectively. $\Delta\theta$ is the amplitude of position offset, i.e., $\Delta\theta = |\Delta\theta_p| = |\Delta\theta_n|$. In addition, it has been proved in Chapter 5 that the estimated PM flux linkage is not affected by the VSI non-linearity.

9.3 Experimental Results

The above-mentioned three estimation methods, i.e., the acceleration/deceleration-based method in Eq. (9.7), the MRAS observer–based method in Eq. (9.10), and the motion equation–based method in Eqs. (9.14–9.16), are evaluated experimentally, which are denoted by Schemes I to III. The prototype Machines 1 and 2 are employed, of which the key parameters have been listed in Table 9.1, and their complete specifications are presented in Appendix B. The machines are driven by the field-oriented control (FOC) system based on a DS1006 control box, and a direct current (dc) load machine is used as a load, of which the nominal rotor moment of inertia is 6.70×10^{-3} kg m^2. In addition, to accurately calculate the electromagnetic torque, the POPE method is employed to obtain the value of the rotor PM flux linkage.

Three moments of inertia will be estimated as follows:

C1: Only the moment of inertia of the PMSM rotor.
C2: The combined moment of inertia including the PMSM rotor and the coupling.
C3: The combined moment of inertia including the PMSM rotor and the dc load machine.

Table 9.1 Main design parameters of Prototype Machines 1 and 2.

Parameters	Prototype Machine 1	Prototype Machine 2
Rated current/rated rotor speed	10 A/400 r/min	4 A/1000 r/min
dc-link voltage	36 V	158 V
Phase resistance (20°C)	0.3 Ω	6.0 Ω
Rotor PM flux linkage	70.7 mWb	236.0 mWb
Moment of inertia	3.41×10^{-4} kg m^2	3.61×10^{-4} kg m^2
Moment of inertia of rotor coupling	2.42×10^{-4} kg m^2	2.42×10^{-4} kg m^2

9.3.1 Comparison Between Schemes I and III

First, the estimation results of Scheme I are tabulated in Table 9.2. It can be easily found that the accuracy of Scheme I is significantly inferior under the tests *C1* and *C2* when the moment of inertia is low. It can be explained that the influence of friction in these two tests is critical, but it is neglected in Scheme I.

In comparison, the influence of friction is considered in Scheme III, which only requires three variables, i.e., $i_q(t_0)$, $i_q(t_p)$, and Ω_p, to calculate J and F. By using the measured signals shown in Figure 9.4, the mechanical parameters J and F can be calculated by Eqs (9.14–9.16), as listed in Table 9.3. In theory, the determination of J in Scheme III is independent of friction term $F\omega_r$ if the detection of zero-crossing of rotor speed is ideal. However, in reality, the speed measurement is not ideal and the real rotor speed at the detected zero-crossing point is usually not zero. Consequently, the influence of $F\omega_r$ will be not zero, but the influence of static friction can be regarded as zero since $\omega_r \neq 0$.

Furthermore, it can be seen from Table 9.3 that the frequency of perturbation signal f plays a dominant role in the estimation of J, i.e., the estimation error increases significantly with f from 3 to 10 Hz, in all three tests. Meanwhile, when f is set to 3 or 5 Hz, the estimation errors in all three tests are much lower than those estimated by Scheme I. By way of example, the minimum estimation errors of the test *C3* in Schemes I and III are 31.5% and 0.5%, respectively. It can be explained that the amplitude of perturbation current is relatively high (1.5 and 2 A) at *C3*, and the term $F\omega_r$ at the detected zero-crossing point is small compared with the transient T_e. Therefore, to ensure a higher ratio of signal versus noise, f should be set smaller than 5 Hz.

Table 9.2 Estimation results of Scheme I.

Tests	Moment of inertia (J) [$\times 10^{-4}$]		Error (%)
	Nominal	Estimated	
C1	3.61	16.0	347.9
C2	6.03	23.0	274.4
C3	80.6	106	31.5

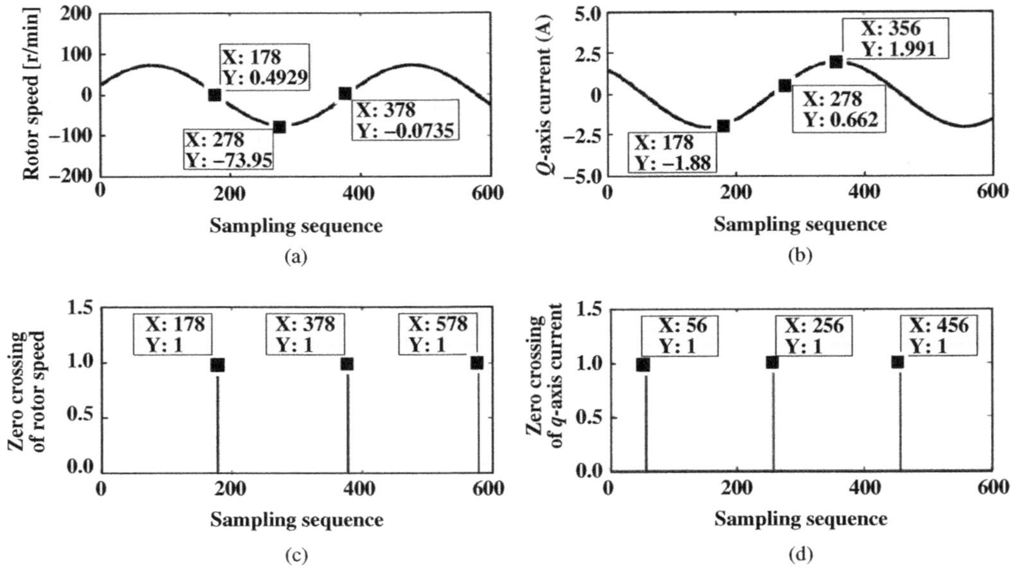

Figure 9.4 Measured rotor speed and q-axis current and detected zero-crossing points [5], IEEE: (a) rotor speed, (b) injected q-axis current, (c) detected zero-crossing of rotor speed, and (d) detected zero-crossing of q-axis current.

Table 9.3 Estimation results of Scheme III.

Tests	i^*_{qp}	f	F [×10⁻⁴]	Moment of inertia (J) [×10⁻⁴]		Error
				Estimated	Nominal	
C1	0.5 A	3 Hz	7.6	4.13	3.61	14.4%
		5 Hz	21	4.37		21.1%
		10 Hz	180	4.61		27.7%
C2	0.3 A	3 Hz	24	7.57	6.03	25.5%
		5 Hz	4.8	7.59		25.9%
	0.5 A	3 Hz	18	7.42		23.1%
		5 Hz	41	7.65		26.9%
		10 Hz	360	8.06		33.7%
C3	1.5 A	3 Hz	170	81.0	80.6	0.5%
		5 Hz	230	80.0		0.7%
		10 Hz	810	109		35.2%
	2.0 A	3 Hz	910	79.0		2.0%
		5 Hz	4200	82.0		1.7%
		10 Hz	4500	124		53.8%

9.3.2 Comparison Between Schemes II and III

Both Schemes II and III can consider the influence of F in their respective estimation model. Therefore, a good MRAS-based estimation of J can be achieved if the transient mechanical angular speed ω_r can be tracked. However, as can be seen from Eq. (9.10), the setup of PI constants of observers is important to ensure the tracking ability of algorithm since the input and output signals of observers are in the form of periodic forward–reverse perturbation. As can be seen from Figure 9.5, under the same PI constant, the convergence speed and the results of the MRAS-based J observer are different under various perturbation signals.

Furthermore, the influences of different PI constants of MRAS observer on the estimation accuracy are illustrated in Figure 9.6 based on the *C2* test. It can be easily found that the setup of PI constants of MRAS observer should match the perturbation signal properly, and thus, the tuning of PI constants becomes quite complex, which usually depends on the expert experience.

In comparison, Scheme III does not have these issues, where J is obtained within one sinusoidal perturbation cycle less than 0.5 s, while the traditional MRAS needs more than 1 s, as shown in Figure 9.5. Therefore, it can be concluded that Scheme III is faster than the traditional MRAS in estimating J. In addition, it can be observed from Table 9.3 that the estimation errors of Schemes II and III are kept within 20% and 3% at tests *C1* and *C3* (if $f \leq 5$ Hz), respectively. It can be explained that both methods have considered the influence of friction coefficient F, while the measured data are of the same ratio of signal versus noise in both methods.

Therefore, the estimation accuracy of Scheme III is competitive to those conventional observer-based methods, e.g., Scheme II, while it consumes much less computational time and no need to tune the convergence speed to match the perturbation signal.

9.3.3 Influence of Variation in Rotor PM Flux Linkage

As aforementioned, the POPE method is employed to estimate the rotor PM flux linkage, where the positive and negative position offsets θ_p and θ_n are injected [21]. The measured dq-axis currents, voltages, rotor speeds, and the estimated rotor PM flux linkages of Prototype Machines 1 and 2 are shown in Figures 9.7 and 9.8, respectively, where θ_u is the unit electrical angle.

It is obvious that the injected rotor position offsets lead to significant changes in the d-axis voltage, while the change in q-axis voltage is quite negligible. In addition, the sampling time for the two sets of data corresponding to $\Delta\theta_p$ and $\Delta\theta_n$ is set to 10 s to ensure that there is enough time for the settling down of voltages and currents. In practice, the sampling time can be less than 1 s due to the fast response of current loop regulators (usually less than 10 ms). Furthermore, the estimated rotor PM flux linkages are compared to the finite element analysis (FEA) results, as shown in Figure 9.9.

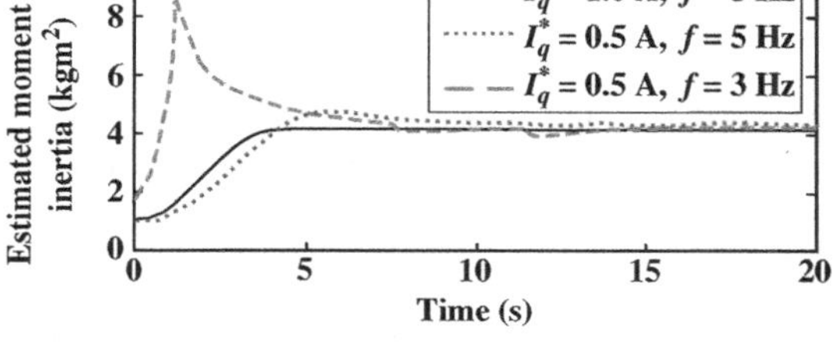

Figure 9.5 Estimation moment of inertia of PMSM rotor by MRAS with different perturbation signals under same PI constants (PI-1: $P_J = 0.01$, $I_J = 1000$, $P_F = 1$, and $I_F = 1$) [5], IEEE.

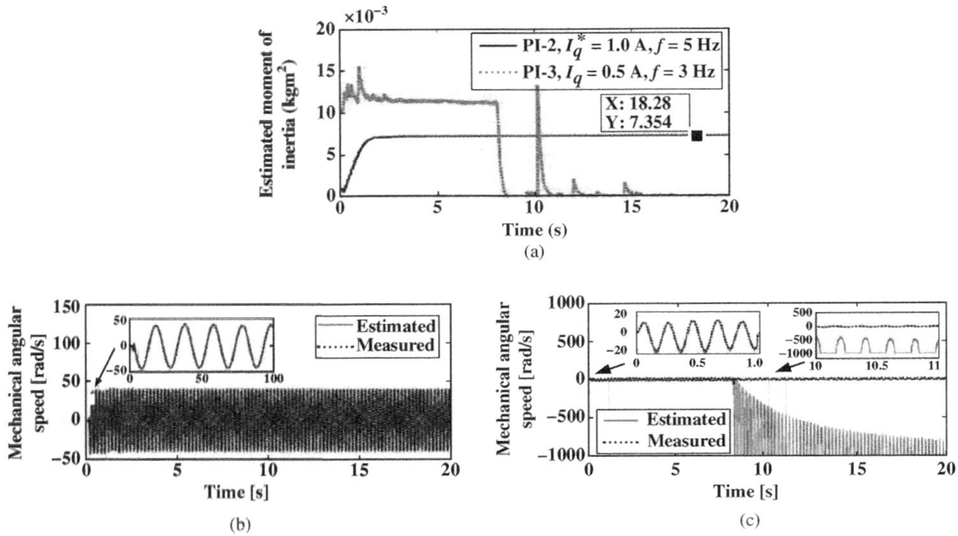

Figure 9.6 MRAS-based estimation of J with different PI constant based on $C2$ test [5], IEEE: (a) estimation of J with two sets of PI constants, (b) estimated and measured ω_r (PI-2: $P_J = 1000$, $I_J = 20000$, $P_F = 10$, and $I_F = 100$), and (c) estimated and measured ω_r (PI-3: $P_J = 1000$, $I_J = 1000$, $P_F = 100$, and $I_F = 1000$).

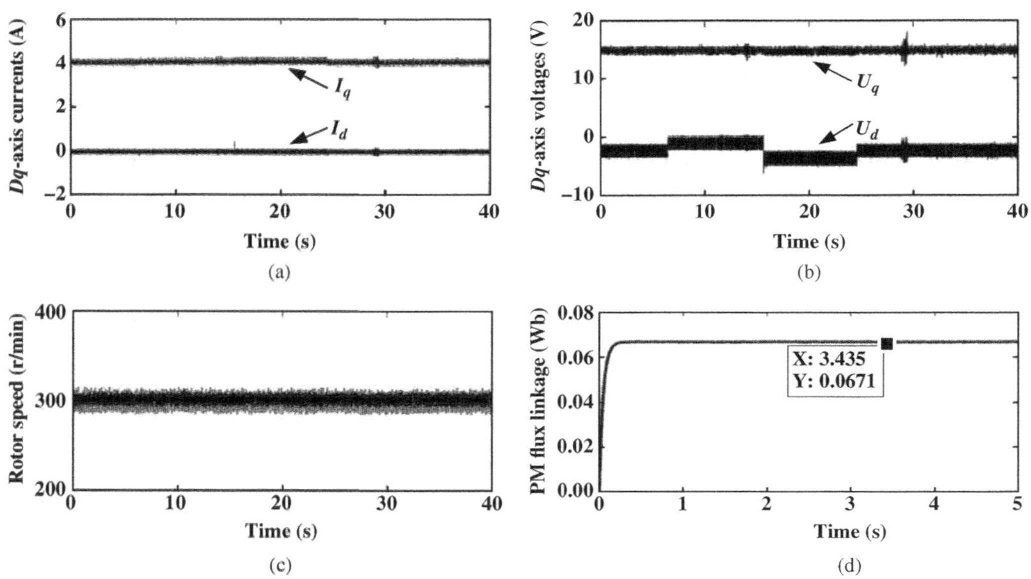

Figure 9.7 Estimation of PM flux linkage based on measured quantities of Prototype Machine 1 at 300 r/min ($|\theta_p| = |\theta_n| = 10\theta_u$) [6], IEEE: (a) measured dq-axis currents, (b) measured dq-axis voltages, (c) Measured rotor speed, and (d) estimated PM flux linkage.

As can be observed, with the increase in the amplitudes of q-axis currents, the rotor PM flux linkages decrease due to magnetic saturation, whether they are estimated by POPE or FEA. To demonstrate their influence on the estimated moment of inertia, the variation in the estimated

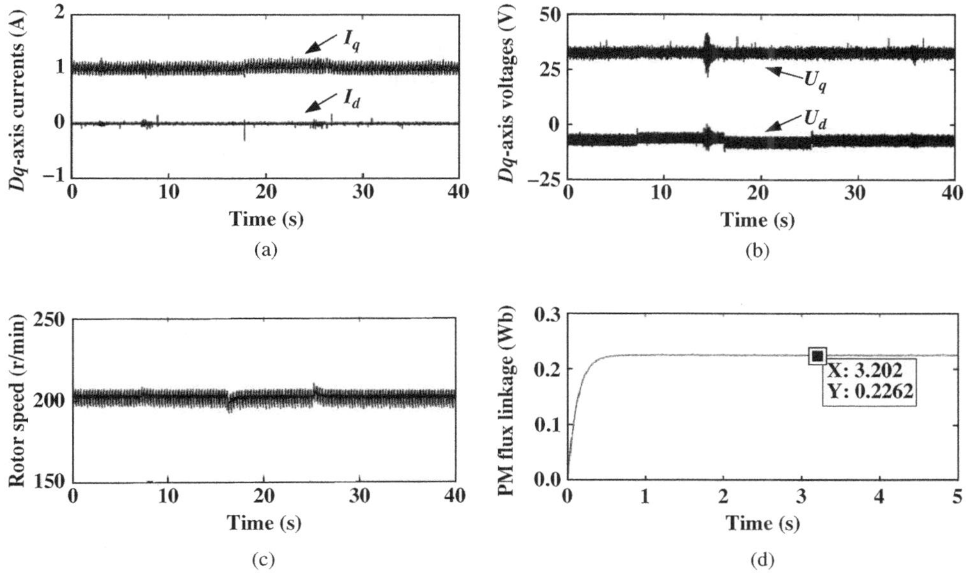

Figure 9.8 Estimation of PM flux linkage based on measured quantities of Prototype Machine 2 at 200 r/min ($|\theta_p| = |\theta_n| = 5\theta_u$) [6], IEEE: (a) measured dq-axis currents, (b) measured dq-axis voltages, (c) measured rotor speed, and (d) estimated PM flux linkage.

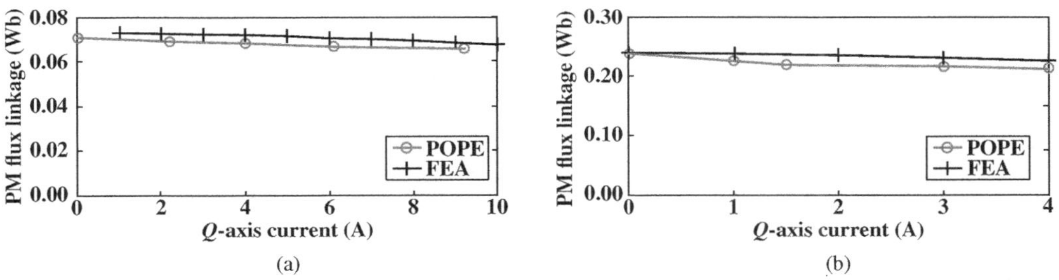

Figure 9.9 Comparison of estimated rotor PM flux linkages and FEA results [6], IEEE: (a) Prototype Machine 1 and (b) Prototype Machine 2.

moment of inertia of the dc load machine and PMSM rotor with different values of rotor PM flux linkage is shown in Figure 9.10, where $\hat{\psi}_{pm}$ is the nominal value of the rotor PM flux linkage.

It can be easily found that the variation in the estimated moment of inertia is almost proportional to the rotor PM flux linkage due to their linear relationship in Eq. (9.14b). Therefore, the accurate knowledge of the rotor PM flux linkage is essential for the estimation of moment of inertia [21].

9.3.4 Sensitivity Analysis of Sinusoidal Perturbation Signal

In this sub-section, the sensitivities of frequency and amplitude of the perturbation signal on the estimated mechanical parameters are analysed based on Prototype Machine 1, as presented in Figure 9.11. The frequency of the injected q-axis current is set to 2.5, 7.5, and 10 Hz, respectively, while the amplitude of the mechanical angular speed decreases from 8 to 1.5 rad/s correspondingly. This is due to the limitation that the mechanical response is usually much slower than the current response.

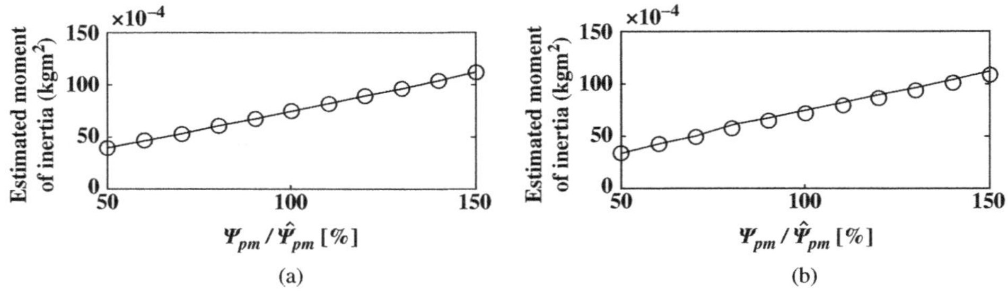

Figure 9.10 Variation of estimated moment of inertia of dc load machine and PMSM rotor with different rotor PM flux linkage [6], IEEE: (a) Prototype Machine 1 with $i_q^* = \sin(10\pi t)$ and (b) Prototype Machine 2 with $i_q^* = 0.8\ \sin(10\pi t)$.

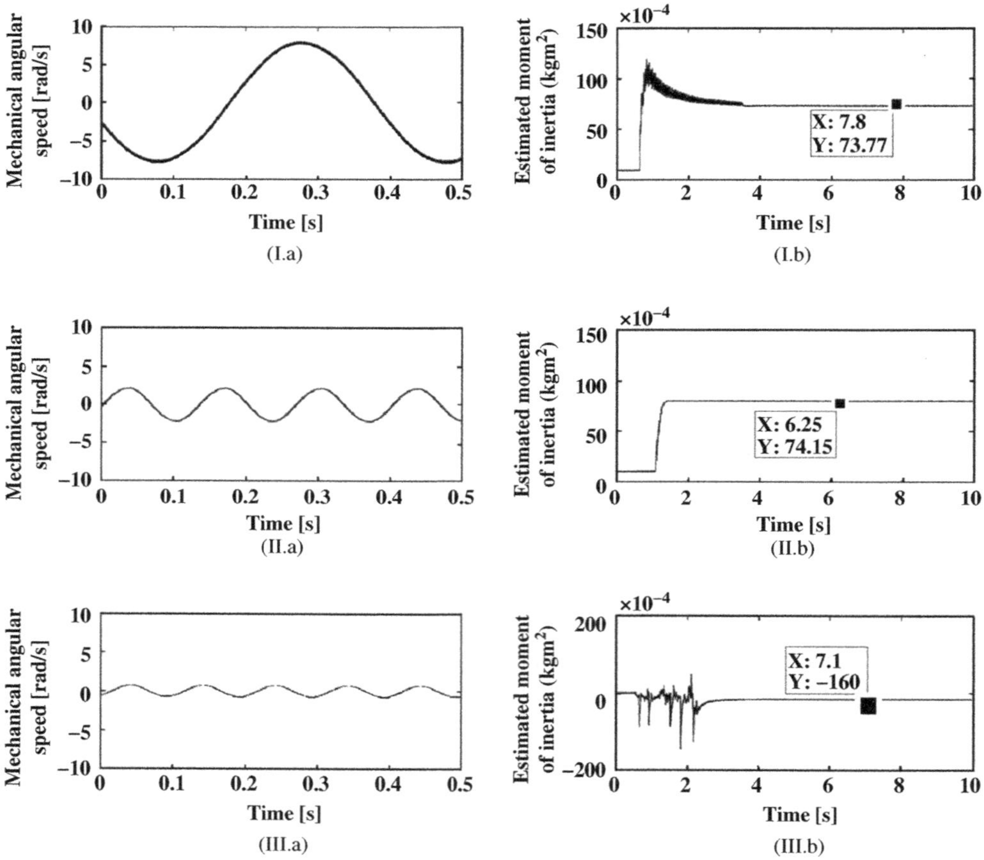

Figure 9.11 Variations of estimated moment of inertia of dc load machine and PMSM rotor with different injected q-axis currents under i_d [6], IEEE: (I) Q-axis current $i_q^* = 2\ \sin(5\pi t)$: (I.a) mechanical angular speed and (I.b) estimated moment of inertia; (II) Q-axis current $i_q^* = 1.5\ \sin(15\pi t)$: (II.a) mechanical angular speed and (II.b) estimated moment of inertia; and (III) Q-axis current $i_q^* = 1.5\ \sin(20\pi t)$: (III.a) mechanical angular speed and (III.b) estimated moment of inertia.

The estimation results in Figures 9.11(I) and (II) are close to the nominal value. However, the result in Figure 9.11(III) converges to a wrong point, since the value of $d\omega_r/dt$ is too small at 10 Hz and the viscous friction becomes dominant in the mechanical system (Eq. 9.2). In addition, when the amplitude of ω_r is too small, the signal versus noise in the measurement may also be deteriorated. Therefore, it is necessary to consider the amplitude of $d\omega_r/dt$ during the estimation of J, which can be maximized by online tuning the frequency and amplitude of perturbation signals manually.

9.4 Design of PI Speed Regulator

Under $i_d = 0$ control, the FOC-based mechanical parameter estimation system is presented in Figure 9.12, and the transfer function of speed loop under FOC is shown in Figure 9.13, where the current loop $C_{iq}(s)$ acts as an inner loop of the whole speed loop.

In Figure 9.13a, ω_r^*, T_ω, $K_e = 1.5p\psi_{pm}$, K_n, and $K_{ni} = K_n/\tau_n$ are the reference speed, the time constant of low-pass filter for measured rotor speed, the torque constant, and the PI constants of speed regulators. Since the time constants of current loop, T_s and T_ω, are usually much smaller than those of the speed loop, a conventional solution to reduce the order of the drive system model is to combine those small time constants together, and thus, the simplified system transfer function can be expressed as Figure 9.13b, and the optimized PI constants for speed loop can be expressed as follows:

$$K_n = \frac{J}{2K_e T_{\text{sum}}} \tag{9.19}$$

$$K_{ni} = \frac{K_n}{\tau_n} = \frac{F}{2K_e T_{\text{sum}}},\ \tau_n = \frac{J}{F} \tag{9.20}$$

where $T_{\text{sum}} = T_\omega + 2(T_s + T_d + T_{if})$. T_d and T_{if} are the dead time of the PWM control and the low-pass filter time constant of the measured dq-axis currents, respectively.

Besides, the PI regulators of the dq-axis current loops can be obtained by substituting the nominal values of L_d, L_q, and R as follows [6]:

$$K_{pd} = \frac{L_d}{2T_{sf}},\ K_{pq} = \frac{L_q}{2T_{sf}} \tag{9.21}$$

$$K_i = \frac{R}{2T_{sf}} \tag{9.22}$$

where $T_{sf} = T_s + T_d + T_{if}$. K_{pd}, K_{pq}, and K_i are the PI constants of the dq-axis current controllers.

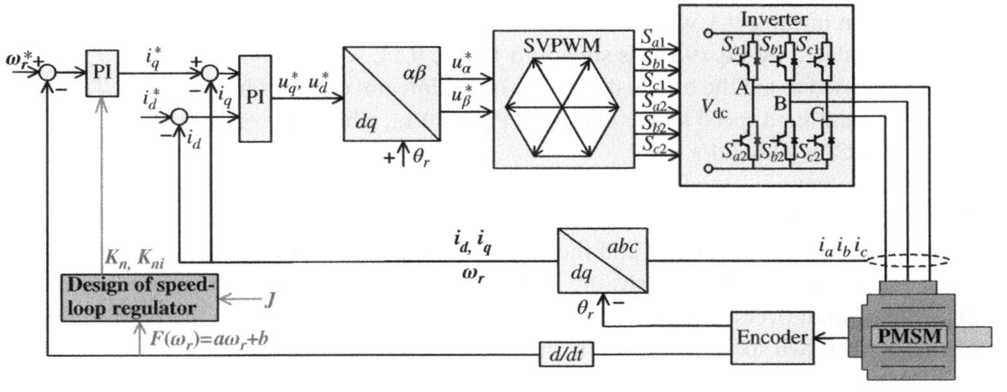

Figure 9.12 Block diagram of FOC-based mechanical parameter estimation system.

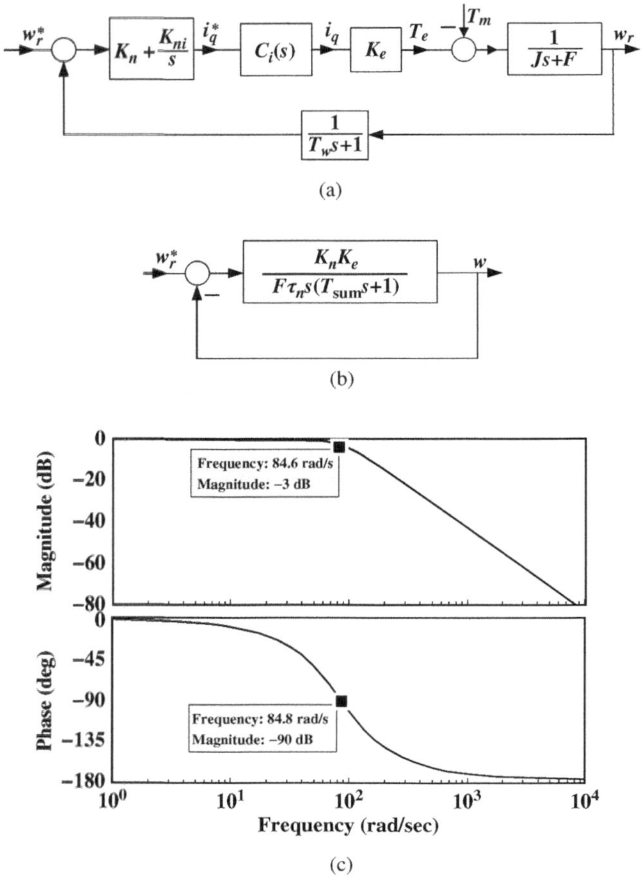

Figure 9.13 Transfer function of speed loop of field-oriented control [5], IEEE: (a) speed loop model, (b) simplified speed loop, and (c) Bode diagram of simplified speed loop for Prototype Machine 2 ($K_{pd} = 104.7$, $K_{pq} = 196.3$, and $K_i = 20805$).

For the determination of F at different rotor speeds, the proportional constant of the speed regulator is derived by Eq. (9.19) using the average value of calculated J (8.05×10^{-3} kg m^2) at $f = 3$ or 5 Hz, as shown in Table 9.3, while the integral constant is set to 0.1. The curve of the friction coefficient F calculated by Eq. (9.16) is shown in Figure 9.14, where $F(\omega_r) = a\,\omega_r + b$. Furthermore, $F(\omega_r)$ is then used to aid the online derivation of the integral constant by Eq. (9.20). The bode diagram of the simplified speed loop is shown in Figure 9.13, which shows that the bandwidth of the speed loop at $C3$ is 84.6 rad/s.

The performance of speed PI regulator derived from Eqs. (9.19) and (9.20) is evaluated in Figure 9.15 based on Prototype Machine 2, where the proportional constant is derived by using the average value of calculated J and the integral constant is derived by using the real-time calculated $F(\omega_r) = a\,\omega_r + b$. Obviously, the PMSM can rotate in the forward and reverse directions with quite a small overshoot and short settling time. Figure 9.16 shows the performance comparison between two speed regulators using PI constants derived with and without online calculation of $F(\omega_r) = a\,\omega_r + b$. The speed regulator without online calculation of $F(\omega_r)$ uses its average value to derive the integral constant by Eq. (9.20).

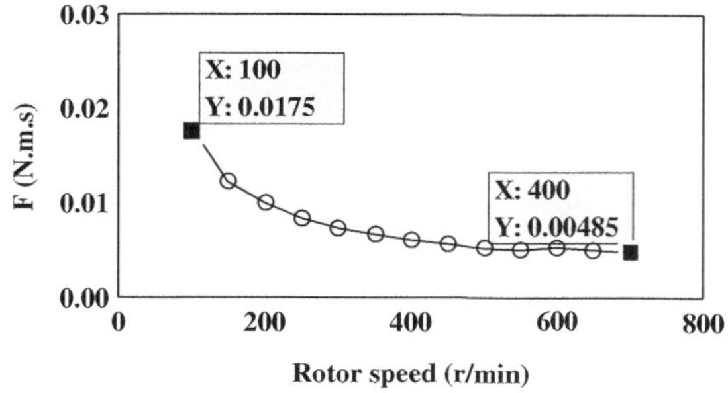

Figure 9.14 Calculated $F(\omega_r)$ at different rotor speeds [5], IEEE.

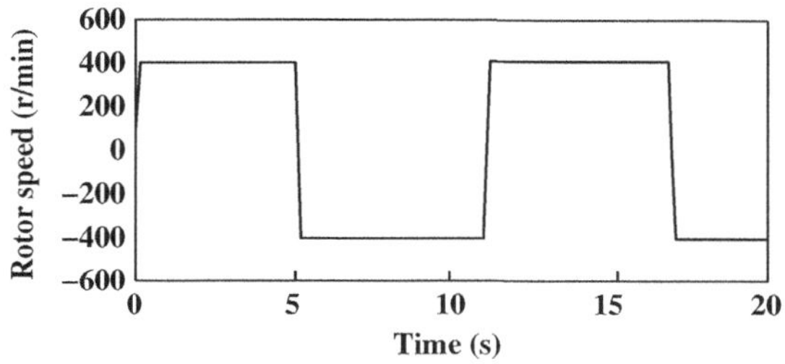

Figure 9.15 Step response of rotor speed in forward and reverse directions [5], IEEE.

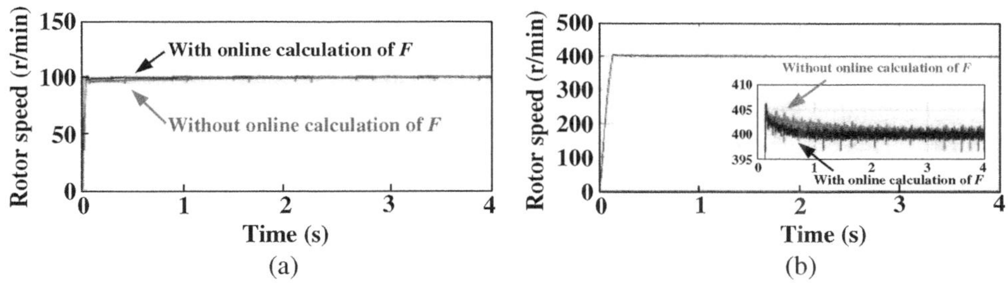

Figure 9.16 Performance comparison between designed speed regulators with and without online calculation of F ($i_d = 0$ and $i_d = 2.5$ A) [5], IEEE: (a) comparison at low-speed region and (b) comparison at high-speed region.

From the comparison results at low and high rotor speeds, it is found that the speed regulator with the online calculation of F exhibits better performance in eliminating the steady-state error. This can be explained that since F decreases with the rotor speed, the derived integral constant calculated by using the average value of F is too small at the low-speed region, while it becomes considerable at the high-speed region. Hence, it can be concluded that the performance of the speed regulator at different speeds can be improved by using online calculated $F(\omega_r)$.

Figure 9.17 compares the performance of two speed PI regulators with different proportional constants, of which both the integral constants are tuned with the real-time calculated F with respect to the transient rotor speed. It is found that i_q of both tests reaches the maximum

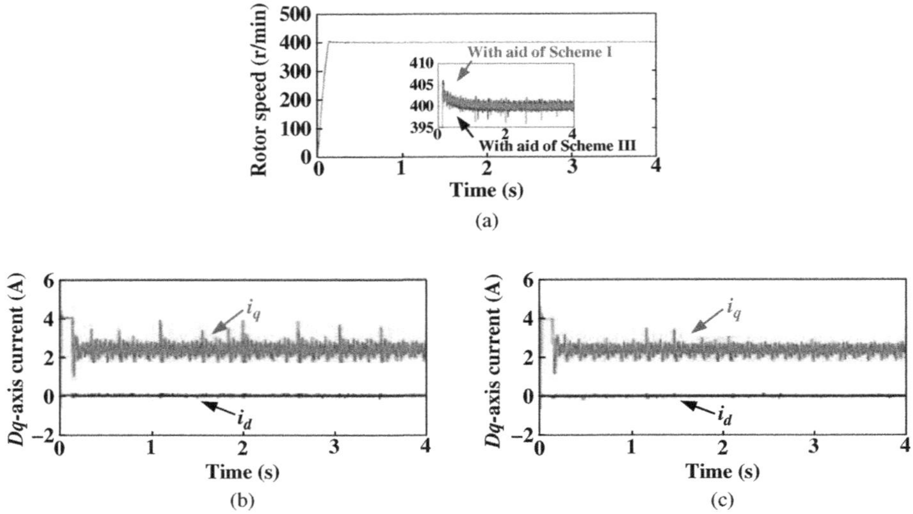

Figure 9.17 Comparison of rotor speeds and *dq*-axis currents by using proportional constants of speed regulators with aids from Schemes I and III ($i_d = 0$ and $i_q = 2.5$ A) [5], IEEE: (a) rotor responses of speed regulators, (b) *Dq*-axis current in Scheme I, and (c) *Dq*-axis current in Scheme III.

limit (4 A) immediately at the beginning of step response, as shown in Figure 9.17b and c, which will result in a similar step response of the rotor speed in Figure 9.17a. However, the speed regulator using the calculated J by Scheme III exhibits better dynamic performance in the q-axis current at the steady state. It can be explained that the calculated J by Scheme I is 31.7% larger than that of Scheme III, which will result in a large peak-to-peak ripple in the q-axis current. Therefore, the identified J by Scheme I can also be employed for the design of speed loop regulator, and the dynamic performance can be further improved if the identified J by Scheme III is employed.

9.5 Summary

This chapter has described three widely used online or offline mechanical parameter estimation techniques, i.e., acceleration/deceleration-based, MRAS observer–based, and fundamental motion equation–based methods. The former method is based on an acceleration/deceleration process, while the latter two methods are implemented under zero-mean sinusoidal rotor variations excited by the q-axis current injection. It has been shown that the injected q-axis current affects the accuracy of calculated electromagnetic torque due to variation of rotor PM flux linkage. Thus, the rotor PM flux linkage is preferred to be known a priori, e.g., by using the POPE method.

Furthermore, the estimated mechanical parameters have been used for the design of the speed-loop PI regulator. It has been demonstrated that the observer-based and fundamental motion equation–based methods can achieve similar results and are more accurate than the acceleration/deceleration-based method since the latter ignores the influence of viscous friction. Moreover, the fundamental motion equation–based method has the advantages of faster determination, easier implementation, and lower computational cost than the MRAS observer. Besides, for fast determination of the moment of inertia used for the speed-loop PI regulator, both the acceleration/deceleration-based and fundamental motion equation–based methods are satisfactory, while the PI regulator designed by the latter exhibits a better dynamic performance.

References

1 S. Ichikawa, M. Tomita, S. Doki, and S. Okuma, "Sensorless control of permanent-magnet synchronous motors using online parameter identification based on system identification theory." *IEEE Trans. Ind. Electron*, vol. 53, no. 2, pp. 363–372, Apr. 2006.

2 S. Morimoto, M. Sanada, and Y. Yakeda, "Mechanical sensorless drives of IPMSM with online parameter identification." *IEEE Trans. Ind. Appl.*, vol. 42, no. 5, pp. 1241–1248, Sep/Oct. 2006.

3 K. W. Lee, D. H. Jung, and, I. J. Ha. "An online identification method for both stator resistance and back-EMF coefficient of PMSMs without rotational transducers." *IEEE Trans. Ind. Electron*, vol. 51, no. 2, pp. 507–510, Apr. 2004.

4 Y. Feng, X. Yu, and, F. Han. "High-order terminal sliding-mode observer for parameter estimation of a permanent-magnet synchronous motor." *IEEE Trans. Ind. Electron*, vol. 60, no. 10, pp. 4272–4280, Oct. 2013.

5 K. Liu and Z. Q. Zhu. "Fast determination of moment of inertia of permanent magnet synchronous machine drives for design of speed loop regulator." *IEEE Trans. Control Syst. Tech.*, vol. 25, no. 5, pp. 1816–1824, Sept. 2017.

6 K. Liu and Z. Q. Zhu. "Parameter estimation of PMSM for aiding PI regulator design of field oriented control." *Proc. The 17th Int. Conf. on Electrical Machines and Systems (ICEMS2014)*, 22-25 Oct. 2014, *Hangzhou, China*, pp. 2705–2711, Oct. 2014.

7 S. Underwood and I. Husain. "On-line parameter estimation and adaptive control of permanent magnet synchronous machines." *IEEE Trans. Ind. Electron*, vol. 57, no. 7, pp. 2435–2443, Jun. 2010.

8 K. Liu and Z. Q. Zhu. "Online estimation of rotor flux linkage and voltage source inverter nonlinearity in permanent magnet synchronous machine drives." *IEEE Trans. Power Electron*, vol. 29, no. 1, pp. 418–427, Jan. 2014.

9 D. Basic, F. Malrait, and P. Rouchon. "Current controller for low frequency signal injection and rotor flux position tracking at low speeds." *IEEE Trans. Ind. Electron*, vol. 58, no. 9, pp. 4010–4022, Sep. 2011.

10 Adegis. *Panasonic instruction manual AC servo motor and driver, MINAS A4 Series*. [online]. Available: https://adegis.com/media/asset/ffee066e40d56bef3c773e4564af5a67b8b5e7a5e8ac4bad8b6954303e71d17c.pdf

11 R. Mizutani, T. Takeshita, and N. Matsui. "Current model-based sensorless drives of salient-pole PMSM at low speed and standstill." *IEEE Trans. Ind. Appl.*, vol. 34, no. 4, pp. 841–846, Jul. 1998.

12 L. Sun, M. Cheng, H. Wen, and L. Song. "Motion control and performance evaluation of a magnetic-geared dual-rotor motor in hybrid powertrain." *IEEE Trans. Ind. Electron*, vol. 64, no. 3, pp. 1863–1872, Mar. 2017.

13 R. Babau, I. Boldea, T. J. E. Miller, and N. Muntean. "Complete parameter identification of large induction machines from no-load acceleration–deceleration tests." *IEEE Trans. Ind. Electron*, vol. 54, no. 4, pp. 1962–1972, Aug. 2007.

14 W. Lin, T. Su, and R. Wu. "Parameter identification of induction machine with a starting no-load low-voltage test." *IEEE Trans. Ind. Electron*, vol. 59, no. 1, pp. 352–360, Jan. 2012.

15 G. S. Buja, R. Menis, and M. I Valla. "Disturbance torque estimation in a sensorless DC drive." *IEEE Trans. Ind. Electron*, vol. 42, no. 4, pp. 351–357, Aug. 1995.

16 F. Lin and H. Su. "A high-performance induction motor drive with on-line rotor time-constant estimation," *IEEE Trans. Energy Convers*, vol. 12, no. 4, pp. 297–303, Dec. 1997.

17 S. Li and Z. Liu, "Adaptive speed control for permanent-magnet synchronous motor system with variations of load inertia," *IEEE Trans. Ind. Electron*, vol. 56, no. 8, pp. 3050–3059, Aug. 2009.

18 S. Yang and J. Lin, "Observer-based automatic control loop tuning for servo motor drives," In *Proc. 2013 IEEE 10th Int. Conf. Power Electron. Drive Syst. (PEDS)*, Kitakyushu, Japan, 2013, pp. 302–305.

19 K.-H. Kim and M.-J. Youn, "A nonlinear speed control for a PM synchronous motor using a simple disturbance estimation technique," *IEEE Trans. Ind. Electron*, vol. 49, no. 3, pp. 524–535, Jun. 2002.

20 J. S. Ko, J. H. Lee, and M. J. Youn, "Robust digital position control of brushless DC motor with adaptive load torque observer," *IEE Proc. Electr. Power Appl*, vol. 141, no. 2, pp. 63–70, Mar. 1994.

21 K. Liu and Z. Q. Zhu, "Mechanical parameter estimation of permanent-magnet synchronous machines with aiding from estimation of rotor PM flux linkage," *IEEE Trans. Ind. Appl.*, vol. 51, no. 4, pp. 3115–3125, July–Aug. 2015.

22 F. Andoh, Moment of inertia identification using the time average of the product of torque reference input and motor position. *IEEE Trans. Power Electron.*, vol. 22, no. 6, pp. 2534–2542, Nov. 2007.

23 K. Liu, C. Hou, and W. Hua, "A novel inertia identification method and its application in PI controllers of PMSM drives," *IEEE Access*, vol. 7, pp. 13445–13454, 2019.

24 S. W. Wang and S. M. Wan, "Estimations of load parameters for PMSM by MRAS," In *Proc. IEEE Int. Conf. Elect. Control Eng. (ICECE)*, Sep. 2011, pp. 657–660.

25 K. J. Åström and B. Wittenmark, Adaptive Control. Reading, MA: Addison-Wesley, 1995.

26 S. Wang and S. Wan, "Estimation of load parameters for PMSM by MRAS," In *Proc. IEEE Int. Conf. Electr. Control Eng.*, pp. 657–660, 2011.

27 X. W. Wang, and J. Yi, "Method for identification of rotational inertia of AC servo," China Patent 101 231 207 A. Jul. 30, 2008.

10

Modern Control and Optimization Theory–Based Parameter Estimation Algorithms

10.1 Introduction

In the previous chapters, several parameter estimation schemes for PMSM have been introduced, while a complete estimation system also requires a dedicated estimation algorithm to solve the established state-space model. Since 1960s, due to vigorous development of modern control and optimization theories, various adaptive algorithm-based methods [1–29] and gradient-based [30–35] methods have been applied to system parameter identification. Since 1990s, high-performance microprocessors further promoted the development of evolutionary algorithm–based methods, which have proven to be highly successful across a wide range of computational tasks in parameter estimation [36–48].

The main modern control theory-based estimation algorithms will be introduced in this chapter, including the following:

1. *Adaptive algorithm-based methods:*
   - Recursive least-squares (RLS) algorithm
   - Model reference adaptive system (MRAS)
   - Extended Kalman filter (EKF)
   - Adaline neural network (ANN)
2. *Gradient-based methods:*
   - Steepest descent method
   - (Quasi) Newton's method
   - Gauss–Newton method
3. *Evolutionary algorithm–based methods:*
   - Particle swarm optimization (PSO)
   - Genetic algorithm (GA)

The principles of all the above algorithms will be described briefly. Moreover, they will be used to estimate PMSM electrical parameters based on a current injection–based scheme. The simulation results will also be presented for illustrative purposes.

10.2 Designed Parameter Estimation Scheme

Generally, the parameter estimation for PMSMs under FOC is shown in Figure 10.1, where the *dq*-axis currents and voltages, as well as the rotor speed, are input quantities for the parameter estimator.

In the adaptive algorithm-based methods, the estimated parameters are updated at each time step according to the deviations of the predicted and measured values at the previous time step.

Parameter Estimation of Permanent Magnet Synchronous Machines, First Edition. Zi Qiang Zhu, Kan Liu, and Dawei Liang.

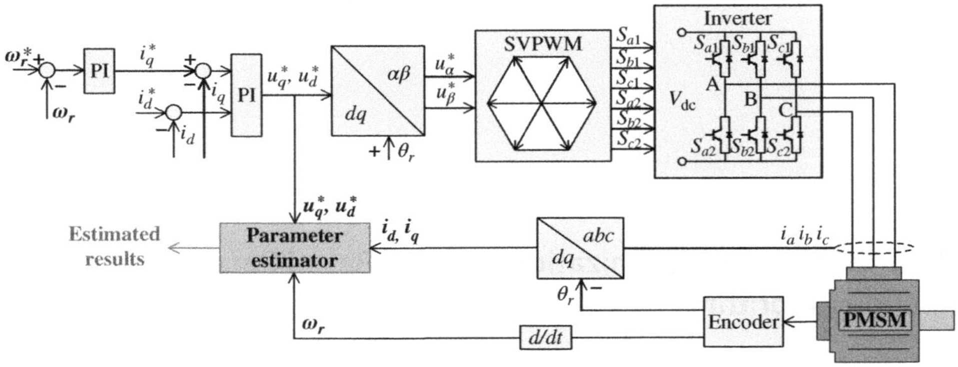

Figure 10.1 Schematic diagram of online parameter estimation.

The adaptive law can be designed by RLS, KF, MRAS, and ANN. In comparison, in the gradient-based and evolutionary algorithms, all measured data are saved in RAM after sampling. Afterwards, the estimated parameters are obtained by finding the optimum from the corresponding objective functions.

A current injected-based parameter estimation scheme is designed based on Figure 10.1 for the following illustration, where the PMSM is controlled under $i_d = 0$ at 600 r/min. A negative d-axis current of 1 A is injected in a short period of 0.5 s to estimate three electrical parameters, i.e., the winding resistance R_s, the dq-axis inductance $L_d = L_q = L_s$, as well as the rotor PM flux linkage ψ_{pm}. Additionally, the VSI non-linearity has also been compensated, as detailed in Chapter 2.

Therefore, the two sets of dq-axis equations are written as follows:

$$u_{d0} = -\omega_r L_s i_{q0} \tag{10.1a}$$

$$u_{q0} = R_s i_{q0} + \omega_r \psi_{pm} \tag{10.1b}$$

$$u_{d1} = R_s i_{d1} - \omega_r L_s i_{q1} \tag{10.1c}$$

$$u_{q1} = R_s i_{q1} + \omega_r (L_s i_{d1} + \psi_{pm}) \tag{10.1d}$$

where the subscripts "0" and "1" denote the dq-axis currents/voltages when $i_d = 0$ and $i_d \neq 0$, respectively.

The complete estimation scheme is presented in Figure 10.2, and the simulated currents, voltages, and the rotor speed are shown in Figure 10.3. The nominal values of R_s, L_s, and ψ_{pm} are 6.0 Ω, 6.0 mH, and 0.25 Wb, respectively.

The above electrical quantities are used for the parameter estimation, which is implemented in three steps. First, the value of L_s is estimated by Eq. (10.1a). Second, R_s is estimated by Eq. (10.1c)

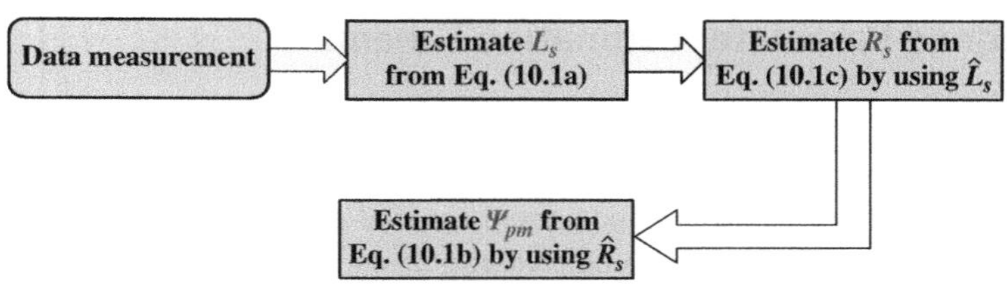

Figure 10.2 Flowchart of three-step parameter estimation.

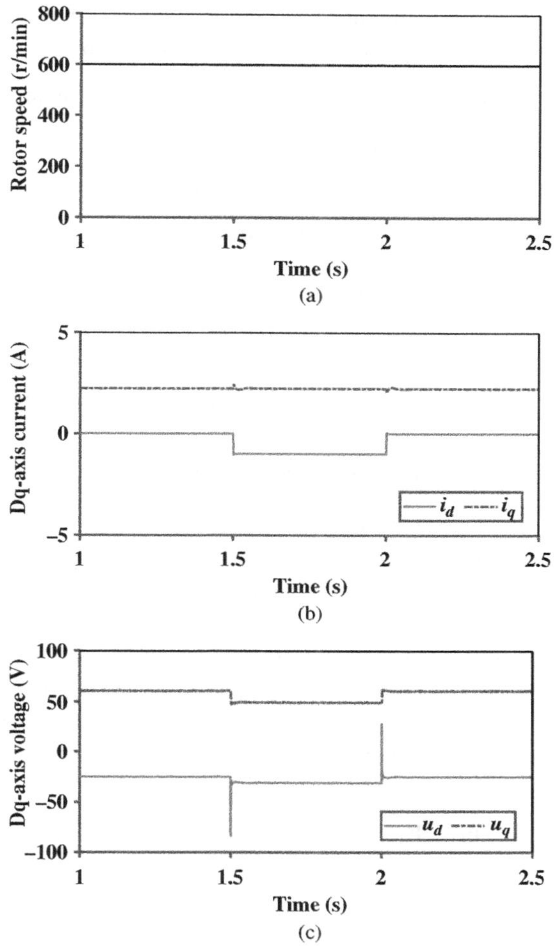

Figure 10.3 Simulated electrical quantities used for parameter estimation: (a) rotor speed, (b) *Dq*-axis currents, and (c) *Dq*-axis voltages.

using the estimated $\widehat{L}_s$. At last, the rotor PM flux linkage ψ_{pm} is estimated with the aid of the estimated $\widehat{R}_s$ from Eq. (10.1b).

10.3 Recursive Least-Squares

10.3.1 Basic Principle

The recursive least-squares (RLS) method is an adaptive algorithm to obtain the parameters by recursively minimizing a linear least-square function. The concept of least-square regression was discovered by Gauss, which is used for statistical treatment [1]. After over 100 years, it was rediscovered by Plackett [2] and has obtained great interest since 1960s because of rapid development of control theory [4]. Nowadays, the RLS algorithm has been widely used in online parameter estimation due to its fast convergence [5–9].

First, it is assumed that the model is expressed as:

$$Y = \Phi\Theta \tag{10.2}$$

where $\boldsymbol{\Phi}$ and $\boldsymbol{Y}$ are the input and output matrices. $\boldsymbol{\Theta}$ is the vector containing unknown parameters.

By minimizing the least-square errors ε_k between the measured $\boldsymbol{y}_k$ and predicted $\hat{\boldsymbol{y}}_k$ in Eq. (10.3), the unknown parameters can be calculated by Eq. (10.4).

$$\varepsilon_k = \left(\boldsymbol{Y}_k - \widehat{\boldsymbol{Y}}_k\right)^2 = \left(\boldsymbol{Y}_k - \boldsymbol{\Phi}_k\widehat{\boldsymbol{\Theta}}_k\right)^T\left(\boldsymbol{Y}_k - \boldsymbol{\Phi}_k\widehat{\boldsymbol{\Theta}}_k\right) \tag{10.3}$$

$$\widehat{\boldsymbol{\Theta}}_k = (\boldsymbol{\Phi}_k^T\boldsymbol{\Phi}_k)^{-1}\boldsymbol{\Phi}_k^T\boldsymbol{Y}_k \tag{10.4}$$

Therefore, the linear recursive estimator can be written as follows:

$$\widehat{\Theta}_k = \widehat{\Theta}_{k-1} + \boldsymbol{K_k}\left(\boldsymbol{Y}_k - \boldsymbol{\Phi}_k\widehat{\Theta}_{k-1}\right) \tag{10.5}$$

where $\boldsymbol{K}_k$ is the estimator gain vector. $\left(\boldsymbol{Y}_k - \boldsymbol{\Phi}_k\widehat{\Theta}_{k-1}\right)$ is the correction term. As can be observed, the new estimated parameters $\widehat{\boldsymbol{\Theta}}_k$ is updated based on the previous step value and the correction term multiplied by a gain vector $\boldsymbol{K}_k$.

Moreover, the gain vector $\boldsymbol{K}_k$ is obtained by minimizing the aggregate variance of the estimation errors and expressed as:

$$K_k = \boldsymbol{P}_{k-1}\boldsymbol{\Phi}_k^T(I + \boldsymbol{\Phi}_k^T P_{k-1}\boldsymbol{\Phi}_k)^{-1} \tag{10.6}$$

$$\boldsymbol{P}_k = (\boldsymbol{I} - \boldsymbol{K}_k\boldsymbol{x}_k^T)\boldsymbol{P}_{k-1} \tag{10.7}$$

where $\boldsymbol{P}_k$ is the covariance matrix. $\boldsymbol{I}$ is the identity matrix.

Nevertheless, the above conventional RLS estimator suffers from the "data saturation" issue in practical applications, i.e., with increasing the number of sampling steps, the gain and covariance matrices $\boldsymbol{K}_k$ and $\boldsymbol{P}_k$ become smaller. Consequently, the newly collected measured output has a minor impact on correcting the estimated parameters, i.e., the RLS cannot track the changes in the parameters.

To mitigate this issue, a forgetting factor λ_{rls} [4] is introduced in the RLS to set less weight to the previously measured data. Therefore, the updating algorithm of RLS with the forgetting factor can be expressed as:

$$\widehat{\boldsymbol{\Phi}}_k = \widehat{\boldsymbol{\Phi}}_{k-1} + \boldsymbol{K}_k\left(y_k - \hat{\boldsymbol{x}}_k^T\widehat{\boldsymbol{\Phi}}_{k-1}\right) \tag{10.8a}$$

$$\boldsymbol{K}_k = \boldsymbol{P}_{k-1}\boldsymbol{x}_k(\lambda_{\text{rls}}I + \boldsymbol{x}_k^T P_{k-1}\boldsymbol{x}_k)^{-1} \tag{10.8b}$$

$$\boldsymbol{P}_k = \lambda_{\text{rls}}{}^{-1}\boldsymbol{P}_{k-1}(\boldsymbol{I} - \boldsymbol{K}_k\boldsymbol{x}_k^T) \tag{10.8c}$$

where λ_{rls} is usually selected between 0.9 and 1.

10.3.2 Application to Parameter Estimation for PMSMs

The RLS algorithm for the parameter estimation is summarized as follows:

Algorithm 1: RLS algorithm for parameter estimation

Step 1: Initialization

$\widehat{\Phi}_0 = 0$, $P_0 = \delta^{-1}\boldsymbol{I}$ (δ=small/large positive constants for high/low SNR)

Step 2: Recursive estimation

Computing Eqs. (10.8a)–(10.8c)

Step 3: Output estimated parameters

Following the estimation scheme in Section 10.2, the RLS algorithm is employed to estimate the stator winding resistance, the rotor PM flux linkage, and the dq-axis inductance, which are presented in Figure 10.4. As can be observed, the RLS has fast convergence capability, and the converged estimated results are almost identical to the nominal results.

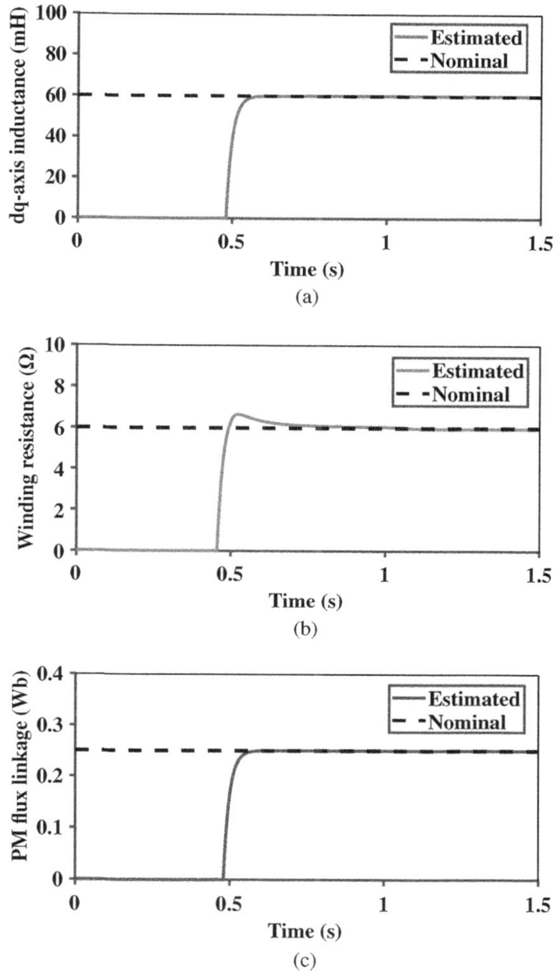

Figure 10.4 Estimated *dq*-axis inductance, stator winding resistance, rotor PM flux linkage by RLS: (a) *dq*-axis inductance, (b) stator winding resistance, and (c) rotor PM flux linkage.

10.4 Kalman Filter

10.4.1 Basic Principle

Kalman filter (KF), proposed by R.E. Kalman in 1960 [10] streaming from RLS algorithm, is developed for time-discrete data linear filtering problem. The Kalman filter provides a computational efficiency solution of the least-square method. Compared to the RLS estimator, which updates the estimation of a static parameter, the Kalman filter can update and estimate an evolving state, i.e., past, present, and future [11], where the noise measurement risks are also considered in the recursive processing.

10.4.1.1 Conventional Kalman Filter

For a time-discrete linear system, the state equation at time k is evolved from the state at $k-1$ and can be written as follows:

$$\boldsymbol{x}_k = \boldsymbol{A}_k \boldsymbol{x}_{k-1} + \boldsymbol{B}_k \boldsymbol{u}_{k-1} + \boldsymbol{w}_{k-1} \tag{10.9}$$

where $\boldsymbol{A}$ is the state transition matrix. $\boldsymbol{B}$ is the control-input model. $\boldsymbol{u}$ is the control vector. $\boldsymbol{w}$ is the zero-mean white Gaussian system noises with process noise covariance $\boldsymbol{Q}$ to indicate the model uncertainty. When $\boldsymbol{Q}$ is small, the model is more reliable with less noise and lower uncertainty level.

The observation (or measurement) output is expressed as:

$$\boldsymbol{y}_k = \boldsymbol{H}_k \boldsymbol{x}_k + \boldsymbol{v}_k \tag{10.10}$$

where $\boldsymbol{H}$ is the observation matrix. $\boldsymbol{v}$ is the observation/measurement noise with measurement noise covariance $\boldsymbol{R}$ to indicate the reliability of measurement. When $\boldsymbol{R}$ is small, the measurement results contain less noise and are more reliable.

The Kalman filter is conceptualized as two distinct phases, i.e., *predict* and *update*, where the state is represented by two variables, i.e., *posteriori* state $\boldsymbol{x}_{k|k}$ and *posteriori* estimate covariance matrix.

1. *Predict Phase*

   In the predict phase, the state estimated from the previous time step is used to predict the state at the current time step. This phase is also known as a *prior* state estimation since the observation information from the current time step is not involved.

   $$\text{Prior state} \quad \hat{\boldsymbol{x}}_{k|k-1} = \boldsymbol{A}_k \boldsymbol{x}_{k-1|k-1} + \boldsymbol{B}_k \boldsymbol{u}_k \tag{10.11a}$$

   $$\text{Prior covariance} \quad \boldsymbol{P}_{k|k-1} = \boldsymbol{A}_k \boldsymbol{P}_{k-1|k-1} \boldsymbol{A}_k^T + \boldsymbol{Q}_k \tag{10.11b}$$

2. *Update Phase*

   In the update phase, the deviation between the prior prediction and current observation is multiplied by the Kalman gain to refine the state estimation in Eq. (10.12). First, the Kalman filter in Eq. (10.12a) is updated. Afterwards, the actual observed/measured output is used to generate the *posterior* state $\boldsymbol{x}_{k|k}$ in Eq. (10.12b). The final step is to obtain the *posterior* error covariance matrix by Eq. (10.12c). Overall, the complete schematic diagram of the Kalman filter is presented in Figure 10.5.

   $$\text{Kalman gain} \quad \boldsymbol{K}_k = \boldsymbol{P}_{k|k-1} \boldsymbol{H}_k^T \left[\boldsymbol{H}_k \boldsymbol{P}_{k|k-1} \boldsymbol{H}_k^T + \boldsymbol{R}_k\right]^{-1} \tag{10.12a}$$

   $$\text{Updated state} \quad \boldsymbol{x}_{k|k} = \hat{\boldsymbol{x}}_{k|k-1} + \boldsymbol{K}_k \left[\boldsymbol{y}_k - \boldsymbol{H} \hat{\boldsymbol{x}}_{k|k-1}\right] \tag{10.12b}$$

   $$\text{Updated covariance matrix} \quad \boldsymbol{P}_{k|k} = (\boldsymbol{I} - \boldsymbol{K}_k \boldsymbol{H}_k) \boldsymbol{P}_{k|k-1} \tag{10.12c}$$

In the actual implementation of the Kalman filter, the covariance matrices of the measurement $\boldsymbol{R}$ and process $\boldsymbol{Q}$ noises need to be determined in a prior, which have a significant impact on the dynamic performance of the Kalman filter, i.e., the transient process of the Kalman filter will be

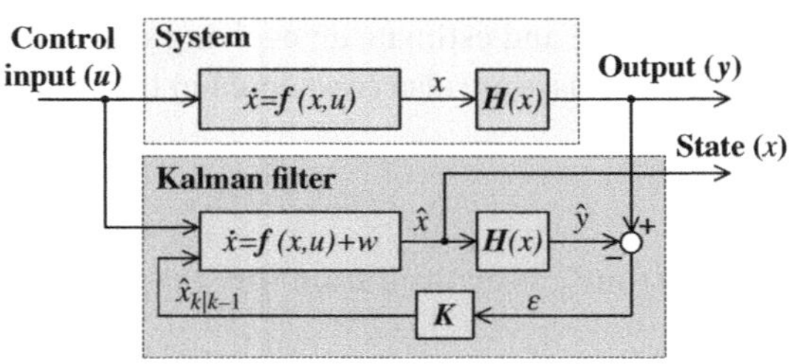

Figure 10.5 Schematic diagram of parameter estimation based on Kalman filter.

faster with a smaller $\boldsymbol{R}$ and larger $\boldsymbol{Q}$. Normally, the measurement noise covariance $\boldsymbol{R}$ can be determined by offline sample measurement. Nevertheless, the determination of process noise covariance $\boldsymbol{Q}$ is generally challenging since the estimation process is usually impossible to be observed directly. Therefore, the trial-and-error method is a common solution [12].

10.4.1.2 Extended Kalman Filters

The above derivation of the Kalman filter is based on a linear system (Eq. 10.9), while it can also be extended to a non-linear system Eqs. (10.13) and (10.14) by using the EKF. For a time-discrete non-linear system:

$$\boldsymbol{y}_k = \boldsymbol{f}(\boldsymbol{x}_{k-1}, \boldsymbol{u}_{k-1}) + \boldsymbol{w}_{k-1} \tag{10.13}$$

$$\boldsymbol{y}_k = \boldsymbol{h}(\boldsymbol{x}_k) + \boldsymbol{v}_k \tag{10.14}$$

The EKF utilizes the first-order Taylor series of the predict function $\boldsymbol{f}(\boldsymbol{x}, \boldsymbol{u})$ and update function $\boldsymbol{h}(\boldsymbol{x})$ to linearize, where $\boldsymbol{f}(\boldsymbol{x}, \boldsymbol{u})$ must be differentiable. Therefore, the state transition $\boldsymbol{A}_k$ and observation $\boldsymbol{H}_k$ matrices in Eqs. (10.9) and (10.10) are calculated by Eqs. (10.15) and (10.16). Besides, both the predict and update phases are the same as the Kalman filter, as shown in Figure 10.5.

$$\boldsymbol{A}_k = \left.\frac{\partial f}{\partial x}\right|_{\hat{\boldsymbol{x}}_{k-1|k-1}, \boldsymbol{u}_k} \tag{10.15}$$

$$\boldsymbol{H}_k = \left.\frac{\partial h}{\partial x}\right|_{\hat{\boldsymbol{x}}_{k-1|k-1}} \tag{10.16}$$

Although the EKF is one of the most widely used algorithms for non-linear systems, it exhibits inferior performance for a highly $\boldsymbol{f}(\boldsymbol{x}, \boldsymbol{u})$ and $\boldsymbol{h}(\boldsymbol{x})$. It can be explained that the EKF utilized linearization on the time scale for the update. To address these issues, the unscented Kalman filter is proposed in Ref. [16]. In the unscented Kalman filter, the approximations of the non-linear process and observation models do not exist, and it utilizes a set of deterministic sampling points to represent the underlying Gaussian distribution. Afterwards, the *posterior* distribution can be obtained by using unscented transformation [16]. The unscented Kalman filter has the same procedure as the classical/EKFs, i.e., *predict* and *update*.

1. *Predict Phase*

   Prior state $\quad \hat{\boldsymbol{x}}_{k|k-1} = \sum_{i=0}^{2L} \boldsymbol{W}_i^a \boldsymbol{x}_i$ (10.17a)

   Prior covariance $\quad \boldsymbol{P}_{k|k-1} = \sum_{i=0}^{2L} \boldsymbol{W}_i^c (\boldsymbol{x}_i - \hat{\boldsymbol{x}}_{k|k-1})(\boldsymbol{x}_i - \hat{\boldsymbol{x}}_{k|k-1})^T + \boldsymbol{Q}_k$ (10.17b)

   where $\boldsymbol{W}_i^a$ and $\boldsymbol{W}_i^c$ are the first- and second-order weights. L is the dimension of the system state.
2. *Update Phase*

   Kalman gain $\quad \boldsymbol{K}_k = \sum_{i=0}^{2L} \boldsymbol{W}_i^c (\boldsymbol{x}_j - \hat{\boldsymbol{x}}_{k|k-1})(\boldsymbol{y}_i - \hat{\boldsymbol{y}})^T \hat{\boldsymbol{S}}_k^{-1}$ (10.18a)

   Updated state $\quad \boldsymbol{x}_{k|k} = \hat{\boldsymbol{x}}_{k|k-1} + \boldsymbol{K}_k[\boldsymbol{y}_k - \hat{\boldsymbol{y}}]$ (10.18b)

   Updated covariance matrix $\quad \boldsymbol{P}_{k|k} = \boldsymbol{P}_{k|k-1} - \boldsymbol{K}_k \hat{\boldsymbol{S}}_k \boldsymbol{K}_k^T$ (10.18c)

   where $\hat{\boldsymbol{S}}_k$ is the covariance of the transformed sigma points.

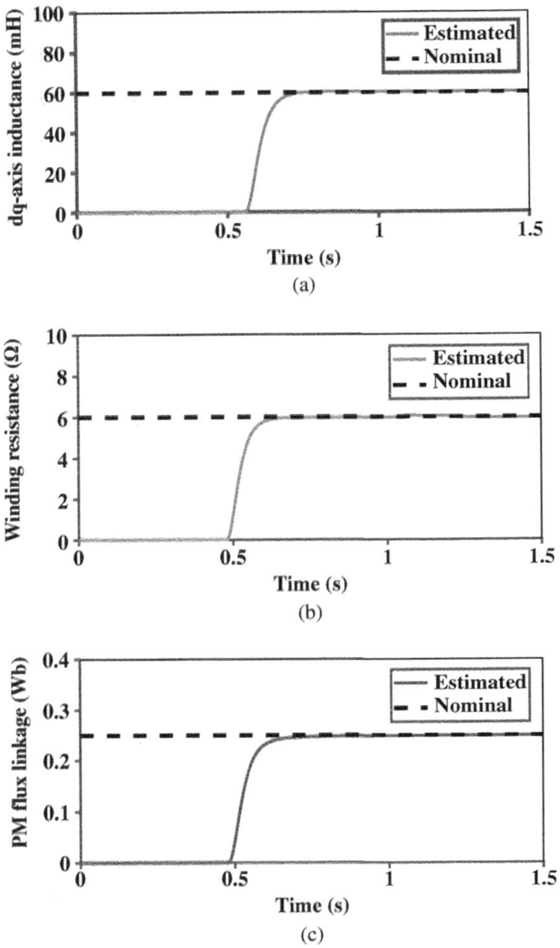

Figure 10.6 Estimated *dq*-axis inductance, stator winding resistance, and rotor PM flux linkage by Kalman filter: (a) *dq*-axis inductance, (b) stator winding resistance, and (c) rotor PM flux linkage.

10.4.2 Application to Parameter Estimation for PMSMs

To date, different types of Kalman filters have been extensively applied to parameter estimation for PMSMs [14–18]. For illustration, the standard Kalman filter is employed to estimate *dq*-axis stator inductance, stator winding resistance, and rotor PM flux linkage, where Eqs. (10.11) and (10.12) are introduced into the *dq*-axis voltage equations in Eq. (10.1). The estimation results are shown in Figure 10.6, where high estimation accuracy and fast convergence capability can be observed.

10.5 Model Reference Adaptive System

10.5.1 Basic Principle

Model reference adaptive system (MRAS) was developed in the 1950s for high-performing aircraft autopilots [19]. Since the state-space model and Lyapunov stability criteria were introduced 10 years later, MRAS has obtained great interest and is successfully applied to diverse fields, including parameter estimation for electrical machines [20–23].

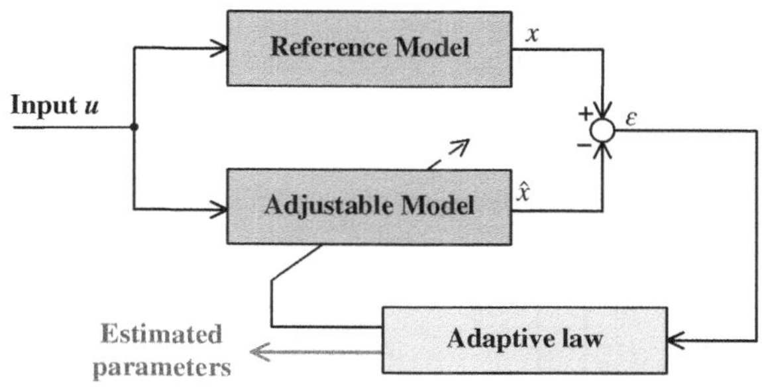

Figure 10.7 Schematic diagram of MRAS estimator.

The structure of MRAS estimator includes the reference model, the independent adjustable model, and the adaptive law, where the estimation error, ε, between the reference model, $\boldsymbol{x}$, and the adjustable model, $\hat{\boldsymbol{x}}$, is calculated and then fed back to an appropriate adaptive law to adjust the estimated parameters, as illustrated in Figure 10.7.

Generally, there are two theories for designing the adaptive mechanism [20], i.e., Lyapunov stability theory and Popov's integral inequality [21–24]. More specifically, to ensure global asymptotic stability, a design of adaptive law based on the Lyapunov stability theorem requires designers' experience. In comparison, the Popov stability criterion is more flexible in obtaining the adaptive law, i.e., Popov's integral inequality in Eq. (10.19):

$$\int_0^{t_1} e^T(-\zeta) \geq \delta_0^2 \tag{10.19}$$

for all $t_1 \geq 0$. δ_0 is a finite positive constant. Usually, the PI technique is employed as the adaptive mechanism.

10.5.2 Application to Parameter Estimation for PMSMs

According to the estimation scheme in Figure 10.2, the transient dq-axis voltage equations can be rewritten as follows:

$$\begin{bmatrix} \frac{di_d}{dt} \\ \frac{di_q}{dt} \end{bmatrix} = \begin{bmatrix} -\frac{R_s}{L_s} & \omega_r \\ -\omega_r & -\frac{R_s}{L_s} \end{bmatrix} \begin{bmatrix} i_d \\ i_q \end{bmatrix} + \begin{bmatrix} \frac{1}{L_s} & 0 \\ 0 & \frac{1}{L_s} \end{bmatrix} \begin{bmatrix} u_d \\ u_q \end{bmatrix} + \begin{bmatrix} 0 \\ -\frac{\psi_{pm}}{L_s}\omega_r \end{bmatrix} \tag{10.20}$$

First, the variable model for the estimation of L_s is constructed based on Eq. (10.1):

$$\frac{d\hat{i}_d}{dt} = \omega_r i_{q0} + \frac{1}{\hat{L}_s} u_d \tag{10.21}$$

By subtracting Eq. (10.20) from Eq. (10.21), the system deviation can be calculated as follows:

$$\frac{di_d}{dt} - \frac{d\hat{i}_d}{dt} = \frac{1}{L_s} u_d + \frac{1}{\hat{L}_s} u_d \tag{10.22}$$

Defining $e_d = i_d - \hat{i}_d$, $b = 1/L_s$, and $\hat{b} = 1/\hat{L}$, Eq. (10.22) can be written as follows:

$$\frac{de_d}{dt} = (b - \hat{b}) u_d \tag{10.23}$$

Therefore, by using Popov's integral inequality (Eq. 10.19), the adaptive law for estimating L_s can be designed as follows:

$$\hat{b} = \left(\frac{k_{b1}}{s} + k_{b2}\right) u_d + b(0) \tag{10.24}$$

Second, the variable model for estimating R_s and ψ_{pm} is written as follows:

$$\begin{bmatrix} \frac{d\hat{i}_d}{dt} \\ \frac{d\hat{i}_q}{dt} \end{bmatrix} = \begin{bmatrix} -\frac{\widehat{R}_s}{L_s} & \omega_r \\ -\omega_r & -\frac{\widehat{R}_s}{L_s} \end{bmatrix} \begin{bmatrix} \hat{i}_d \\ \hat{i}_q \end{bmatrix} + \begin{bmatrix} \frac{1}{L_s} & 0 \\ 0 & \frac{1}{L_s} \end{bmatrix} \begin{bmatrix} u_d \\ u_q \end{bmatrix} + \begin{bmatrix} 0 \\ -\frac{\widehat{\psi}_{pm}}{L_s}\omega_r \end{bmatrix} \tag{10.25}$$

where $\widehat{R}_s$ and $\widehat{\psi}_{pm}$ are the identified values of stator winding resistance and rotor PM flux linkage, respectively.

By subtracting Eq. (10.25) from Eq. (10.24), the deviation matrix can be written as follows:

$$\frac{d}{dt}\boldsymbol{e} = K\begin{bmatrix} -\frac{R_s}{L_s} & \omega_r \\ -\omega_r & -\frac{R_s}{L_s} \end{bmatrix}\boldsymbol{e} + \begin{bmatrix} -\frac{R_s - \widehat{R}_s}{L_s} & 0 \\ 0 & -\frac{R_s - \widehat{R}_s}{L_s} \end{bmatrix} \begin{bmatrix} \hat{i}_d \\ \hat{i}_q \end{bmatrix} + \begin{bmatrix} 0 \\ -\frac{\psi_{pm} - \widehat{\psi}_{pm}}{L_s}\omega_r \end{bmatrix} \tag{10.26}$$

where $\boldsymbol{e} = \boldsymbol{i} - \hat{\boldsymbol{i}} = [e_d \quad e_q]$.

To satisfy Popov's inequality criterion (Eq. 10.19), the PI adaptive law can be expressed as follows:

$$\widehat{R}_s = -\left(k_{a1} + \frac{k_{a2}}{s}\right)\left(\frac{\hat{i}_d}{L_s}e_d + \frac{\hat{i}_q}{L_s}e_q\right) \tag{10.27a}$$

$$\widehat{\psi}_{pm} = -\left(k_{b1} + \frac{k_{b2}}{s}\right)(e_d\omega_r) \tag{10.27b}$$

where k_{a1} and k_{a2} are PI constants for identifying R_s. k_{b1} and k_{b2} are PI constants for identifying ψ_{pm}.

Finally, when the MRAS estimators (Eqs. 10.24 and 10.27) are employed in the estimation scheme in Section 10.2, the estimated stator winding resistance, rotor PM flux linkage, and *dq*-axis stator inductance are presented in Figure 10.8.

10.6 Adaline Neural Network

10.6.1 Basic Principle

ANN is a single-layer artificial neural network, which was developed by Widrow and Holf [25]. The mathematical model of an ANN is expressed in Eq. (10.28), of which the structure is shown in Figure 10.9:

$$O(W_i, X_i) = \sum_{i=0}^{n} W_i X_i \tag{10.28}$$

where W_i is the net weight and X_i is the input signal.

The activation function $O(W_i, X_i)$ of the network output node is a linear function. The objective function is defined as minimizing the least mean square (LMS), and thus, the updated W_i in each sampling period is presented as follows:

$$OW_{i,k+1} = W_{i,k} + 2\eta X_i(\widehat{O}_k - O_k) \tag{10.29}$$

where η is the convergence factor adjusting the convergence speed. $\widehat{O}_k$ is the predicted output. k and $k+1$ are the sampling indexes.

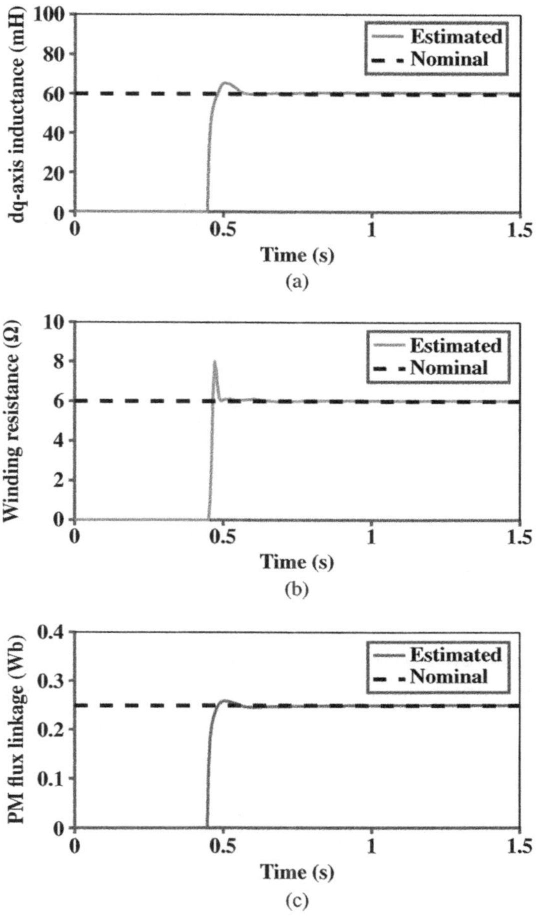

Figure 10.8 Estimated *dq*-axis inductance, stator winding resistance, and rotor PM flux linkage by MRAS estimators: (a) *dq*-axis inductance, (b) stator winding resistance, and (c) rotor PM flux linkage.

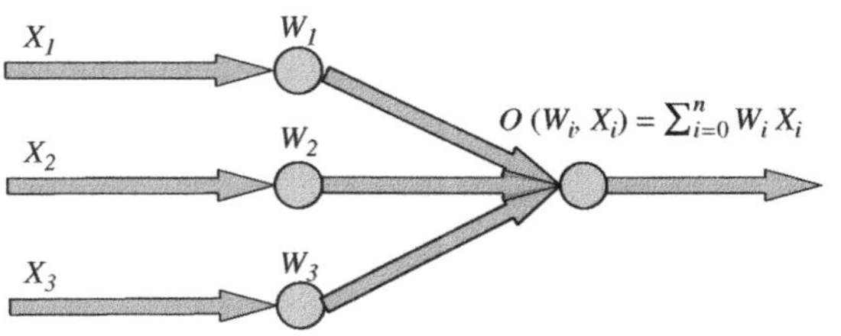

Figure 10.9 Structure of an ANN.

10.6.2 Application to Parameter Estimation for PMSMs

When the ANN is employed to estimate electrical parameters for PMSMs [15–29], its complete parameter estimation system is presented in Figure 10.10. In the estimation system, the ANN estimator follows the variation of PMSM parameters based on the reference model, i.e., the real PMSM. The network is composed of different single-node subnets, and the parameters to be estimated, e.g., winding resistance, inductance, and flux linkage, are regarded as the weights W_i of subnets. The dq-axis voltages u_d and u_q of the PMSM will be compared to the outputs from the ANN estimator, i.e., $\hat{u}_d$ and $\hat{u}_q$. Afterwards, the output difference will be fed back to adjust the weights by using the LMS learning rule.

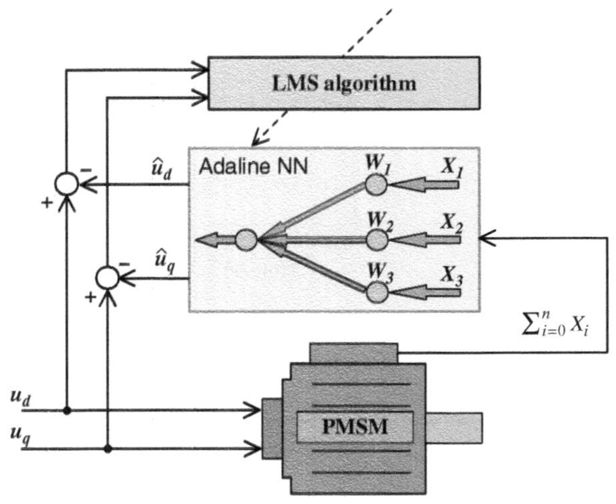

Figure 10.10 ANN-based estimation system for PMSM.

According to Eq. (10.29), if $d(k)$ is the sampled target output, the weight adjustment can be obtained as follows:

$$W_i(k+1) = W_i(k) + 2\eta X_i[d(k) - O(W_i, X_i)] \tag{10.30}$$

According to the estimation scheme in Section 10.2, the ANN estimators for estimating stator winding resistance, dq-axis inductance, and rotor PM flux linkage are designed as follows, and the subnet structures are shown in Figure 10.11.

First, for the estimation of L_s, it is assumed that $X_i = \omega_r(k) i_{q0}(k)$, $O = -\hat{u}_{d0}(k)$, and $d(k) = -u_{d0}(k)$, and thus

$$\hat{L}_s(k+1) = \hat{L}_s(k) + 2\eta\omega_r(k) i_{q0}(k)[\hat{u}_{d0}(k) - u_{d0}(k)] \tag{10.31}$$

Second, the estimation of R_s is derived from Eq. (10.1c) with the aid of the estimated $\hat{L}_s$ at the first step. In this case, the input signal X_i and the predicted output O are designed as $X_i = i_{d1}(k)$ and $O = \hat{u}_{d1}(k) + \omega_r(k) L_s i_{q1}(k)$.

$$\hat{R}_s(k+1) = \hat{R}_s(k) + 2\eta i_{d1}(k)[\hat{u}_{d1}(k) - u_{d1}(k)] \tag{10.32}$$

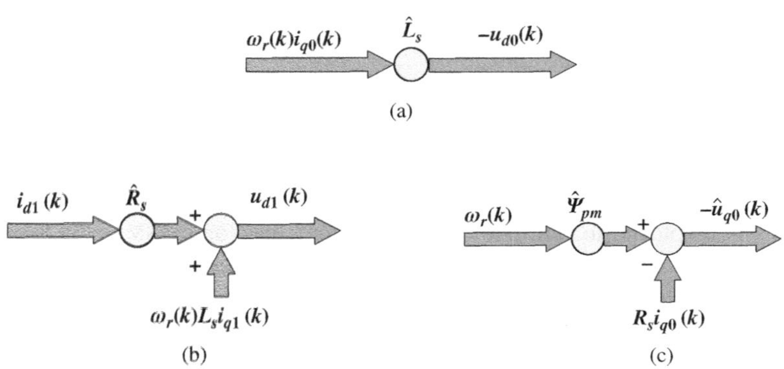

Figure 10.11 Subnet structures of designed ANN estimators: (a) dq-axis stator inductance, (b) stator winding resistance, and (c) rotor PM flux linkage.

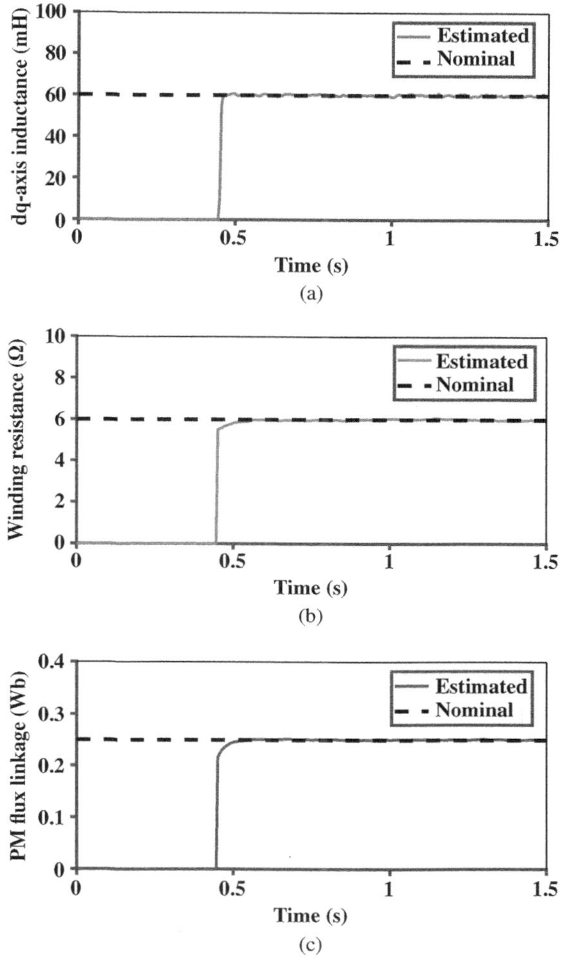

Figure 10.12 Estimated (a) *dq*-axis inductance, (b) stator winding resistance, and (c) rotor PM flux linkage by ANN estimators.

Third, with the aid of the estimated $\hat{R}_s$, the rotor PM flux linkage can be estimated as follows:

$$\hat{\psi}_{pm}(k+1) = \hat{\psi}_{pm}(k) + 2\eta\omega_r(k)\left[\hat{u}_{q0}(k) - u_{q0}(k)\right] \tag{10.33}$$

where $X_i = \omega_r(k)$ and $O = \hat{u}_{q0}(k) - R_s i_{q0}(k)$.

Finally, when the above three ANN estimators are employed in the prototype PMSM described in Section 10.2, the estimated parameters are presented in Figure 10.12. More experimental results employed in parameter estimation for PMSM can also be found in Chapters 3–5.

10.7 Gradient-Based Methods

The previous sections 10.3–10.6 have introduced several adaptive estimation methods, which are based on a similar principle, i.e., the identified parameter is updated in each time step according to the measured information. This section introduces another type of estimation algorithm based on the gradient of the least-square objective function, including steepest descent, Newton, and Newton–Gauss methods. In this case, the estimated parameters are obtained by solving an optimization problem, and the optimum is adjusted stepwise according to the first-order or second-order gradients of the least-square objective function [30–33].

10.7.1 Basic Principle

10.7.1.1 Steepest Descent Method

For an objective differentiable function $f(x)$ with the first-order gradient $g(x_k) = \nabla f(x_k)$ at step x_k, the next-step x_{k+1} can be written as:

$$x_{k+1} = x_k + a_k d_k \tag{10.34}$$

where $d_k = -\nabla f(x_k)$. a_k is the step length, which is subject to:

$$a_k = \text{argmin } f(x_k + a_k d_k) \tag{10.35}$$

where argmin refers to the argument of the minimum for $f(x)$.

Therefore, the algorithm of the steepest descent method can be implemented by:

Algorithm 2: Steepest descent method

Step 1: Initialization

Initialize x_0 and $d_0 = -\nabla f(x_0)$

Step 2: Iterative computation

Calculate gradient $\nabla f(x_k)$

Update step location by Eq. (10.34)

Recalculate gradient $\nabla f(x_{k+1})$

Step 3: Satisfice terminal criterion?

If not, back to Step 2.

Step 4: Output

Although the steepest descent is the simplest and best-known method for optimization problem, it is not commonly used in practice due to its extremely low convergence. Therefore, the Newton and Gauss–Newton methods are more practical.

10.7.1.2 Newton's Method

For the same optimization function, i.e., min $f(x)$, Newton's method solves this problem by constructing a sequence $\{x_k\}$ from the initial point x_0 to converge towards the minimum with the aid of second-order Taylor approximation (Eq. 10.36) in each iteration.

$$f(x_k + \Delta x_k) \approx f(x_k) + f'(x_k)t + \frac{1}{2}f''(x_k)\Delta x_k^2 \tag{10.36}$$

where Δx_k is the step variation, which is determined as follows:

$$\frac{d\left[f(x_k) + f'(x_k)\Delta x_k + \frac{1}{2}f''(x_k)\Delta x_k^2\right]}{d\Delta x_k} = 0 \tag{10.37a}$$

$$\Delta x_k = -\frac{f'(x_k)}{f''(x_k)} \tag{10.37b}$$

In addition, for multivariate functions, the above derivation is also applicable, where the first-order and second-order partial derivatives, i.e., named by Jacobian and Hessian matrices are calculated for iteration. Overall, the algorithm of Newton's method can be summarized as follows:

Algorithm 3: Newton's method

Step 1: Initialization
Initialize $x_0, f'(x_0,)$, and $f''(x_0)$

Step 2: Iterative computation
Calculate step by Eq. (10.37)
Update step: $x_{k+1} = x_k + \Delta x_k$

Step 3: Satisfice terminal criterion?
If not, back to Step 2.

Step 4: Output

Newton's method is a highly efficient method and has fast convergence capability. However, it is sensitive to the setting of initial values, especially when the objective function is not non-convex problem. Meanwhile, the iterative calculation of the Jacobian and Hessian matrices brings a heavy computation burden, and sometimes, they are unavailable. Therefore, an alternative way is using the quasi-Newton method, where the Hessian matrix is approximated by a positive-definite matrix $\boldsymbol{B}$ that satisfies the quasi-Newton condition in Eq. (10.38).

$$\boldsymbol{B}_{k+1}[x_{k+1} - x_k] = \nabla f(x_{k+1}) - \nabla f(x_k) \tag{10.38}$$

There are some popular formulas to update the approximate Hessian matrix $\boldsymbol{B}_{k+1}$, e.g., BFGS, DFP, etc., as detailed in Refs [32–34].

10.7.1.3 Gauss–Newton Method

The Gauss–Newton method is an extension of Newton's method for finding a minimum, particularly for non-linear least-squares problems (Eq. 10.39), which is widely used in the field of parameter estimation [35].

$$\min F(x) = \|f(x)\|^2 \tag{10.39}$$

When $f(x)$ is expressed by Taylor series, the objective function $F(x)$ can be approximated as follows:

$$F(x_k + \Delta x_k) \approx \|f(x_k) + \nabla f(x_k)\Delta x_k\|^2 \tag{10.40}$$

To minimize the objective function (Eq. 10.40), Δx_k is calculated as follows:

$$\frac{d\|f(x_k) + \nabla f(x_k)\Delta x_k\|^2}{d\Delta x_k} = 0 \tag{10.41a}$$

$$\Delta x_k = -\frac{f(x_k)f'(x_k)}{[f'(x_k)]^2} \tag{10.41b}$$

In the comparison of iterative step calculation in Eqs. (10.37b) and (10.41b), the denominator term in Eq. (10.41b), $[f'(x_k)]^2$ is an approximation of the Hessian matrix of Eq. (10.37b). The algorithm of the Gauss–Newton method is presented as follows.

Algorithm 4: Gauss–Newton method

Step 1: Initialization
Initialize x_0 and $f'(x_0,)$

Step 2: Iterative computation
Calculate step by (10.41)
Update step: $x_{k+1} = x_k + \Delta x_k$

Step 3: Satisfice terminal criterion?
If not, back to Step 2.

Step 4: Output

10.7.2 Application to Parameter Estimation for PMSMs

According to the estimation scheme described in Section 10.2, the least-squares objective functions, P_1, P_2, and P_3, are constructed to estimate L_s, R_s, and ψ_{pm}, respectively.

$$P_1(\hat{L}_s) = \sum_{k=1}^{n} \left[K_1 \left\| u_{d0}(k) - \omega_r(k) \hat{L}_s i_{q0}(k) \right\|^2 \right] \tag{10.42}$$

$$P_2(\hat{R}_s) = \sum_{k=1}^{n} \left[K_2 \left\| u_{d1}(k) - \hat{R}_s i_{d1}(k) + \omega_r(k) L_s i_{q1}(k) \right\|^2 \right] \tag{10.43}$$

$$P_3(\hat{\psi}_{pm}) = \sum_{k=1}^{n} \left[K_3 \left\| u_{q0}(k) - R_s i_{q0}(k) + \omega_r(k) \hat{\psi}_{pm} \right\|^2 \right] \tag{10.44}$$

where the superscript "^" denotes the estimated parameters. K_1, K_2, and K_3 are the gains of the objective functions. n is the measured data length.

The above three gradient-based methods, i.e., steepest descent, Newton, and Gauss–Newton methods are applied to minimize the above objective functions, where the initial value of each parameter is set to the same values. The estimated results are shown in Figure 10.13. As can be seen, the Newton and Gauss–Newton methods have much faster convergence speed than the steepest descent method since they consider the curvature information of objective functions. Meanwhile, after iterative calculation, all estimation results can be converged to the nominal values.

10.8 Particle Swarm Optimization

10.8.1 Basic Principle

Particle swarm optimisation (PSO) is a nature-inspired algorithm proposed by Kennedy and Eberhart in 1995 and was originally intended to simulate the social behaviour in a bird flock or fish school [36]. Nowadays, the PSO algorithm has been extensively applied in computation science to solve optimization problems by iteratively improving a candidate solution amongst a given measure of quality [37–39]. In addition, the PSO algorithm is a metaheuristic method, and thus, it does not require that the optimization problem must be differentiable, as it is required by classical gradient-based optimization methods described in Section 10.7.

Figure 10.13 Estimated (a) dq-axis inductance, (b) stator winding resistance, and (c) rotor PM flux linkage by gradient-based estimators.

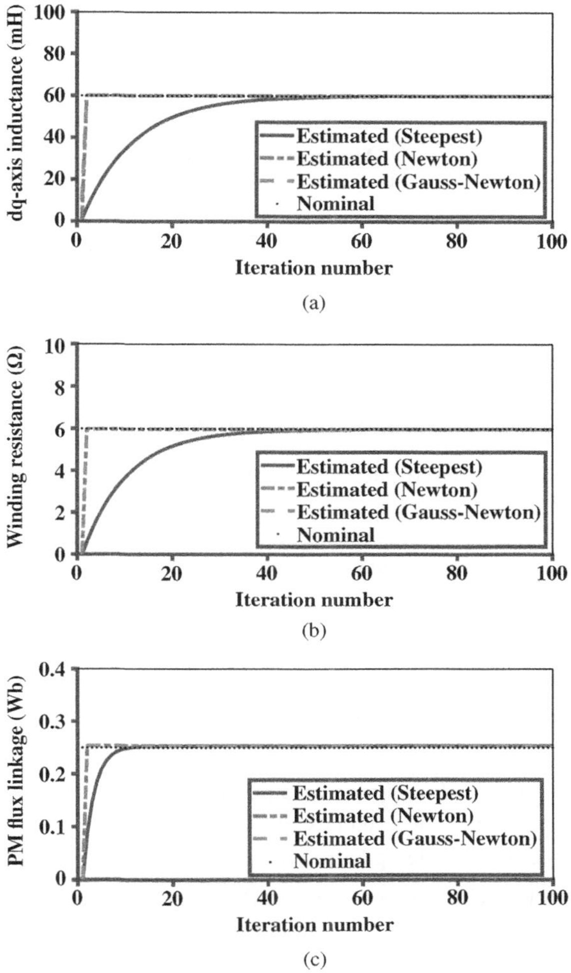

In the PSO algorithm, the estimation problem is solved based on a population of candidate solutions, i.e., particles. In a determined search space, all particles move toward the best-known positions by simple mathematical formulae (Eq. 10.45), which will be updated as better positions are found by other particles. Finally, the swarm will move toward the best solutions, i.e., the sequence of the estimated parameters is converged as follows:

$$v_{i+1} = wv_i + c_1r_1(x_p - x_i) + c_2r_2(x_g - x_i) \tag{10.45a}$$

$$x_{i+1} = x_i + v_i \tag{10.45b}$$

where w is the inertia weight; c_1, c_2 are the cognitive and social acceleration constants, respectively; x_p, x_g are the particle best and overall best locations, respectively; x_i is the ith particle location; v_i represents the particle velocity; r_1 and r_2 are the random numbers between 0 and 1.

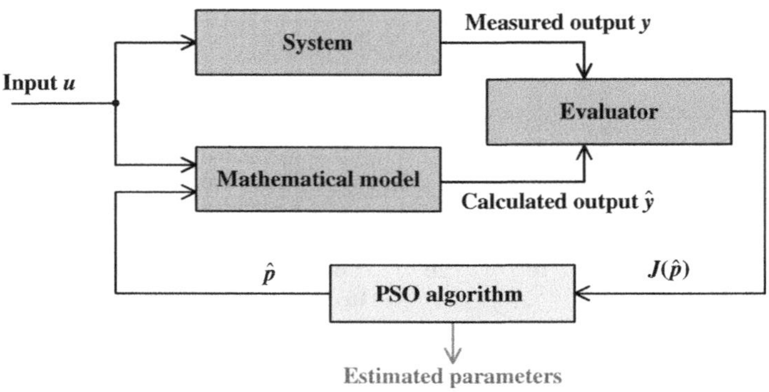

Figure 10.14 PSO-based parameter estimation.

10.8.2 Application to Parameter Estimation for PMSMs

The PSO-based parameter estimation system can be illustrated in Figure 10.14 and explained as follows.

Algorithm 5: PSO algorithm for parameter estimation

Step 1: Initialization
Initialize parameters, optimization settings (number of particles, maximum iterative number, terminal conditions).

Step 2: Data loading
Load measured data for parameter estimation.

Step 3: Update speed and position
Update particle's velocity in Eq. (10.45a)
Update particle's position in Eq. (10.45b)

Step 4: Update best personal and global best positions
If Fit(x_i) < Fit(x_p), then update $x_p \leftarrow x_i$
If Fit(x_p) < Fit(x_g), then update $x_g \leftarrow x_p$

Step 5: Evaluation of cost function
Evaluate the fitness of the global best position and the iterative number, if not satisfy the terminal condition, return to Step 3, or else go to Step 6.

Step 6: Output estimated parameters

According to the estimation scheme described in Section 10.2, the objective functions, P_4, P_5, and P_6, for estimating L_s, R_s, and ψ_{pm} are designed to minimize the respective absolute values.

$$P_4(\widehat{L}_s) = \sum_{k=1}^{n} \left[K_4 |u_{d0}(k) + \omega_r(k)\widehat{L}_s i_{q0}(k)|\right] \tag{10.46}$$

$$P_5(\widehat{R}_s) = \sum_{k=1}^{n} \left[K_5 |u_{d1}(k) - \widehat{R}_s i_{d1}(k) + \omega_r(k) L_s i_{q1}(k)|\right] \tag{10.47}$$

$$P_6(\widehat{\psi}_{pm}) = \sum_{k=1}^{n} \left[K_6 |u_{q0}(k) - R_s i_{q0}(k) + \omega_r(k)\widehat{\psi}_{pm}|\right] \tag{10.48}$$

where the superscript "^" denotes the estimated parameters. K_4, K_5, and K_6 are the gains of the objective functions. n is the measured data length.

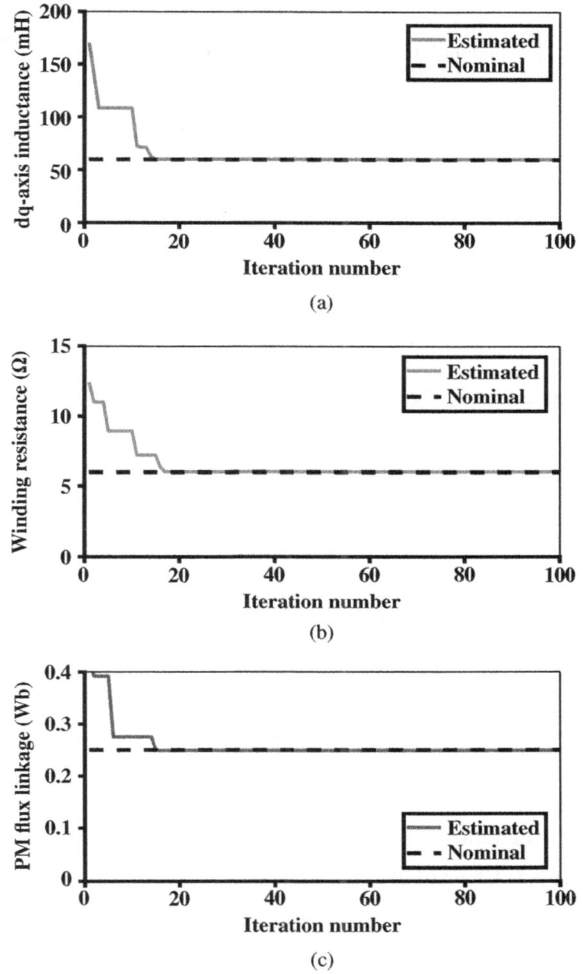

Figure 10.15 Estimated (a) *dq*-axis inductance, (b) stator winding resistance, and (c) rotor PM flux linkage by PSO estimators.

The PSO algorithm is applied to minimize the objective functions (Eq. 10.46)–(Eq. 10.48), and the estimated results are presented in Figure 10.15. As can be seen, all estimated values are converged to the nominal values within 20 iterations. It is worth mentioning that the convergence speed of the PSO algorithm is affected by the selected search space.

10.9 Genetic Algorithm

10.9.1 Basic Principle

Genetic algorithm (GA) is another popular metaheuristic method inspired by Charles Darwin's natural evolution theory. This algorithm reflects the process of natural selection, where the best individuals are selected for reproduction to reproduce for a new generation [40]. In computation science, GA is widely used to find the optimum based on biologically inspired operations, i.e., mutation, crossover, and selection, which will be repeated on the current population until the new

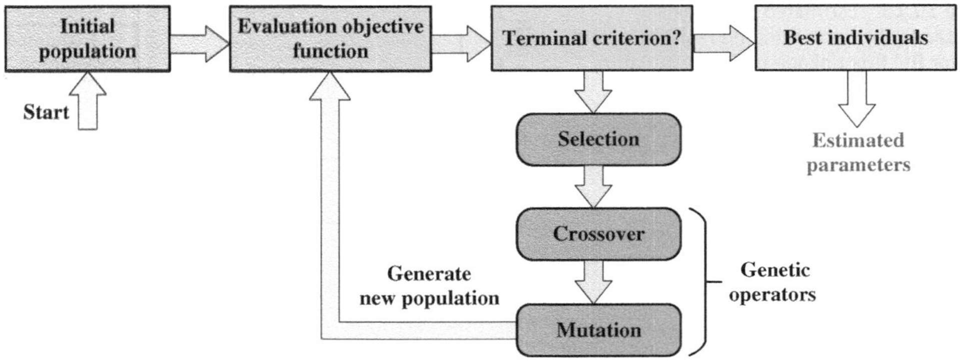

Figure 10.16 Flowchart of GA algorithm.

population is reproduced. It can be expected that the average fitness will be increased by the iterative reproduction of population.

For the classical GA, its optimization procedure is shown in Figure 10.16 and explained as follows.

Algorithm 6: Classical GA algorithm for parameter estimation

Step 1: Initialization

A group of solutions for objective function is usually generated randomly, which is designated as the gene pool of a population.

Step 2: Selection

In each successive generation, a portion of the existing population is selected to reproduce for a new generation by a fitness-based process.

Step 3: Crossover

The crossover operator is used to generate a "child" solution by combining the genetic information of a pair of "parent" solutions.

Step 4: Mutation

The mutation operator is applied to new generations with a certain possibility.

Step 5: Terminal criterion

The evolution is terminated when the offspring has no significant difference from the previous populations, or the maximum generation number is reached. Otherwise, go back to Step 2.

Step 6: Output estimated parameters

More recently, the quantum genetic algorithm (QGA) developed in Ref. [43] is the product of the combination of quantum computation and GA and has the advantages of faster convergence speed, without premature convergence, and enhanced exploration capability. The conventional GA uses binary strings or other classical representations encode solutions, while the solutions in QGA are represented by quantum bits. More specifically, every bit of gene is encoded in the form of Q-bit, i.e., $\begin{bmatrix}\alpha\\ \beta\end{bmatrix}$, where $|\alpha|^2 + |\beta|^2 = 1$. $|\alpha|^2$ and $|\beta|^2$ represent the probabilities that Q-bit can be found in the binary "0" and "1" states. The algorithm is presented as follows, as detailed in Ref. [44].

Algorithm 7: QGA algorithm for parameter estimation

Step 1: Initialization
Initialize a group of random solutions.

Steps 2–4: Selection, crossover, and mutation
Similar as the conventional GA in Algorithm 6.

Step 5: Quantum rotation gate operator
The quantum rotation gate operator makes the Q-bit genes converge to fitter states.

Step 6: Terminal criterion
Once the termination condition is satisfied, the evolutionary process will terminate or else will repeat Step 2–5.

Step 7: Output estimated parameters

10.9.2 Application to Parameter Estimation for PMSMs

The GA has also been widely used in parameter estimation for PMSMs [45–48]. For illustration, the GA algorithm is employed in the designed estimation scheme as described in Section 10.2. The constructed objective functions are the same as those in PSO in Eqs. (10.46–10.48). Figure 10.17 shows the variations of the estimated dq-axis inductance, winding resistance, and PM flux linkage with iteration number, which converge to the nominal values within 20 iterations.

10.10 Summary

In this chapter, several most popular algorithms have been introduced, including the adaptive algorithm-based methods (RLS, KF, MRAS, and ANN), the gradient-based methods (Steepest descent, Newton's and Gauss–Newton methods), and evolutionary algorithms (PSO and GA). All the above-mentioned methods have been applied to current injection–based parameter estimation schemes for a PMSM in the MATLAB platform.

Some brief remarks are presented as follows.

1. *Adaptive Algorithm–Based Methods*
   RLS is simple for implementation, while it only has one tuning parameter, i.e., forgetting factor, to compromise the convergence capability with the robustness to noise. Kalman and EKFs, used for linear and non-linear systems, are robust against measurement noise. However, the EKF has an inferior performance for highly non-linear systems. The unscented Kalman filter is tailored for non-linear problems, while its construction is more complex, and the computation burden is also increased accordingly. MRAS has a simple estimation structure, while its design is complicated, especially for multivariable estimation, and MRAS is also sensitive to measurement noise. ANN estimator has a simple structure, only consisting of a single adaptive linear neuron, and thus, it has the advantages of easy implementation and fast training process. However, due to its linear activation function, ANN is inappropriate for non-linear problems, which can only be solved by more advanced NN containing hidden layers. Meanwhile, if the input data is not scaled or normalized properly, the estimated results from the ANN estimator may be deteriorated.

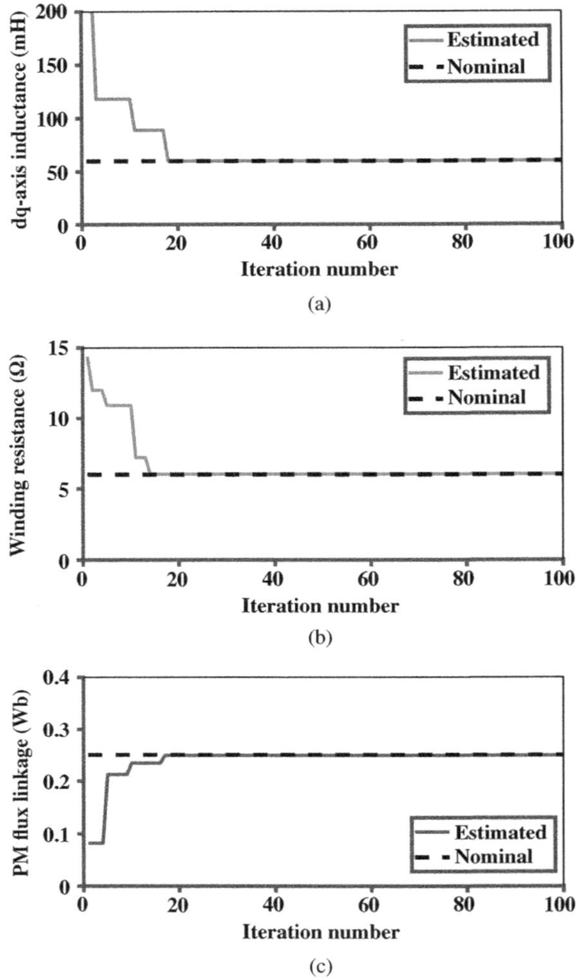

Figure 10.17 Estimated (a) *dq*-axis inductance, (b) stator winding resistance, and (c) rotor PM flux linkage by GA estimators.

2. *Gradient-Based Methods*
   The gradient-based methods are reliable and fast converged when the number of unknown parameters is small. However, it is hard to implement large-scale multivariable optimization due to mathematical difficulty, i.e., the Jacobian and Hessian matrices are not always available and computable. Moreover, when the optimization problem is non-convex, the gradient-based methods are easily stuck in the stationary points, e.g., local minimum and saddle points.

3. *Evolutionary Algorithm–Based Methods*
   The evolutionary algorithms, e.g., PSO and GA, have better compatibilities for multi-parameter global optimization since they maintain diverse solutions. However, the evolutionary algorithms converge more slowly than the conventional gradient-based methods.

Table 10.1 summarizes the merits/demerits, the absolute percentage errors, and the computation time of all algorithms. In addition, the computation time is recorded via a standard PC (hardware: Intel(R) Core(TM) i5-10300H CPU @ 2.50GHz).

It can be easily found that all estimation algorithms have satisfactory estimation performance. The absolute percentage errors of all algorithms are kept within 1%, while the evolutionary algorithms,

Table 10.1 Comparison of various estimation algorithms.

Algorithm	Absolute percentage error	Computation time	Merits/demerits
RLS	0.40%	3.2 ms	✔ Simple implementation and fast convergence ✘ Sensitive to measurement noise/outliers
(Extended) Kalman filters	0.31%	5.1 ms	✔ Robust against measurement noise ✘ Sensitive to model setup ✘ Not applicable for highly non-linear system (extended KF)
MRAS	0.34%	1.6 ms	✔ Simple structure and high computational efficiency ✘ Complicated design of adaptive law
ANN	0.36%	2.8 ms	✔ Simple structure and low memory requirement ✘ Not applicable for non-linear system ✘ Sensitive to input scaling
Gauss–Newton method	0.58%	1.4 ms	✔ Fast computation and good local convergence capability ✘ Sensitive to initial condition and number of identified parameters
PSO	0.23%	27 ms	✔ Easy implementation with fewer tuning parameters ✘ Inferior searchability for local optimum
GA	0.21%	35 ms	✔ Large scale of search space and parallelism capability ✘ Risk of premature convergence ✘ Difficulty in tuning parameters

i.e., SPO and GA, exhibit the lowest errors due to their excellent global optimization capabilities. Clearly, the appropriate establishment of an estimation scheme plays a critical role in accuracy.

In terms of computation time, all adaptive and gradient-based methods exhibit similar computational efficiency, whereas the MRAS and Gauss–Newton methods consume the least computation time due to their simple update equations. In comparison, the evolutionary algorithms, i.e., SPO and GA, are typically computationally expensive since they require not only a selection of variations but also a fitness function to evaluate variants. To ease the computation burden, the search space needs to be reduced. Besides, the multi-step estimation scheme, which are described in Section 10.2 and employed in Chapters 5 and 6, reduces the unknown parameter number at each estimation procedure, which could also significantly ease the computation burden.

Overall, it can be concluded that although the estimation algorithms are obtained based on different simple or complex principles, the precision of the estimation is less related to the nature of the estimator once it has been established appropriately.

References

1 Z. Q. Zhu, D. Liang, and K. Liu, "Online parameter estimation for permanent magnet synchronous machines: An overview," *IEEE Access*, vol. 9, pp. 59059–59084, 2021.

2 C. F. Gauss, *Theoria Motus Corporum Celestium: In Sectionibus Conicis Solem Ambientium*. Bruxelles: Culture et Civilisation, 1968.

3 R. L. Plackett, "Some theorems in least squares," Biometrika, vol. 37, no. 1/2, pp. 149–157, June 1950.

4 A. H. Sayed and T. Kailath, "A state-space approach to adaptive RLS filtering," *IEEE Signal Process. Mag.*, vol. 11, no. 3, pp. 18–60, Jul. 1994.

5 S. Kallio, J. Karttunen, P. Peltoniemi, P. Silventoinen, and O. Pyrhönen, "Online estimation of double-star IPM machine parameters using RLS algorithm," *IEEE Trans. Ind. Electron.*, vol. 61, no. 9, pp. 4519–4530, Sept. 2014.

6 G. Feng, C. Lai, and N. C. Kar, "A novel current injection-based online parameter estimation method for PMSMs considering magnetic saturation," *IEEE Trans. Magn.*, vol. 52, no. 7, pp. 1–4, Jul. 2016.

7 S. J. Underwood and I. Husain, "Online parameter estimation and adaptive control of permanent-magnet synchronous machines," *IEEE Trans. Ind. Electron.*, vol. 57, no. 7, pp. 2435–2443, July 2010.

8 C. Lian, F. Xiao, J. Liu, and S. Gao, "Parameter and VSI nonlinearity hybrid estimation for PMSM drives based on recursive least square," *IEEE Trans. Transport. Electrif.*, vol. 9, no. 2, pp. 2195–2206, June 2023.

9 S. Nalakath, M. Preindl, and A. Emadi, "Online multi-parameter estimation of interior permanent magnet motor drives with finite control set model predictive control," *IET Electr. Power Appl.*, vol. 11, no. 5, pp. 944–951, 5 2017.

10 R. E. Kalman, "A new approach to linear filtering and prediction problems," *J. Basic Eng.*, vol. 82, no. 1, pp. 35–45, 1960.

11 G. Welch and G. Bishop, "An introduction to the Kalman filter," Uni. North Carolina at Chapel Hill, Chapel Hill, NC, USA, 95–41, 1995.

12 S. Bolognani, L. Tubiana, and M. Zigliotto, "Extended Kalman filter tuning in sensorless PMSM drives," *IEEE Trans. Ind. Appl.*, vol. 39, no. 6, pp. 1741–1747, Nov.–Dec. 2003.

13 S. J. Julier, J. K. Uhlmann, and H. F. Durrant-Whyte, "A new approach for filtering nonlinear systems," In *Proc. 1995 American Control Conf. - ACC'95*, Seattle, WA, USA, 1995, pp. 1628–1632 vol. 3.

14 Z.Q. Zhu, X. Zhu, P.D. Sun, and D. Howe, "Estimation of winding resistance and PM flux-linkage in brushless AC machines by reduced-order extended Kalman filter," In *Proc. 2007 IEEE Int. Conf. Network., Sens. Contr.*, UK, 15–17 April 2007, pp. 740–745.

15 F. Auger, M. Hilairet, J. M. Guerrero, E. Monmasson, T. Orlowska-Kowalska, and S. Katsura, "Industrial applications of the Kalman filter: A review," *IEEE Trans. Ind. Electron.*, vol. 60, no. 12, pp. 5458–5471, Dec. 2013.

16 S. J. Julier, J. K. Uhlmann, and H. F. Durrant-Whyte, "A new approach for filtering nonlinear systems," In *Proc. 1995 American Control Conf. (ACC-95)*, Seattle, USA, 1995, pp. 1628–1632 vol. 3.

17 S. Jafarzadeh, C. Lascu and M. S. Fadali, "Square root unscented Kalman Filters for state estimation of induction motor drives," *IEEE Trans. Ind. Appl.*, vol. 49, no. 1, pp. 92–99, Jan./Feb. 2013.

18 T. Boileau, N. Leboeuf, B. Nahid-Mobarakeh, and F. Meibody-Tabar, "Online identification of PMSM parameters: parameter identifiability and estimator comparative study," *IEEE Trans. Ind. Appl.*, vol. 47, no. 4, pp. 1944–1957, July-Aug. 2011.

19 H. Whitaker, J. Yamron, and A. Kezer "*Design of Model-Reference Adaptive Control Systems for Aircraft*," Massachusetts Inst. Tech., Instrumentation Laboratory, 1958.

20 K. Liu, Q. Zhang, Z.Q. Zhu, J. Zhang, A.W. Shen, and P. Stewart, "Comparison of two novel MRAS strategies for identifying parameters in permanent magnet synchronous motors," *Int. J. Automat. Comput.*, vol. 7, no. 4, pp. 516–524, 2010.

21 T. Boileau, B. Nahid-Mobarakeh, and F. Meibody-Tabar, “On-line identification of PMSM parameters: Model-reference vs EKF,” In *Proc. IEEE Ind. Appl. Soc. Annu. Meeting*, Edmonton, AB, Canada, Oct. 2008, pp. 1–8.

22 P. Cao, X. Zhang, and S. Yang, “A unified-model-based analysis of MRAS for online rotor time constant estimation in an induction motor drive,” *IEEE Trans. Ind. Electron.*, vol. 64, no. 6, pp. 4361–4371, June 2017.

23 O. C. Kivanc and S. B. Ozturk, “Sensorless PMSM drive based on stator feedforward voltage estimation improved with MRAS multiparameter estimation,” *IEEE/ASME Trans. Mechatron.*, vol. 23, no. 3, pp. 1326–1337, June 2018.

24 C. Attaianese, A. Damiano, G. Gatto, I. Marongiu, and A. Perfetto, “Induction motor drive parameters identification,” *IEEE Trans. Power Electron.*, vol. 13, no. 6, pp. 1112–1122, Nov. 1998.

25 B. Widrow and M. Lehr, “30 years of adaptive neural networks: Perceptron, Madaline, and backpropagation,” *Proc. IEEE*, vol. 78, no. 9, pp. 1415–1442, Sep. 1990.

26 K. Liu, Q. Zhang, J. Chen, Z.Q. Zhu, and J. Zhang, “Online multiparameter estimation of nonsalient-pole PM synchronous machines with temperature variation tracking,” *IEEE Trans. Ind. Electron.*, vol. 58, no. 5, pp. 1776–1788, May 2011.

27 K. Liu, Z.Q. Zhu, and D. Stone, “Parameter estimation for condition monitoring of PMSM stator winding and rotor permanent magnets,” *IEEE Trans. Ind. Electron.*, vol. 60, no. 12, pp. 5902–5913, Apr. 2013.

28 K. Liu, Z.Q. Zhu, and D.A. Stone, “Online estimation of rotor flux linkage and voltage source inverter nonlinearity in permanent magnet synchronous machine drives,” *IEEE Trans. Power Electron.*, vol. 29, no. 1, pp. 418–427, 2014.

29 K. Liu and Z. Q. Zhu, “Position-offset-based parameter estimation using the Adaline NN for condition monitoring of permanent-magnet synchronous machines,” *IEEE Trans. Ind. Electron.*, vol. 62, no. 4, pp. 2372–2383, April 2015.

30 A. Björck, *Numerical Methods for Least Squares Problems*. Philadelphia, PA: SIAM, Soc. Ind. Appl. Math., 1996.

31 J. Meza, “Steepest *Descent*,” Lawrence Berkeley National Laboratory, USA, 2010.

32 P. R. Adby and D. M. A. H., *Introduction to Optimization Methods*. London: Chapman and Hall, 1974.

33 R. Fletcher, *Practical Methods of Optimization*. Hoboken: Wiley, 2013.

34 J. Nocedal and S. J. Wright, *Numerical Optimization*. Berlin: Springer, 2006.

35 A. Perera and R. Nilsen, “Recursive prediction error gradient-based algorithms and framework to identify PMSM parameters online,” *IEEE Trans. Ind. Appl.*, vol. 59, no. 2, pp. 1788–1799, 2023.

36 J. Kennedy, and R. Eberhart, Particle swarm optimization. In *Proc. IEEE Int. Conf. Neural Networks*, Perth, Australia, 1995, vol. 4, pp. 1942–1948.

37 Z. Liu, H. Wei, X. Li, K. Liu, Q. Zhong, Global identification of electrical and mechanical parameters in PMSM drive based on dynamic self-learning PSO. *IEEE Trans. Power Electron.*, vol. 33, no. 12, pp. 10858–10871, 2018.

38 Z.H. Liu, X.H. Li, L.H. Wu, S.W. Zhou, and K. Liu, GPU-accelerated parallel coevolutionary algorithm for parameters identification and temperature monitoring in permanent magnet synchronous machines. *IEEE Trans. Ind. Informat*, vol. 11, no. 5, pp. 1220–1230, 2015.

39 O. Sandre-Hernandez, R. Morales-Caporal, J. Rangel-Magdaleno, H. Peregrina-Barreto, and J.N. Hernandez-Perez, Parameter identification of PMSMs using experimental measurements and a PSO algorithm. *IEEE Trans. Instrum. Meas.* vol. 64, no. 8, pp. 2146–2154, 2015.

40 M. Mitchell, *An Introduction to Genetic Algorithms*. Cambridge, MA: MIT, 1998.

41 S. Katoch, S.S. Chauhan, and V. Kumar, A review on genetic algorithm: Past, present, and future. *Multimed. Tools Appl.* vol. 80, no. 5, pp. 8091–8126, 2020.

42 D. Liu, Mathematical modeling analysis of genetic algorithms under schema theorem. *J. Comput. Methods Sci. Eng.* vol. 19, no. S1, pp. 131–137, 2019.

43 R. Ibarrondo, G. Gatti, and M. Sanz, Quantum genetic algorithm with individuals in multiple registers. [Online]. Available at https://arxiv.org/abs/2203.15039.

44 K.H. Han, and J.H. Kim, Quantum-inspired evolutionary algorithm for a class of combinatorial optimization. *IEEE Trans. Evol. Comput.* vol. 6, no. 6, pp. 580–593, 2002.

45 P. Pillay, R. Nolan, and T. Haque, Application of genetic algorithms to motor parameter determination for transient torque calculations. *IEEE Trans. Ind. Appl.* vol. 33, no. 5, pp. 1273–1282, 1997.

46 F. Alonge, F. DIppolito, G. Ferrante, and F.M. Raimondi, Parameter identification of induction motor model using genetic algorithms. In: *Proc. Inst. Elect. Eng. Contr. Theory Appl.* vol. 145, no. 6, pp. 587–593, 1998.

47 K. Liu, and Z.Q. Zhu, Quantum genetic algorithm-based parameter estimation of PMSM under variable speed control accounting for system identifiability and VSI nonlinearity. *IEEE Trans. Ind. Electron.* vol. 62, no. 4, pp. 2363–2371, 2015.

48 D.C. Aliprantis, S.D. Sudhoff, and B.T. Kuhn, Genetic algorithm-based parameter identification of a hysteretic brushless exciter model. *IEEE Trans. Energy Convers.* vol. 21, no. 1, pp. 148–154, 2006.

11

Applications of Parameter Estimation

11.1 Introduction

Generally, the application of online parameter estimation techniques is pivotal for addressing the challenges posed by complex electromagnetic, drive control, and thermal interactions within PMSM drive systems. By real-time refining internal models, a PMSM can be regarded as a non-linear white-box model to some extent, and thus, the machine can be driven to the optimal performance accounting for varying load conditions and temperature changes.

Under a sensorless drive system, accurate knowledge of machine parameters according to the actual machine characteristics can significantly enhance the precision of rotor position and speed estimation together with accurate torque and flux information. This is crucial for achieving reliable machine operation without the need for a physical encoder, especially for applications where high control performance is required, such as robotics, electric vehicles, and industrial automation.

Furthermore, the online parameter estimation methods contribute eminently to thermal monitoring techniques. By continuously monitoring and updating critical machine parameters, the potential risks of thermal overheating can be detected at an early stage. Therefore, the reliability and lifespan of PMSMs can be enhanced, and the full potential of PMSMs across a spectrum of applications can be released.

In addition, accurate machine parameters provide a baseline of the healthy mode, and thus, any deviations from the baseline can indicate different faults, e.g., inter-turn short circuit, PM demagnetization, and rotor eccentricity.

Therefore, this chapter describes several important applications of the online parameter estimation techniques, including the following:

1. Improvement in control performance:
   - Design of PI regulators for field-oriented control (FOC)
   - Determination of maximum torque per Ampere (MTPA) current trajectory
2. Performance improvement and application under sensorless control
   - Performance improvement in sensorless control
   - Applications of parameter estimation methods under sensorless control
3. Machine torque estimation by HF signal injection considering cross-coupling effect
4. Thermal condition monitoring of winding and PMs
5. Fault diagnosis
   - Inter-turn short-circuit fault
   - Demagnetization fault

Parameter Estimation of Permanent Magnet Synchronous Machines, First Edition. Zi Qiang Zhu, Kan Liu, and Dawei Liang.

The principles of all the above applications and the importance of the machine parameters on their performance will be described and discussed. The experimental results for most applications will also be provided for illustration.

11.2 Improvement in Control Performance

Parameter estimation techniques are now essential for high-performance PMSM drive systems and have been embodied in commercial industrial drives for PMSMs. The accurate knowledge of machine electrical and mechanical parameters plays a critical role in the control performance, especially for the FOC and model predictive control (MPC). More specifically, the FOC system requires several PI regulators to control the current and rotor speed, whereas the design of these PI regulators requires accurate machine parameters [1–4]. Meanwhile, as widely recognized, parameter dependency is the main drawback of the MPC, since the machine's parameters directly affect the selection of voltage vectors [5] and [6], and thus, affect the control performance, e.g., steady-state current/torque errors, current distortion, and torque ripple, etc. In addition, the PMSM parameters are also used to determine the optimal current trajectory under MTPA/MTPV strategies. If the parameter variations caused by the magnetic saturation and temperature rise are not taken into account, the PMSM will be under-utilized, and the efficiency will also be deteriorated.

This section presents two examples of the applications of parameter estimation in the improvements in control performance, i.e., the design of PI regulators for the FOC and the determination of MTPA current trajectory. The experimental results are also presented.

11.2.1 Design of PI Regulators for FOC

FOC is the most popular control strategy and is widely used in various applications, which include several PI regulators for the control of current and speed loops. Therefore, the PI constants play a critical role in the control performance. In general, they can be determined based on the machine parameters according to the classical control theorem [1–3], i.e., the values of winding resistance R_s and dq-axis inductances L_d and L_q, are used to design current PI regulators, whereas the values of rotor PM flux linkage ψ_{pm} and mechanical parameters, i.e., viscous friction coefficient F and combined moment of inertia J, are used to design speed PI regulator [2, 3]. However, due to inaccessibility or uncertainties of machine parameters, the PI constants can also be tuned by the Ziegler–Nichols method [4]. However, the Ziegler–Nichols method requires a lot of trial-and-error processes, and cannot guarantee the tuned PI constants are optimal, i.e., the unrobust and unstable risks still exist [7].

11.2.1.1 Parameter-Based PI Regulator Design

The block diagram of FOC is shown in Figure 11.1, and the transfer functions of current and speed loops are shown in Figure 11.2, where K_p, $K_i = K_p/\tau_c$ and K_n and $K_{ni} = K_n/\tau_n$ are the PI constants of the current and speed regulators, respectively. In the current loop, T_s, T_d, and T_{if} are the time delays of the inverter, PWM, and measurement, respectively. In the speed loop, K_e is the torque constant and T_ω is the time delay of the rotor position measurement.

First, the open- and closed-loop transfer functions, i.e., G and C, of the dq-axis current loop and the speed loop can be written as follows:

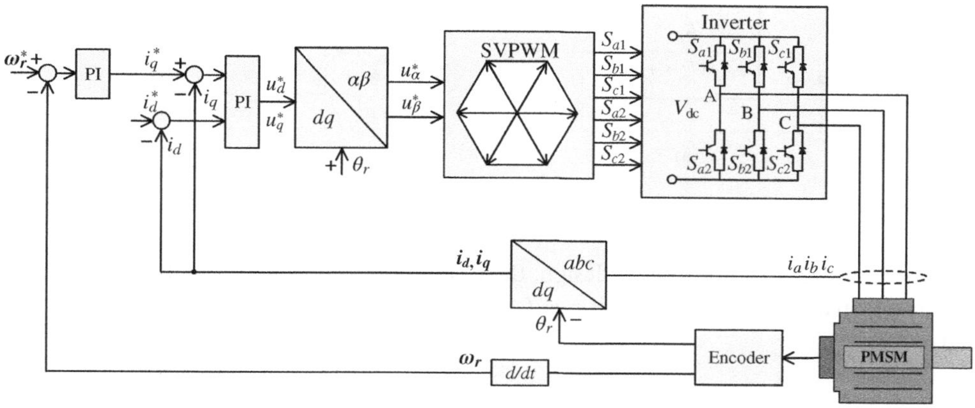

Figure 11.1 Block diagram of FOC for PMSM.

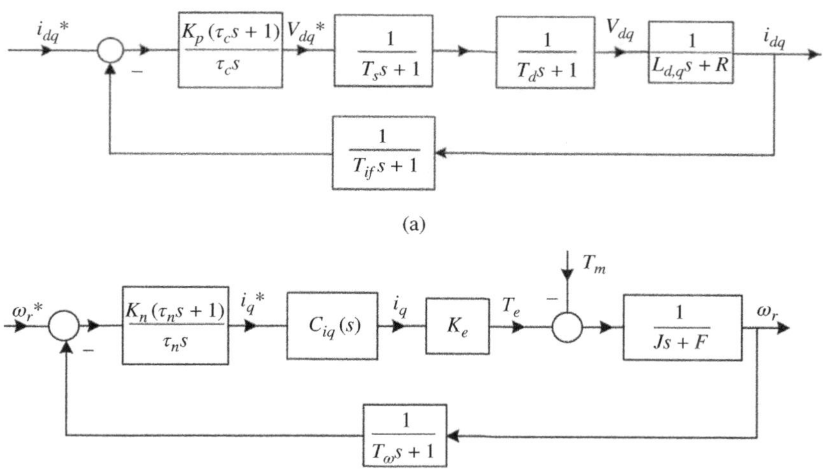

(a)

(b)

Figure 11.2 Complete transfer function of (a) *dq*-axis current loop and (b) speed loop in FOC [3], IEEE.

Current loop:

$$G_i(s) = \frac{K_p(\tau_c s + 1)}{\tau_c \left(L_{d,q} s + R_s\right)(T_s s + 1)(T_d s + 1)\left(T_{if} s + 1\right)} \tag{11.1a}$$

$$C_i(s) = \frac{G_i(s)}{1 + G_i(s)} \tag{11.1b}$$

Speed loop:

$$G_s(s) = \frac{K_e K_n(\tau_n s + 1)\, C_{iq}(s)}{\tau_n s(T_\omega s + 1)(Js + F)} \tag{11.2a}$$

$$C_s(s) = \frac{G_s(s)}{1 + G_s(s)} \tag{11.2b}$$

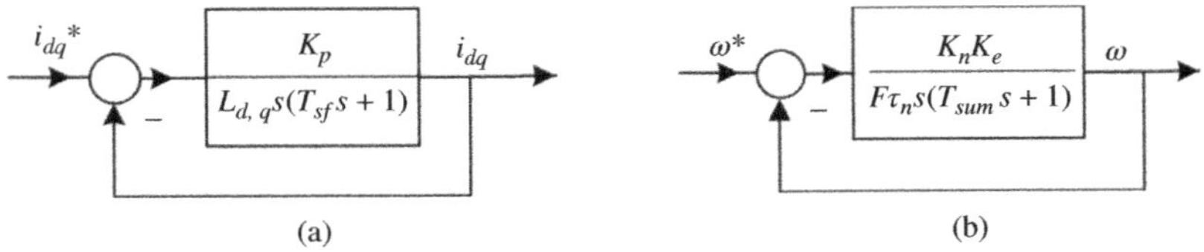

Figure 11.3 Simplified transfer functions [3], IEEE: (a) dq-axis current loop and (b) speed loop.

Furthermore, to facilitate the system design, the first-order systems with small time constants in the current loop are combined. Thus, if $\tau_c = L_{d,q}/R_s$ and $T_{sf} = T_s + T_d + T_{if}$, Eq. (11.2) can be simplified as follows:

$$G_i(s) = \frac{K_p}{L_{d,q}s(T_{sf}s+1)} \tag{11.3a}$$

$$C_i(s) = \frac{K_p}{L_{d,q}T_{sf}} \cdot \frac{1}{s^2 + \frac{1}{T_{sf}}s + \frac{K_p}{L_{d,q}T_{sf}}} = \frac{\omega_n^2}{s^2 + 2\xi\omega_n s + \omega_n^2} \tag{11.3b}$$

As can be observed, the closed-loop transfer function of current loop can be simplified as a second-order system, as shown in Figure 11.3a, of which the $\omega_n = \sqrt{K_p/(L_{d,q}T_{sf})}$ is the natural frequency with the damping ratio $\xi = 0.5\sqrt{L_{d,q}/(K_pT_{sf})}$.

For the transfer function of speed loop, the q-axis current loop term is simplified as a first-order system due to its small time constant, Figure 11.3b, which is given as follows:

$$G_{iq}(s) = \frac{K_p}{L_qT_{sf}}\frac{1}{s^2 + \frac{1}{T_{sf}}s + \frac{K_p}{L_{d,q}T_{sf}}} \approx \frac{K_p}{L_qs + K_p} \tag{11.4}$$

Afterwards, by substituting Eq. (11.4) into Eq. (11.2), the open-loop transfer function of the speed loop is rewritten as follows:

$$G_s(s) = \frac{K_eK_n(\tau_ns+1)}{F\tau_ns(T_\omega s+1)\left(\frac{L_q}{K_q}s+1\right)\left(\frac{J}{F}s+1\right)} \approx \frac{K_eK_n}{F\tau_ns(T_{\text{sum}}+1)} \tag{11.5}$$

where $T_{\text{sum}} = T_\omega + 2T_{sf} \cdot \tau_n = J/F$.

Therefore, the optimized PI constants for dq-axis current and speed regulators can be derived by using the optimal damping ratio $\xi = 1/\sqrt{2}$ [8] and expressed as follows:

$$\textit{Current regulators:} \left\{ K_p = \frac{L_{d,q}}{2T_{sf}} \quad K_i = \frac{K_p}{\tau_c} = \frac{R_s}{2T_{sf}} \right. \tag{11.6}$$

$$\textit{Speed regulator:} \left\{ K_n = \frac{J}{2K_eT_{\text{sum}}} \quad K_{ni} = \frac{K_n}{\tau_n} = \frac{F}{2K_eT_{\text{sum}}} \right. \tag{11.7}$$

Therefore, the machine parameters, including dq-axis inductances, stator winding resistance, rotor flux linkage, combined inertia, and friction coefficient, need to be known for calculating optimized PI constants of the current or speed regulator.

11.2.1.2 Parameter Estimation

An exemplary parameter estimation procedure is shown in Figure 11.4, where Prototype Machine 1 is employed for illustration.

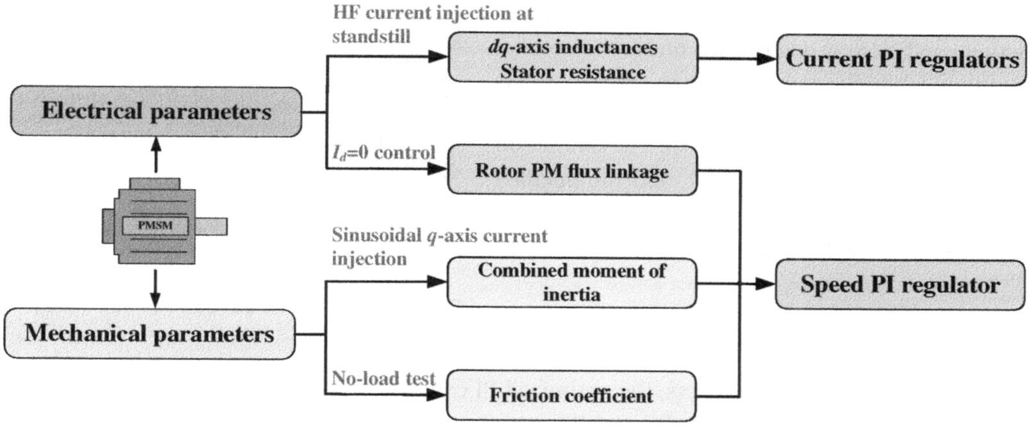

Figure 11.4 Exemplary estimation procedure for design of PI regulators.

First, in terms of electrical parameters, the high-frequency sinusoidal current injection–based offline estimation method (see Chapter 7) is used to estimate the stator winding resistance, the distorted voltage due to VSI non-linearity, and the dq-axis inductances, as shown in Figure 11.5. The injected dq-axis currents are $i_d = \sin(80\pi t)$ and $i_q = \cos(160\pi t)$. Afterwards, the rotor PM flux linkage is estimated during operation under $i_d = 0$ control, where the stator resistance is fixed to the estimated value, as shown in Figure 11.6.

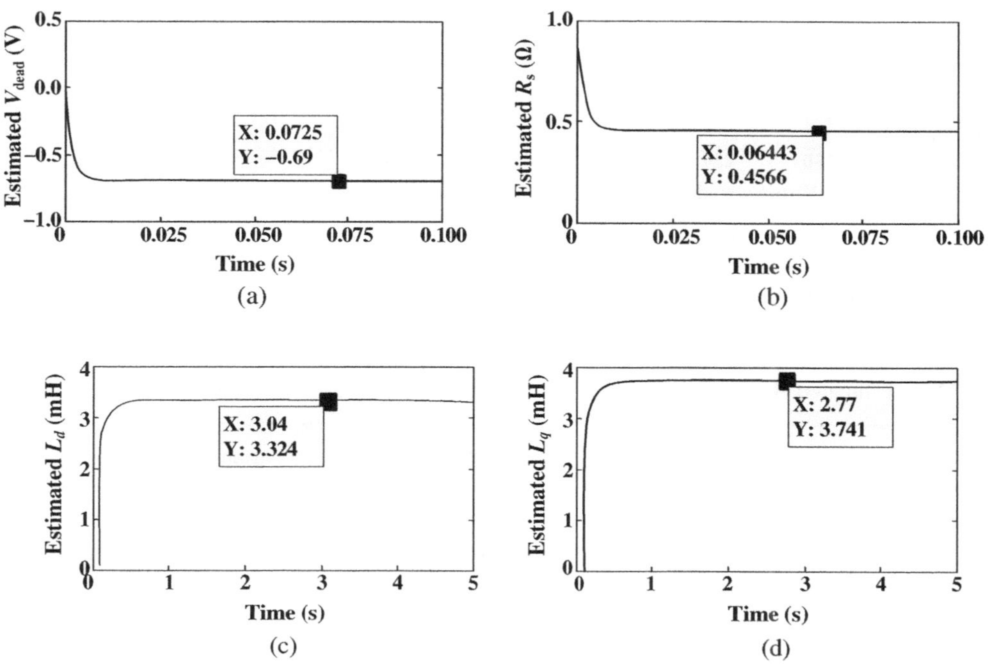

Figure 11.5 Estimated results based on HF current injection–based offline estimation at standstill [3], IEEE: (a) estimated distorted voltage V_{dead}, (b) estimated stator resistance R_s, (c) estimated d-axis inductance L_d, and (d) estimated q-axis inductance L_q.

Second, in terms of mechanical parameters, the friction coefficient is determined under the no-load condition, while the combined moment of inertia is estimated based on the injection of a sinusoidal q-axis current, i.e., $i_q = 1.5\sin(10\pi t)$. The estimation results are shown in Figure 11.7, and the detailed description of the mechanical parameter estimation method is detailed in Chapter 9.

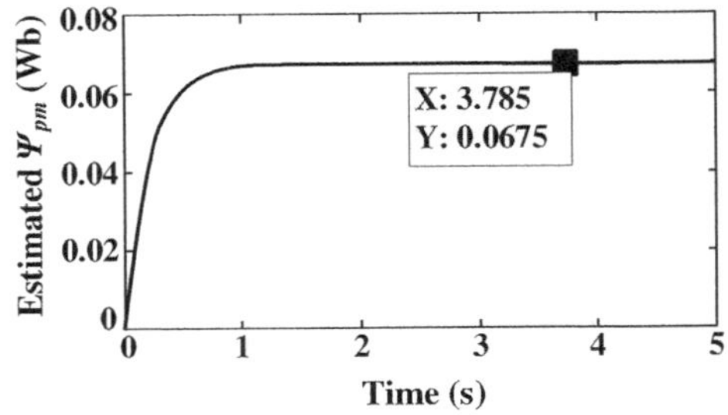

Figure 11.6 Estimated rotor PM flux linkage by fixing resistance ($i_d = 0$ A, $i_q = 2$ A) [3], IEEE.

11.2.1.3 Designed PI Regulators and Experimental Performance

Based on the estimated parameters, the optimized PI constants are listed in Table 11.1, which are further implemented experimentally, and the performance of the designed current and speed PI regulators are shown in Figures 11.8 and 11.9, respectively.

In Figure 11.8a, the q-axis current is fixed to 0A by the PI regulator, while there is an abrupt change in the d-axis reference current from 0A to 1A. It is obvious that the settling time of the d-axis current is smaller, and a slight overshoot at the beginning is acceptable. Besides, the step response in q-axis current is presented in Figure 11.8b, which could achieve a similar dynamic performance.

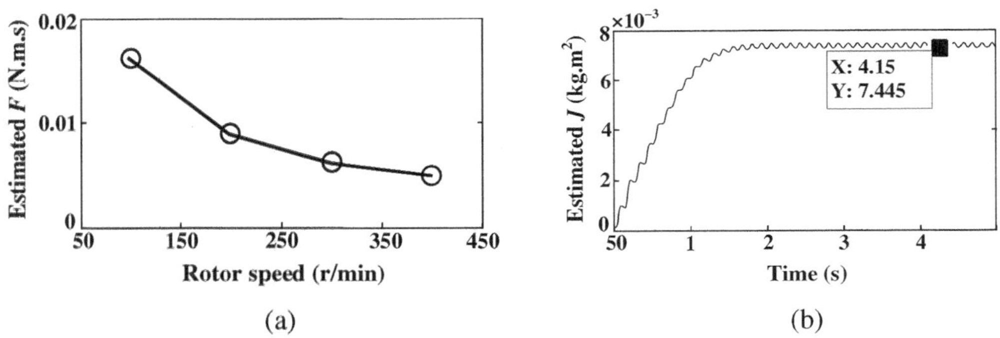

Figure 11.7 Estimated mechanical parameters [3], IEEE: (a) estimated friction coefficient and (b) estimated combined moment of inertia.

Table 11.1 Estimated electrical/mechanical parameters and optimized PI constants.

Estimated parameter	Value	PI regulator	Constant	Value
Stator resistance	0.457Ω	d-axis current regulator	$K_{p,d}$	11.45
d-axis inductance	3.74mH		$K_{i,d}$	1575.90
q-axis inductance	3.32mH	q-axis current regulator	$K_{p,q}$	12.90
Rotor PM flux linkage	67.5mWb		$K_{i,q}$	1575.90
Combined moment of inertia	0.00745kg m^2	Speed regulator	K_n	1.11
Friction coefficient	0.0092N ms		K_{ni}	1.365

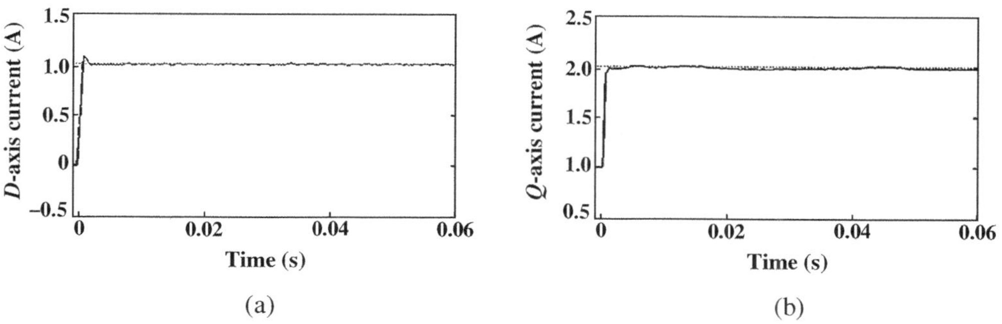

Figure 11.8 Step response of dq-axis currents by using designed current PI regulators [3]: (a) D-axis current and (b) Q-axis current.

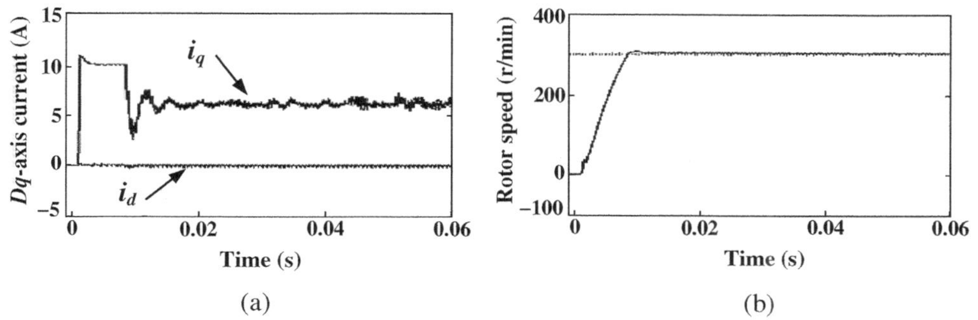

Figure 11.9 Step response of rotor speed by using designed speed PI regulator [3], IEEE: (a) dq-axis current responses and (b)rotor speed change.

In terms of the speed PI regulator, the step response of rotor speed is shown in Figure 11.9. It is obvious that a quite small overshoot will be achieved. In addition, as can be seen from the q-axis currents shown in Figures 11.9a, the q-axis current will reach the upper limit (10 A) immediately at the beginning of step change due to the fast response of the designed q-axis current and speed regulators, which will generate a maximum torque to accelerate the machine rotor to reference speed during the start-up.

11.2.2 Determination of MTPA Current Trajectory

For maximum torque production below the base speed region [9–14], the optimal current trajectory, i.e., MTPA strategy, can be obtained as follows:

$$T_e = \frac{3}{2}p\left[\psi_{pm} i_q + (L_d - L_q) i_d i_q\right] \tag{11.8a}$$

$$\text{with} \begin{cases} \dfrac{\partial T_e}{\partial i_d} = 0 \quad \dfrac{\partial T_e}{\partial i_q} = 0 \\ i_d^2 + i_q^2 \le I_{\max}^2 \end{cases} \tag{11.8b}$$

Therefore, the optimized dq-axis currents, i.e., i_d^* and i_q^*, are expressed as follows:

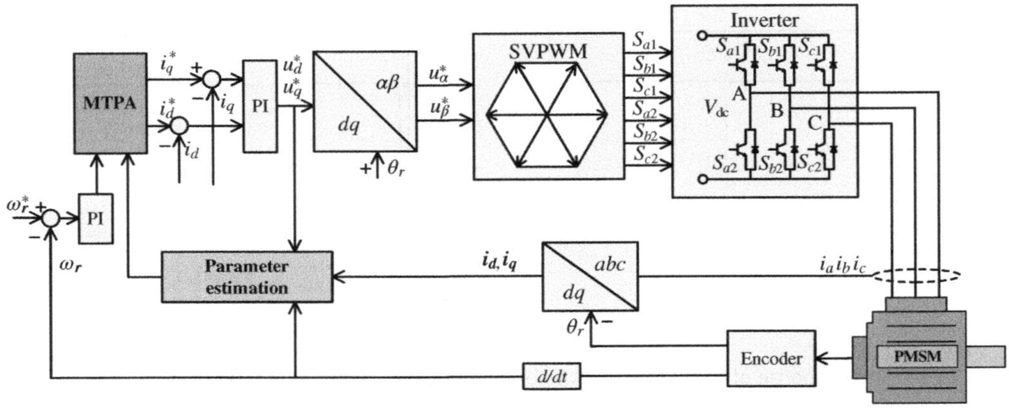

Figure 11.10 Block diagram of FOC-based MTPA control.

$$i_d^* = \frac{-\psi_{pm} + \sqrt{\psi_{pm}^2 + 4(L_d - L_q)^2 i_q^2}}{2(L_d - L_q)} \tag{11.9a}$$

$$i_q^* = \sqrt{I_{max}^2 - i_d^{*2}} \tag{11.9b}$$

It is evident that the optimal current trajectory is highly related to the electrical parameters, including L_d, L_q, and ψ_{pm}. Normally, the MTPA scheme utilizes the fixed nominal parameters to determine the optimal trajectory [9] and [10]. However, as described in the previous chapters, the above parameters are strongly affected by the magnetic saturation and working points. Therefore, it has been reported in Refs [11–16] that those parameters need to be updated according to different load conditions whether they are measured offline or estimated online, as shown in Figure 11.10. Otherwise, the optimal current trajectory will deviate from the actual optimal one, and thus, the torque production will be smaller than the maximum value and the efficiency will deteriorate.

Taking Prototype Machine 2 as an example, when the dq-axis inductances and the rotor PM flux linkage are estimated based on the POPE method in Chapter 5, Figure 11.11 shows the derived MTPA trajectories under ±20% errors in L_d, L_q, and ψ_{pm} and corresponding output torques [8]. As can be observed, the parameter errors in L_d and L_q deviate from the optimal trajectory. Meanwhile, when the estimated parameters are in high accuracy, the output torque agrees well with the theoretical maximum torque. In contrast, if there are ±20% errors in L_d, L_q, and ψ_{pm}, the output torques are 5%, 5%, and 3% lower than the actual maximum torque, respectively.

Moreover, based on Prototype Machine 2, Figure 11.12 compares the MTPA trajectories obtained by fixing the parameters to the nominal values (see Appendix B) and by the estimated dq-axis flux linkages accounting for the magnetic saturation in Chapter 7. As can be seen, the latter MTPA trajectory can achieve a 3.2% higher maximum output torque [16]. Therefore, it can be concluded that the accurate knowledge of machine parameters could benefit maximum torque production and improve efficiency.

11.3 Improvement in Sensorless Control

Sensorless PMSM drives have been receiving increased interest for use in industrial applications since they do not need position sensors, and thus, reduce the cost and size of the drive system while increasing the overall system reliability [17]. Generally, the sensorless control

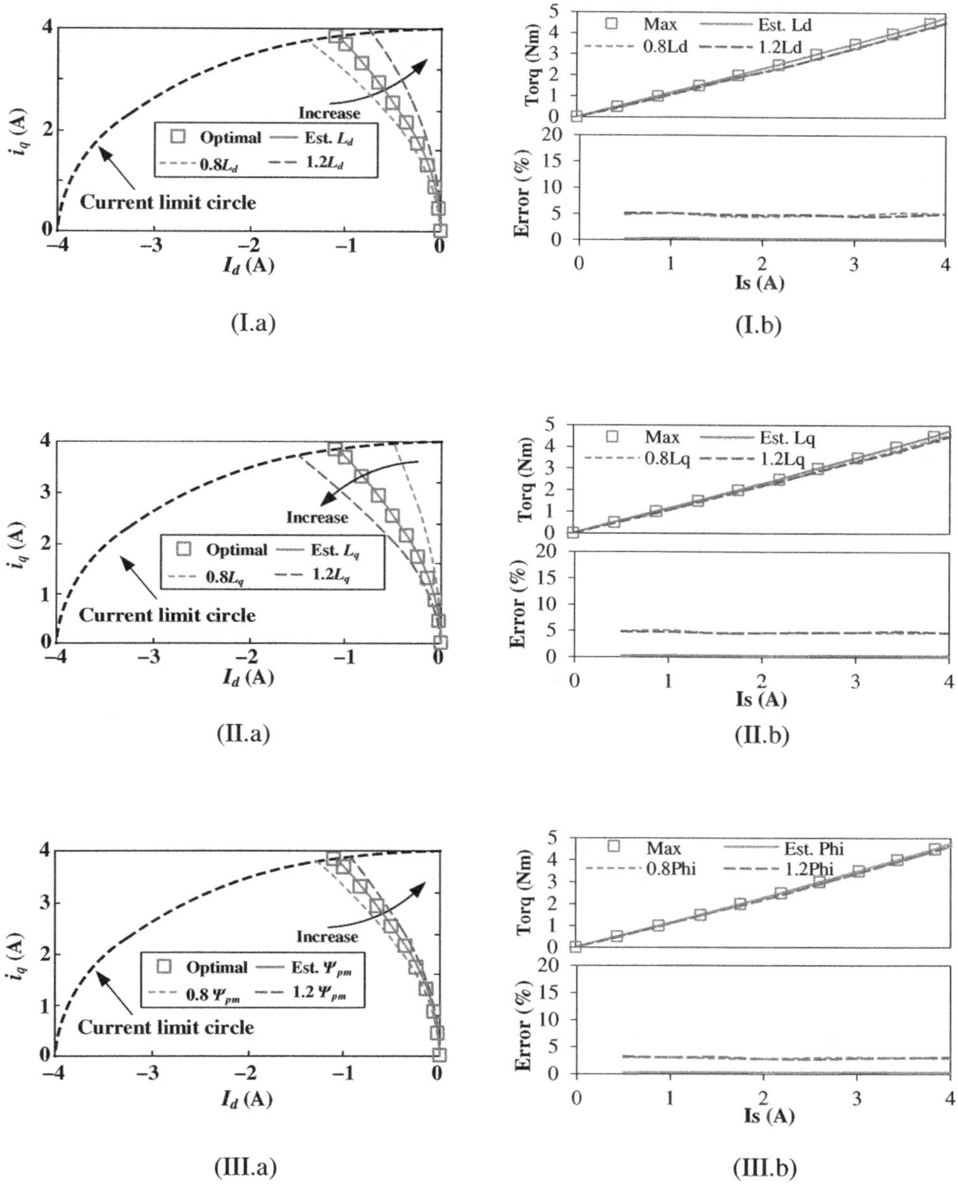

Figure 11.11 Derived MTPA trajectories and corresponding output torques considering parameter deviations [8], IEEE: (I.a) derived MTPA trajectories in ±20% error in L_d, (I.b) output torques in ±20% error in L_d, (II.a) derived MTPA trajectories in ±20% error in L_q, (II.b) output torques in ±20% error in L_q, (III.a) derived MTPA trajectories in ±20% error in ψ_{pm}, and (III.b) output torques in ±20% error in ψ_{pm}.

can be classified into two basic categories: the PMSM model-based and the rotor-saliency-based techniques.

The PMSM model-based sensorless control method utilizes the back-EMF phasor or the rotor flux phasor to estimate the rotor position information. However, the model-based method is sensitive to the variation of electrical parameters and degraded at both standstill and low speed.

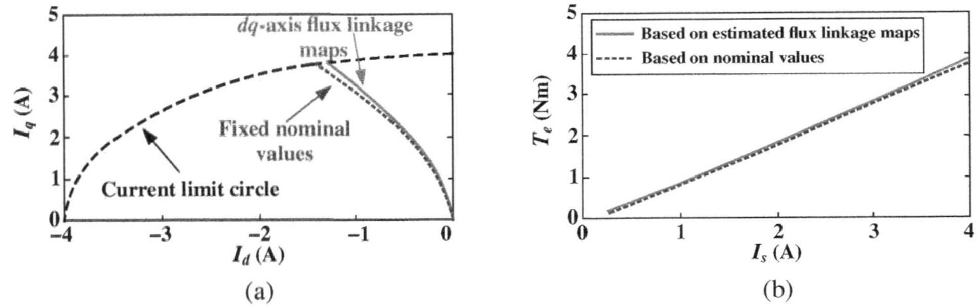

Figure 11.12 Comparison of MTPA trajectories and output torques based on fixed nominal values and estimated *dq*-axis flux linkage maps for Prototype Machine 2 [16], IEEE: (a) derived MTPA trajectories and (b) output torques.

First, the VSI non-linearity and the parameter variations [18] have a great impact on the estimation performance of the back-EMF-based method [19]. Second, the stator resistance is the most sensitive parameter at the low-speed range, and the inaccurate value will result in large position error and instability risks [20–22]. Third, the variation of dq-axis inductances caused by self- and mutual-magnetic saturation is the other critical issue for sensorless control regardless of the rotor speed [19], i.e., the decrease of stator inductances due to magnetic saturation will give rise to a lagged estimation error of the rotor position [23, 24].

On the other hand, the rotor-saliency-based method usually requires the injection of the HF signal, which is often used for starting and low-speed operation [17]. However, both the VSI non-linearity and the cross-coupling magnetic saturation could cause a significant estimation error of the rotor position [24–27].

This section demonstrates the importance of parameter estimation on sensorless control performance with reference to the self- and mutual-magnetic saturation effects in Section 11.3.1. Meanwhile, the applications of the parameter estimation methods under sensorless control are also described in Section 11.3.2.

11.3.1 Improvement in Sensorless Control Performance

11.3.1.1 Extended Back-EMF-Based Method

The block diagram of the extended back-EMF-based sensorless control is presented in Figure 11.13. Accounting for the influences of self- and mutual-magnetic saturation on dq-axis inductances, the dq-axis voltage equations can be written as follows:

$$\begin{aligned}\begin{bmatrix} u_d \\ u_q \end{bmatrix} &= \begin{bmatrix} R_s + L_d^{\text{inc}}\rho - \omega_r L_{qd} & -\omega_r L_q + L_{dq}^{\text{inc}}\rho \\ \omega_r L_d + L_{qd}^{\text{inc}}\rho & R_s + L_q^{\text{inc}}\rho + \omega_r L_{dq} \end{bmatrix} \begin{bmatrix} i_d \\ i_q \end{bmatrix} + \begin{bmatrix} 0 \\ \omega_r \psi_{pm} \end{bmatrix} \\ &= \begin{bmatrix} R_s + L_d^{\text{inc}}\rho - \omega_r L_{qd} & -\omega_r L_q + L_{dq}^{\text{inc}}\rho \\ \omega_r L_q - L_{dq}^{\text{inc}}\rho & R_s + L_d^{\text{inc}}\rho - \omega_r L_{qd} \end{bmatrix} \begin{bmatrix} i_d \\ i_q \end{bmatrix} + \begin{bmatrix} 0 \\ E_{\text{ex}} \end{bmatrix}\end{aligned} \tag{11.10a}$$

with

$$E_{\text{ex}} = \omega_r \psi_{pm} + \left[\omega_r (L_d - L_q) + (L_{qd} + L_{dq})\rho\right] i_d + \left[\omega_r (L_{dq} + L_{qd}) + (L_q - L_q)\rho\right] i_q \tag{11.10b}$$

where E_{ex} is the extended back-EMF. L_d, L_q, L_{dq}, and L_{qd} are the dq-axis self- and mutual-inductances, and the superscript "inc" denotes the incremental type. ρ is the differential operator, i.e., $\rho = d/dt$.

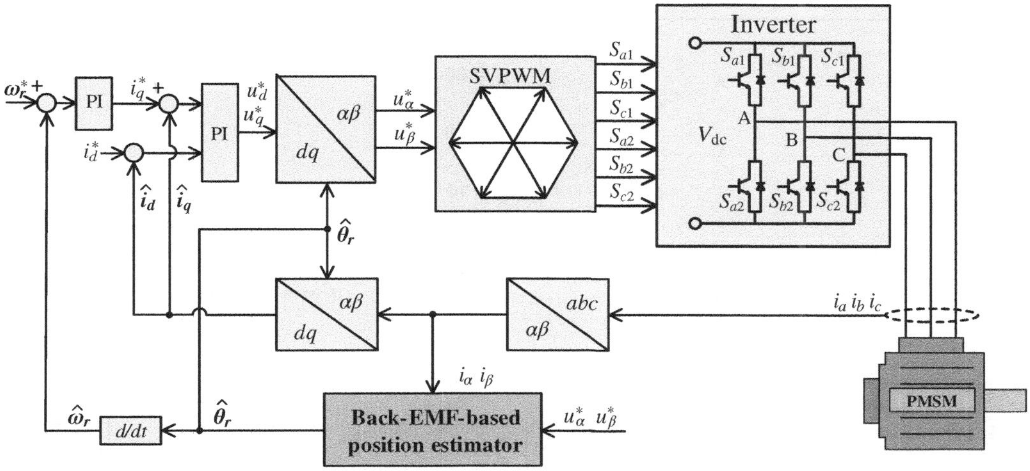

Figure 11.13 Block diagram of extended back-EMF-based sensorless control.

Therefore, the voltage equations in the estimated dq-axis reference frame can be written as follows:

$$\begin{bmatrix} u_d^e \\ u_q^e \end{bmatrix} = \begin{bmatrix} \cos(\Delta\theta) & \sin(\Delta\theta) \\ -\sin(\Delta\theta) & \cos(\Delta\theta) \end{bmatrix} \begin{bmatrix} U_d \\ U_q \end{bmatrix} \tag{11.11}$$

where the superscript "e" denotes the estimated quantities. The rotor position error $\Delta\theta = \theta_r^e - \theta_r$.

By substituting Eq. (11.11) into Eq. (11.10), the dq-axis extended back-EMF is calculated as follows:

$$\begin{cases} E_{\text{ex},d} = u_d^e - \left(R_s + L_d^{\text{inc}}\rho - \omega_r L_{qd}\right) i_d^e + \left(\omega_r L_q - L_{dq}^{\text{inc}}\rho\right) i_q^e \\ E_{\text{ex},q} = u_q^e - \left(R_s + L_d^{\text{inc}}\rho - \omega_r L_{qd}\right) i_q^e - \left(\omega_r L_q - L_{dq}^{\text{inc}}\rho\right) i_d^e \end{cases} \tag{11.12a}$$

$$\Delta\theta = \arctan\left(\frac{E_{\text{ex},d}}{E_{\text{ex},q}}\right) \approx \frac{E_{\text{ex},d}}{E_{\text{ex},q}} \tag{11.12b}$$

At steady state, i.e., $\rho i_d^e \approx 0$ and $\rho i_q^e \approx 0$, Eq. (11.12) can be simplified as follows:

$$\begin{cases} E_{\text{ex},d} = u_d^e - (R_s - \omega_r L_{qd}) i_d^e + \omega_r L_q i_q^e \\ E_{\text{ex},q} = u_q^e - (R_s - \omega_r L_{qd}) i_q^e - \omega_r L_q i_d^e \end{cases} \tag{11.13}$$

Clearly, when the d-axis extended back-EMF term is forced to zero, the estimated rotor position is equal to the actual rotor position, i.e., $\Delta\theta = 0$.

However, if the mutual inductance terms, i.e., $\omega_r L_{qd} i_{d,q}^e$, are neglected, it can be expected that the estimated rotor position will exhibit an obvious error. By way of example, the FEA method is employed to calculate the dq-axis self- and mutual-inductances for Prototype Machine 2, as shown in Figure 11.14. A detailed description of the FEA model can be found in Appendix A.

Furthermore, Figure 11.15 compares the estimation performance of the extended back-EMF-based sensorless control with and without considering self- and mutual-magnetic saturation. When the magnetic saturation is not considered in Figure 11.15a, the q-axis inductance L_q is fixed to 43 mH,

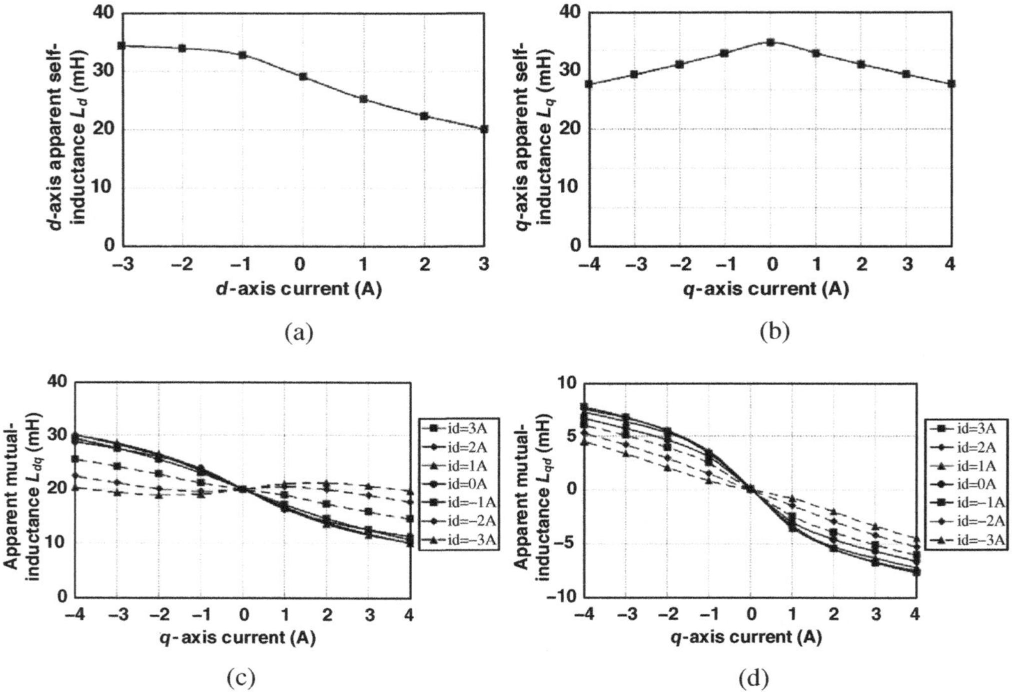

Figure 11.14 Variations of dq-axis apparent self- and mutual-inductances with currents [23], IEEE: (a) d-axis apparent self-inductance L_d, (b) q-axis apparent self-inductance L_q, (c) apparent mutual-inductance L_{dq}, and (d) apparent mutual-inductance L_{qd}.

and the mutual inductance L_{qd} is neglected. In this case, the largest rotor position errors can be observed, i.e., >15°. Furthermore, when only the self-saturation is considered in Figure 11.15b, the variation of L_q with i_q is approximated as a linear equation fitted based on the FEA results, i.e., $L_q = 51.2 - 2i_q^e$. However, the estimated positions in both cases exhibit obvious errors, which vary with the q-axis current from approximately 15° to −15°. In contrast, when both the self- and mutual-magnetic saturation are considered in Figure 11.15c, the estimation errors are reduced significantly and kept within 5°.

It is worth noting that the self- and mutual-magnetic saturation effects not only affect the estimation performance on the extended back-EMF-based method but it also exists in the HF signal injection–based sensorless control, which will be presented in the next subsection.

11.3.1.2 High-frequency Signal Injection–Based Method

When the HF pulsating signal is injected into the dq-axis reference frame, the dq-axis voltage equations can be expressed as follows [24]:

$$\begin{bmatrix} u_{dh}^r \\ u_{qh}^r \end{bmatrix} = \begin{bmatrix} L_d^{\text{inc}} & L_{dq}^{\text{inc}} \\ L_{qd}^{\text{inc}} & L_q^{\text{inc}} \end{bmatrix} \frac{d}{dt} \begin{bmatrix} i_{dh}^r \\ i_{qh}^r \end{bmatrix} \tag{11.14}$$

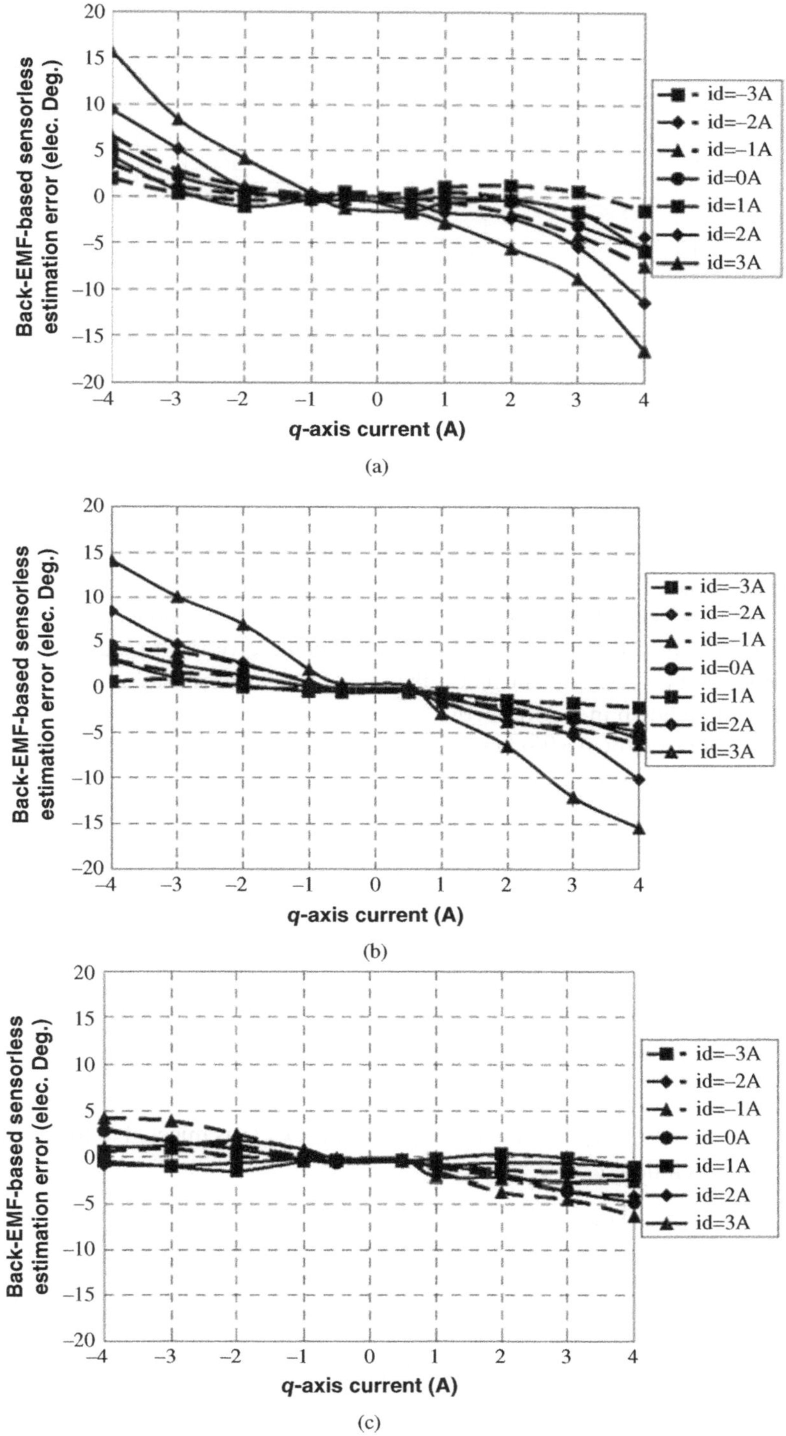

Figure 11.15 Estimation errors of the extend back-EMF-based sensorless control with/without considering self- and mutual-magnetic saturation [23], IEEE: (a) without considering magnetic saturation, (b) only considering self-magnetic saturation, and (c) considering both self- and mutual-magnetic saturation.

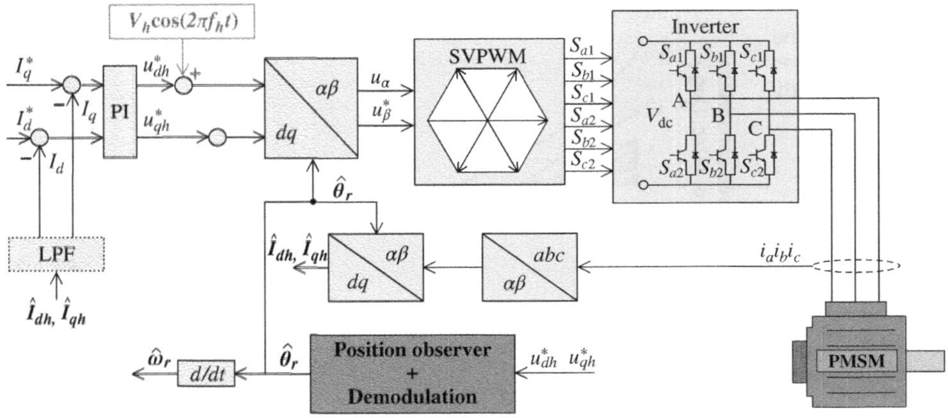

Figure 11.16 Block diagram of d-axis pulsating signal injection-based sensorless control.

where u_{dh}^r, u_{qh}^r, i_{dh}^r, and i_{qh}^r are the HF voltages and currents. The superscript "r" denotes the synchronous rotating reference frame. Meanwhile, the speed-dependent terms and the resistance voltage drop can be neglected compared to the HF reactance [19, 28].

In the HF signal injection–based sensorless control methods, the d-axis signal injection is more common since the q-axis current injection could give rise to higher torque ripple and more loss generation [17, 29, 30]. The block diagram of the d-axis pulsating signal injection–based sensorless control is shown in Figure 11.16.

Considering the position error $\Delta\theta$, the dq-axis voltage Eq. (7.6) can be written as follows:

$$\begin{bmatrix} u_{dh}^e \\ u_{qh}^e \end{bmatrix} = \begin{bmatrix} \cos(\Delta\theta) & \sin(\Delta\theta) \\ -\sin(\Delta\theta) & \cos(\Delta\theta) \end{bmatrix} \begin{bmatrix} u_{dh}^r \\ u_{qh}^r \end{bmatrix} = \begin{bmatrix} L_{\text{avg}} - \hat{L}_{\text{dif}}\cos(2\Delta\theta + \theta_m) & \hat{L}_{\text{dif}}\sin(2\Delta\theta + \theta_m) \\ \hat{L}_{\text{dif}}\sin(2\Delta\theta + \theta_m) & L_{\text{avg}} + \hat{L}_{\text{dif}}\cos(2\Delta\theta + \theta_m) \end{bmatrix} p \begin{bmatrix} i_{dh}^e \\ i_{qh}^e \end{bmatrix} \tag{11.15a}$$

$$\text{with} \begin{cases} L_{\text{dif}} = \dfrac{L_{qh} + L_{dh}}{2} & L_{\text{avg}} = \dfrac{L_{qh} - L_{dh}}{2} \\ \theta_m = \arctan\left[2L_{dqh}/(L_{dh} - L_{qh})\right] & \hat{L}_{\text{dif}} = \sqrt{L_{\text{dif}}^2 + L_{dqh}^2} \end{cases} \tag{11.15b}$$

When the HF voltage signal, i.e., $u_{\text{sig}} = V_h\sin(2\pi f_h t)$, is injected to the d-axis, the dq-axis voltage is written as follows:

$$\begin{bmatrix} u_{\text{sig}} \\ 0 \end{bmatrix} = \begin{bmatrix} L_{\text{avg}} - \hat{L}_{\text{dif}}\cos(2\Delta\theta + \theta_m) & \hat{L}_{\text{dif}}\sin(2\Delta\theta + \theta_m) \\ \hat{L}_{\text{dif}}\sin(2\Delta\theta + \theta_m) & L_{\text{avg}} + \hat{L}_{\text{dif}}\cos(2\Delta\theta + \theta_m) \end{bmatrix} p \begin{bmatrix} i_{dh}^e \\ i_{qh}^e \end{bmatrix} \tag{11.16}$$

The dq-axis HF currents in the estimated rotor-position reference frame are obtained as follows:

$$\begin{cases} i_{dh}^e = \dfrac{u_{\text{sig}}}{p\left(L_{\text{avg}}^2 - \hat{L}_{\text{dif}}^2\right)}\left[L_{\text{avg}} + \hat{L}_{\text{dif}}\cos(2\Delta\theta + \theta_m)\right] \\ i_{qh}^e = \dfrac{-u_{\text{sig}}}{p\left(L_{\text{avg}}^2 - \hat{L}_{\text{dif}}^2\right)}\hat{L}_{\text{dif}}\sin(2\Delta\theta + \theta_m) \end{cases} \tag{11.17}$$

If the cross-coupling effect is not considered, i.e., $L_{dqh} = 0$, the rotor position can be simply estimated by setting $i_{qh}^{e} = 0$, which is calculated as follows:

$$i_{qh}^{e} = \frac{-u_{\text{sig}}}{p\left(L_{\text{avg}}^{2} - \widehat{L}_{\text{dif}}^{2}\right)} \widehat{L}_{\text{dif}} \sin(2\Delta\theta + \theta_m) = 0 \tag{11.18}$$

However, the presence of the mutual inductance will give rise to potentially large position error. Therefore, the rotor position error $\Delta\theta$ should be minimized by considering both dq-axis HF currents, which is derived as follows:

$$i_{qh}^{e} + \lambda i_{dh}^{e} \approx \frac{-u_{\text{sig}}}{p\left(L_{\text{avg}}^{2} - \widehat{L}_{\text{dif}}^{2}\right)} (2L_{\text{dif}} + 2\lambda L_{dqh})\Delta\theta = 0 \tag{11.19}$$

where $\lambda = L_{dqh}/L_{qh}$, which is defined as the coupling factor and can be either calculated by FEA or estimated by the parameter estimation technique described in Chapter 7, as compared in Figure 11.17.

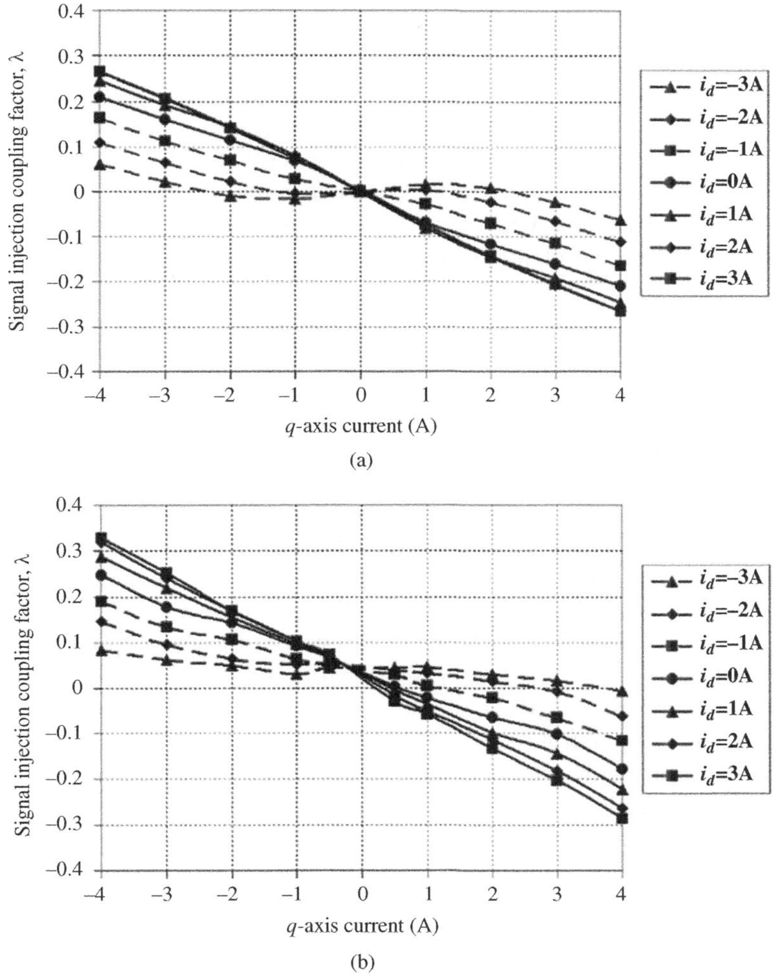

Figure 11.17 Comparison of FEA calculated and estimated coupling factors (λ) [27], IEEE: (a) FEA results and (b) estimated results based on HF signal injection.

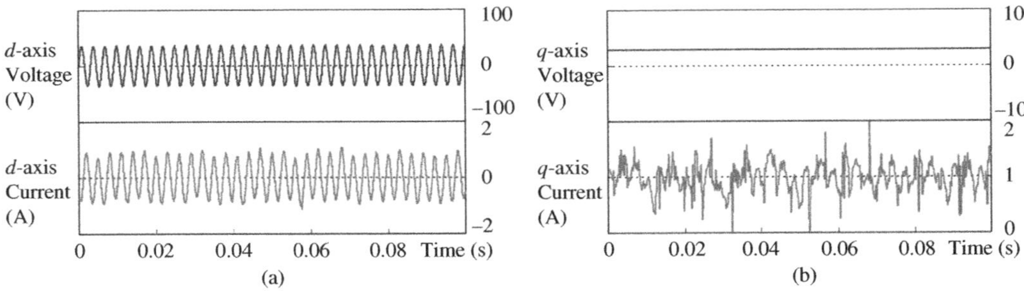

Figure 11.18 Waveforms of voltage and current in signal injection-based sensorless control [24]: (a) *D*-axis voltage and current waveforms and (b) *Q*-axis voltage and current waveforms.

For Prototype Machine 2, the amplitude and frequency of the injected pulsating voltage are 35 V and 330 Hz, respectively. The typical voltage and current waveforms during the HF signal injection are shown in Figure 11.18. By considering the influence of the mutual magnetic saturation, the estimation errors of the rotor position are compared in Figure 11.19. As can be observed, when the mutual inductance is not considered in Eq. (11.18), the position error increases as both the absolute value of i_q and the positive value of i_d are increased, due to the increased saturation and the influence of cross-coupling between the dq-axes, and the maximum error is 45°. In comparison, by using the mutual inductance calculated by FEA, the maximum error can be significantly reduced to approximately 6.1°. Therefore, when the cross-coupling effect is accounted for in the signal injection–based sensorless control, the estimation accuracy can be also improved eminently.

11.3.2 Application of Parameter Estimation Under Sensorless Control

Most parameter estimation methods are developed under sensored control, whereas the accurate parameters under sensorless control are equally important to ensure optimal control performance and safe operation conditions, etc.

Under fundamental model–based position sensorless control, the challenge in parameter estimation arises since the number of fundamental model equations is fixed, while an additional unknown parameter is introduced under sensorless control, the position error. Consequently, the "error transfer issue" commonly exists in the online parameter estimation methods under sensorless control [31]. As shown in Figure 11.20, the error transfer issue denotes that online parameter estimation and position sensorless control are interactive, i.e., the error of position estimation will cause the parameter estimation error, and in turn, the parameter estimation error also deteriorates the position estimation. Consequently, it is difficult to guarantee the convergence of the designed estimator under sensorless control.

It is found in Ref. [19] that the identified stator resistance varies from 1.1 to 1.8 Ω as the position error increases to 10°. To address this issue, this section presents an online parameter estimation method for SPMSM [31] based on position-offset injection tailored for sensorless control, in which the influence of position error on parameter estimation can be eliminated.

11.3.2.1 Position-offset-Based Online Parameter Estimation Under Sensorless Control

The square wave rotor position-offset (SWRPO) with the same amplitude of $\Delta\theta_\delta$ and $-\Delta\theta_\delta$ is injected into the estimated position under sensorless control [31]. During the period of SWRPO injection, it is found that the amplitude response of q-axis voltage fluctuation decreases with

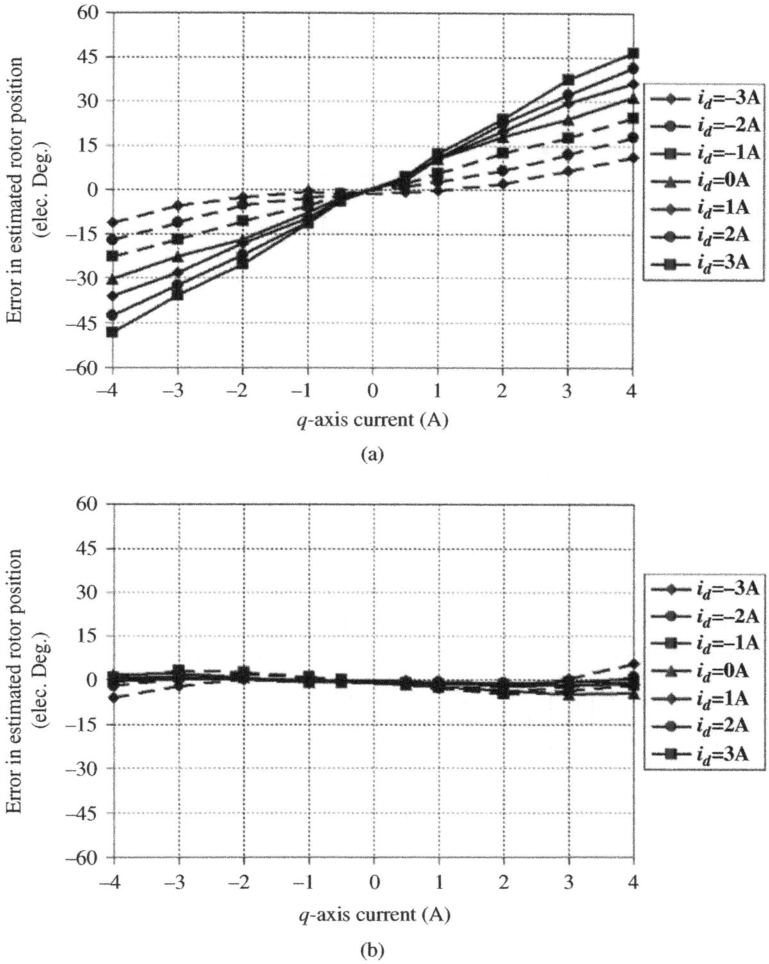

Figure 11.19 Estimation performance of HF signal injection-based sensorless control with and without considering mutual-inductance [27], IEEE (a) without considering mutual-inductance and (b) considering mutual-inductance.

Figure 11.20 Error transfer issue of online parameter estimation under sensorless control.

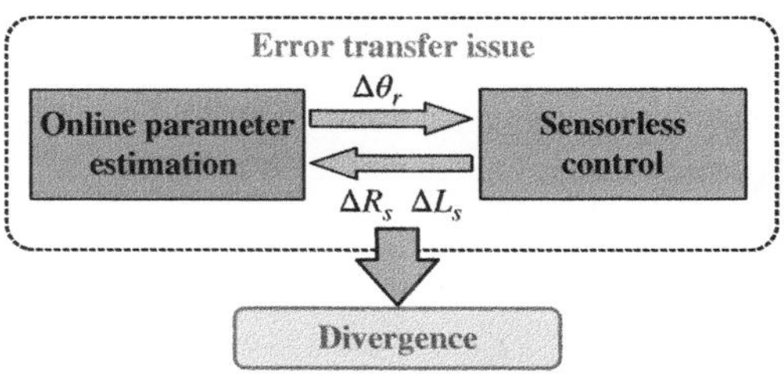

the increasing accuracy of parameters, and thus, both the stator inductance and the resistance can be estimated by controlling the q-axis voltage fluctuation to be zero. Subsequently, the PM flux linkage is estimated from the observed back-EMF. The overall block diagram is shown in Figure 11.21.

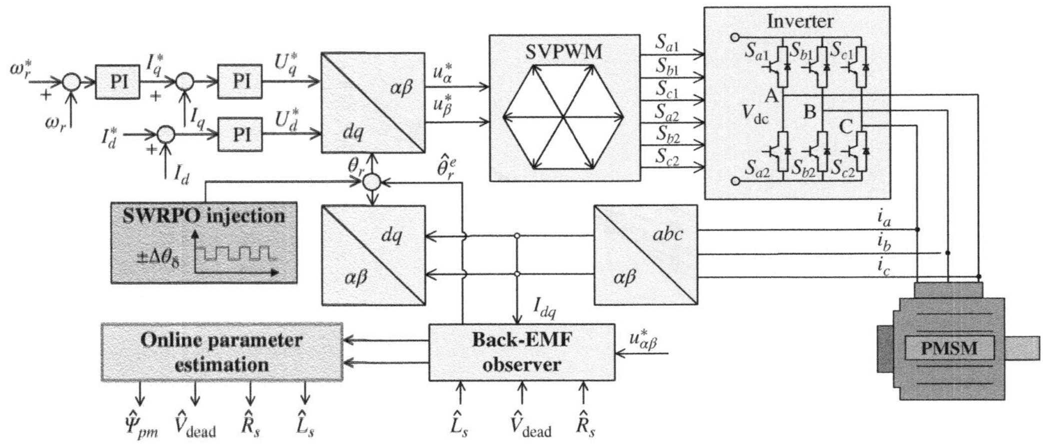

Figure 11.21 Block diagram of SWRPO-based online parameter estimation under sensorless control.

The detailed estimation process is designed as follows.

1. Estimation of stator inductance

Under $i_d = 0$ control, when there is no stator inductance error but stator resistance error, the steady-state estimation model after SWRPO injection can be written as follows:

$$\begin{bmatrix} u_{d1}^e \\ u_{q1}^e \end{bmatrix} = \begin{bmatrix} R_s & -\omega_r^e L_s \\ \omega_r^e L_s & R_s \end{bmatrix} \begin{bmatrix} 0 \\ i_q^e \end{bmatrix} + E_m \begin{bmatrix} \sin(\Delta\theta_\delta + \Delta\theta_{r\_R}) \\ \cos(\Delta\theta_\delta + \Delta\theta_{r\_R}) \end{bmatrix} \tag{11.20a}$$

$$\begin{bmatrix} u_{d2}^e \\ u_{q2}^e \end{bmatrix} = \begin{bmatrix} R_s & -\omega_r^e L_s \\ \omega_r^e L_s & R_s \end{bmatrix} \begin{bmatrix} 0 \\ i_q^e \end{bmatrix} + E_m \begin{bmatrix} \sin(-\Delta\theta_\delta - \Delta\theta_{r\_R}) \\ \cos(-\Delta\theta_\delta - \Delta\theta_{r\_R}) \end{bmatrix} \tag{11.20b}$$

where $\pm\Delta\theta_{r\_R}$ is caused by the component of the current vector in d^e-axis after SWRPO injection.

It is worth noting that under SWRPO injection, $\Delta\theta_{r\_R1} = \Delta\theta_{r\_R2}$, as can be explained by the phasor diagram in Figure 11.22, the components of the current vectors I_{s1} and I_{s2} in the d-axis are $I_{s1}\sin(\Delta\theta_\delta)$ and $I_{s2}\sin(-\Delta\theta_\delta)$, respectively. The d- and q-axis currents are set to constants, thus $I_s = I_{s1} = I_{s2}$, and the two current components will cause opposite position errors ($\pm\Delta\theta_{r\_R}$). Thus, according to Eqs. (11.20a) and (11.20b), the error between $\Delta\bar{u}_{q1}^e$ and $\Delta\bar{u}_{q2}^e$ can be presented as follows:

$$\Delta\bar{u}_q^e = \Delta\bar{u}_{q1}^e - \Delta\bar{u}_{q2}^e = 0 \tag{11.21}$$

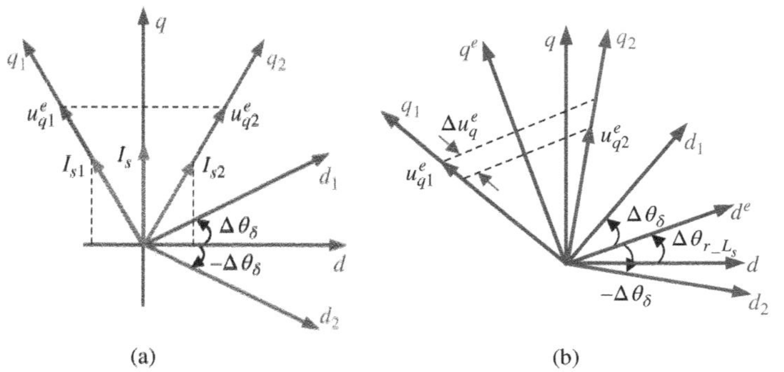

Figure 11.22 Phasor diagram of SPMSM with stator resistance error under $i_d = 0$ control: (a) accurate stator inductance and (b) with stator inductance error.

where the accent '-' represents the average value after filtering.

It can be found from Figure 11.22a that if stator inductance used in back-EMF observer is accurate, the average u_q^e will not fluctuate after injecting SWRPO regardless of stator resistance error in position observer.

Nevertheless, if there is a stator inductance error, as shown in Figure 11.22b, after injecting SWRPO to the estimated position, the voltage equations can be written as follows:

$$\begin{bmatrix} u_{d1}^e \\ u_{q1}^e \end{bmatrix} = \begin{bmatrix} R_s & -\omega_r^e L_s \\ \omega_r^e L_s & R_s \end{bmatrix} \begin{bmatrix} 0 \\ i_q^e \end{bmatrix} + E_m \begin{bmatrix} \sin(\Delta\theta_{r\_R} + \Delta\theta_\delta + \Delta\theta_{r\_L_s}) \\ \cos(\Delta\theta_{r\_R} + \Delta\theta_\delta + \Delta\theta_{r\_L_s}) \end{bmatrix} \tag{11.22a}$$

$$\begin{bmatrix} u_{d2}^e \\ u_{q2}^e \end{bmatrix} = \begin{bmatrix} R_s & -\omega_r^e L_s \\ \omega_r^e L_s & R_s \end{bmatrix} \begin{bmatrix} 0 \\ i_q^e \end{bmatrix} + E_m \begin{bmatrix} \sin(-\Delta\theta_{r_R} - \Delta\theta_\delta + \Delta\theta_{r\_L_s}) \\ \cos(-\Delta\theta_{r_R} - \Delta\theta_\delta + \Delta\theta_{r\_L_s}) \end{bmatrix} \tag{11.22b}$$

Then, subtracting Eq. (11.22b) from Eq. (11.22a), $\Delta \bar{u}_q^e$ is presented as follows:

$$\begin{aligned} \bar{u}_{q1}^e - \bar{u}_{q2}^e &= \bar{E}_m[\cos(\Delta\theta_{r_{L_s}} + \Delta\theta_{r_R} + \Delta\theta_\delta) - \cos(\Delta\theta_{r_{L_s}} - \Delta\theta_{r_R} - \Delta\theta_\delta)] \\ &= -2\bar{E}_m \sin(\Delta\theta_{r_{L_s}}) \sin(\Delta\theta_{r_R} + \Delta\theta_\delta) \\ &= K_{L_s} \sin(\Delta\theta_{r\_L_s}) \end{aligned} \tag{11.23}$$

where $K_{L_s} = -2\bar{E}_m \sin(\Delta\theta_{r\_R} + \Delta\theta_\delta)$. It is evident that the inductance error will give rise to a q-axis voltage fluctuation.

Therefore, the stator inductance L_s can be estimated by controlling the q-axis voltage fluctuation ($\Delta \bar{u}_q^e$) to be zero based on a PI controller. The estimated L_s is calculated as follows:

$$\hat{L}_s = \left(K_{p\_L_s} + \frac{K_{i\_L_s}}{s}\right) \Delta \bar{u}_q^e \tag{11.24}$$

where $K_{p\_L_s}$ and $K_{i\_L_s}$ are the coefficients of the proportional and integral terms, respectively.

2. Estimation of stator resistance

The stator resistance can be estimated under $i_d \neq 0$ control. After injecting SWRPO to the estimated position, the voltage equations can be derived as follows:

$$\begin{bmatrix} u_{d1}^e \\ u_{q1}^e \end{bmatrix} = \begin{bmatrix} R_s & -\omega_r^e \hat{L}_s \\ \omega_r^e \hat{L}_s & R_s \end{bmatrix} \begin{bmatrix} i_d^e \\ i_q^e \end{bmatrix} + E_m \begin{bmatrix} \sin(\Delta\theta_{r\_R} + \Delta\theta_1 + \Delta\theta_\delta) \\ \cos(\Delta\theta_{r\_R} + \Delta\theta_1 + \Delta\theta_\delta) \end{bmatrix} \tag{11.25a}$$

$$\begin{bmatrix} u_{d2}^e \\ u_{q2}^e \end{bmatrix} = \begin{bmatrix} R_s & -\omega_r^e \hat{L}_s \\ \omega_r^e \hat{L}_s & R_s \end{bmatrix} \begin{bmatrix} i_d^e \\ i_q^e \end{bmatrix} + E_m \begin{bmatrix} \sin(\Delta\theta_{r\_R} - \Delta\theta_1 - \Delta\theta_\delta) \\ \cos(\Delta\theta_{r\_R} - \Delta\theta_1 - \Delta\theta_\delta) \end{bmatrix} \tag{11.25b}$$

The generation mechanism of $\Delta\theta_1$ is similar to the stator inductance estimation analysis in Eqs. (11.22a) and (11.22b). Thus, by subtracting Eq. (11.25b) from Eq. (11.25a), $\Delta \bar{u}_q^e$ is expressed as follows:

$$\begin{aligned} \bar{u}_{q1}^e - \bar{u}_{q2}^e &= \bar{E}_m[\cos(\Delta\theta_{r\_R} + \Delta\theta_1 + \Delta\theta_\delta) - \cos(\Delta\theta_{r\_R} - \Delta\theta_1 - \Delta\theta_\delta)] \\ &= -2\bar{E}_m \sin(\Delta\theta_{r\_R}) \sin(\Delta\theta_1 + \Delta\theta_\delta) \\ &= K_{r\_R} \sin(\Delta\theta_{r\_R}) \end{aligned} \tag{11.26}$$

where $K_{r\_R} = -2\bar{E}_m \sin(\Delta\theta_1 + \Delta\theta_\delta)$. Similarly, R_s can be estimated by controlling the q-axis voltage fluctuation to be zero by using a PI controller.

3. Estimation of rotor PM flux linkage

Finally, since the position error due to parameter errors can be significantly minimized by using the estimated L_s and R_s, the rotor PM flux linkage ψ_{pm} can be obtained subsequently from the observed back-EMF. Moreover, it can be concluded that the complete parameter estimation process is immune to position error under sensorless control.

11.3.2.2 Experimental Results

For illustration, Figure 11.23 shows the parameter estimation results of stator inductance and resistance at 30 r/min based on Prototype Machine 3. When the parameter estimation algorithm is enabled, the q-axis voltage fluctuations are controlled, which leads to stator inductance and resistance quickly converging to the steady-state value (21.5 mH and 4.15 Ω) with 2.3% inductance error and 1.2% resistance error, respectively, and the position error also converges to zero. It indicates that the stator inductance and resistance estimations are immune to the position error.

Moreover, Figure 11.24 presents the variations of the position offsets with the estimated parameters, in which the estimation results based on the POPE method in Chapter 6 under sensored control (without considering position error) are also compared.

As can be observed, the errors of estimated values are increased with the position error when the POPE method developed under sensored control is directly applied under the sensorless control due to the significance of position error. In comparison, the described method is tailored for parameter estimation under sensorless control, and the estimated parameters are immune to position error and overcome the error transfer issue.

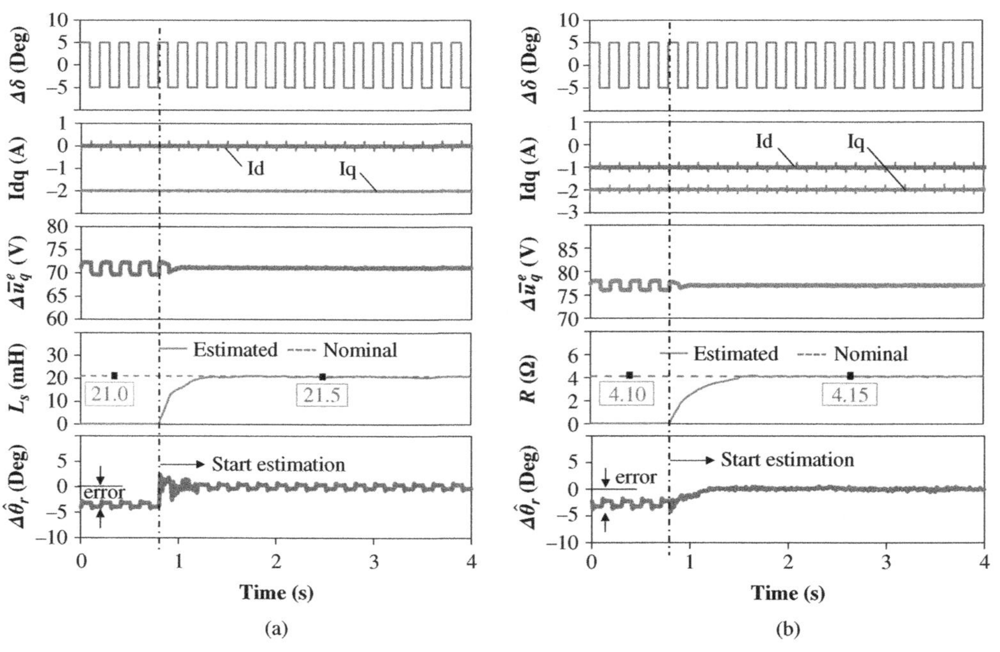

Figure 11.23 Parameter estimation results based on Prototype Machine 3 at 30 r/min: (a) stator inductance estimation and (b) stator resistance estimation.

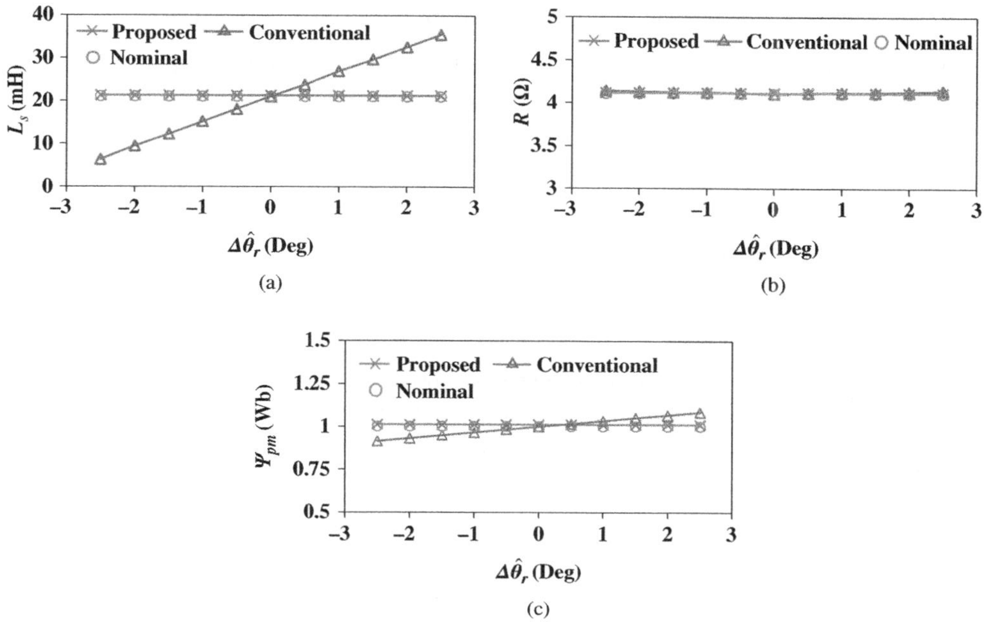

Figure 11.24 Comparison of position-offset-based parameter estimation method with and without considering influence of different position errors: (a) estimated L_s, (b) estimated R_s, and (c) estimated ψ_{pm}.

11.4 Precise Torque Estimation

Precise torque control of PMSM is widely adopted in many applications, e.g., robotics, elevators, and electric vehicles [32], and accurate torque feedback of the machine is essential for both control and monitoring purposes. Although the torque can be measured by transducers such as strain gauges and torsional displacement, they are usually expensive and require additional installation space and cables [33]. Therefore, torque estimation without relying on sensors is preferred.

To date, the machine torque can be estimated by two methods, i.e., the voltage/current observers based on the fundamental electrical models [34–36] and the HF signal injection-based methods [37–40]. Generally, the fundamental electrical model-based method is appropriate at medium- and high-speed ranges but may lose effectiveness at standstill and low-speed ranges due to reduced amplitudes of the back EMF signal. In contrast, the HF signal injection method is appropriate for the estimation at standstill and low-speed range, while not appropriate for occasions with high fundamental electrical frequency.

Evidently, the accurate machine parameters have a great impact on the accuracy of the estimated torque in both methods [37–42]. For the accurate knowledge of the fundamental electrical model, it has been extensively described in the previous chapters. This section presents an example of torque estimation based on an HF square save voltage injection method, where the influence of the cross-coupling effect is considered.

11.4.1 HF Square Wave Voltage Injection Considering Cross-coupling Effect

When the machine torque is estimated under sensorless control, the voltage equations in the estimated synchronous rotating reference frame are expressed as follows:

$$\begin{bmatrix} i_{dh}^{\hat{r}} \\ i_{qh}^{\hat{r}} \end{bmatrix} = R(\Delta\theta)^{-1} \begin{bmatrix} L_{dh} & 0 \\ 0 & L_{qh} \end{bmatrix}^{-1} RR(\Delta\theta)^{-1} \int \begin{bmatrix} u_{dh}^{\hat{r}} \\ u_{qh}^{\hat{r}} \end{bmatrix} dt \tag{11.27a}$$

$$R(\Delta\theta) = \begin{bmatrix} \cos(\Delta\theta) & \sin(\Delta\theta) \\ -\sin(\Delta\theta) & \cos(\Delta\theta) \end{bmatrix} \tag{11.27b}$$

where $\Delta\theta$ is the estimated rotor position error.

The injection of the HF square wave voltages is shown in Figure 11.25, where each injection cycle consists of four injection stages, i.e., A, B, C, and D, and expressed as follows:

$$\begin{bmatrix} u_{dh}^{\hat{r}}(t) \\ u_{qh}^{\hat{r}}(t) \end{bmatrix} = \begin{bmatrix} \overset{t\in A}{V_h} & \overset{t\in B}{0} & \overset{t\in C}{-V_h} & \overset{t\in D}{0} \\ 0 & V_h & 0 & V_h \end{bmatrix} \tag{11.28}$$

where $t \in A, B, C, D$ denotes the time in stages A, B, C, and D, respectively. U_h is the amplitude of the HF voltage.

As can be seen, the positive and negative square wave voltages are injected to the d-axis at stages A and C, while at stages B and D, the positive and negative square wave voltages are injected to the estimated q-axis. Therefore, the injection sequence can be regarded as an equivalent HF rotating voltage $V_h e^{j(\omega_h t + \hat{\theta}_r)}$ with frequency $\omega_h = \omega_h/2T_s$ injected into the stator winding at four specific sampling points, i.e., $\omega_h t = 0$, $\pi/2$, π, and $3\pi/2$.

The current variations in each injection interval can be obtained from Eqs. (11.27) and (11.28):

$$\begin{bmatrix} \Delta i_{dh}^{\hat{r}}(kT_s) \\ \Delta i_{qh}^{\hat{r}}(kT_s) \end{bmatrix} = \begin{bmatrix} \overset{t\in A}{\Delta I_d} & \overset{t\in B}{\Delta I_\varepsilon} & \overset{t\in C}{-\Delta I_d} & \overset{t\in D}{-\Delta I_\varepsilon} \\ \Delta I_\varepsilon & \Delta I_q & -\Delta I_\varepsilon & -\Delta I_q \end{bmatrix} \tag{11.29a}$$

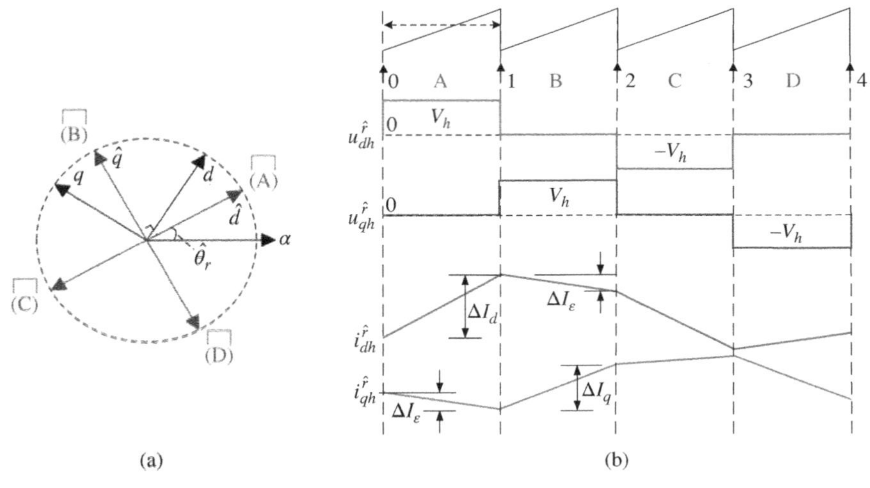

Figure 11.25 Square wave voltage injection [39], IEEE: (a) vector diagram and (b) injection sequence.

$$\begin{cases} \Delta I_{dh} = \dfrac{V_h T_s}{2L_{dh}L_{qh} - L_{dq}^2}\left\{ L_{qh} + L_{dh} + \left(L_{qh} - L_{dh}\right)\cos[2\Delta\theta]\right\} \\ \Delta I_{qh} = \dfrac{V_h T_s}{2L_{dh}L_{qh} - L_{dq}^2}\left\{ L_{qh} + L_{dh} + \left(L_{qh} - L_{dh}\right)\cos[2\Delta\theta]\right\} \\ \Delta I_{dqh} = \dfrac{V_h T_s}{2L_{dh}L_{qh}L_{dq}^2}\left[\left(L_{qh} - L_{dh}\right)\sin[2\Delta\theta]\right] \end{cases} \tag{11.29b}$$

where T_s is the duration for the positive/negative voltage.

Afterwards, the HF dq-axis self- and mutual inductances can be obtained based on Eq. (11.29) as follows:

$$L_{dh} = \frac{1}{2}\left(f_1 + f_2 - \frac{f_2 - f_1}{\sqrt{1 + \tan^2(2\Delta\theta)}}\right) \tag{11.30a}$$

$$L_{qh} = \frac{1}{2}\left(f_1 + f_2 + \frac{f_2 - f_1}{\sqrt{1 + \tan^2(2\Delta\theta)}}\right) \tag{11.30b}$$

$$L_{dqh} = \frac{\tan(2k_c i_q)(f_2 - f_1)}{2\sqrt{1 + \tan^2(2\Delta\theta)}} \tag{11.30c}$$

$$\text{with:}\begin{cases} f_1 = \dfrac{V_h T_s}{(\Delta I_d + \Delta I_q)/2 + \sqrt{[(\Delta I_d - \Delta I_q)/2]^2 + (\Delta I_\varepsilon)^2}} \\ f_2 = \dfrac{V_h T_s}{(\Delta I_d + \Delta I_q)/2 - \sqrt{[(\Delta I_d - \Delta I_q)/2] + (\Delta I_\varepsilon)^2}} \end{cases} \tag{11.30d}$$

Meanwhile, the dq-axis flux linkages can be approximated as follows:

$$\psi_d = \psi_{pm} + L_{dh} i_d \tag{11.31}$$

$$\psi_q = k_{q1} + 3k_{q3} i_q^3 \tag{11.32}$$

where $k_{q1} = L_{qh0}$ and $k_{q3} = (L_{qh} + 2L_{qh0})\, i_q/3$ are the coefficients of the fitted curve.

11.4.2 Torque Estimation Based on Estimated HF Inductances

The obtained HF inductances are used to estimate the machine torque, where the self- and mutual-saturation can be reflected. By substituting Eqs. (11.31) and (11.32) into the general torque Eq. (11.33), the torque can be further estimated based on the estimated dq-axis HF inductances as Eq. (11.34).

$$T_e = \frac{3p}{2}(\psi_d i_q - \psi_q i_d) \tag{11.33}$$

$$T_e = \frac{3p}{2}\left[\psi_{pm} i_q + \left(L_{dh} - \frac{L_{qh} + 2L_{qh0}}{3}\right) i_d i_q\right] \tag{11.34}$$

Meanwhile, the additional torque component T_{cross} caused by the cross-coupling effect is calculated as follows [43]:

$$T_{\text{cross}} = \frac{3p}{2} L_{dqh}\left(i_q^2 - i_d^2\right) \tag{11.35}$$

Therefore, accounting for the cross-coupling effect, the estimated total torque $T_{\text{est}} = T_e + T_{cross}$, which is expressed as follows:

$$T_{\text{est}} = \frac{3p}{2}\left[\psi_{pm} i_q + \left(L_{dh} - \frac{L_{qh} + 2L_{qh0}}{3}\right) i_d i_q + L_{dqh}\left(i_q^2 - i_d^2\right)\right] \tag{11.36}$$

Furthermore, the torque estimation performance is demonstrated based on Prototype Machine 2. When the machine is operated at 60 r/min, and the reference currents are determined by the MTPA strategy (see Section 11.2), the HF square wave voltages are injected into the real dq-axes to estimate the HF inductances under different load torque, where the rotor PM flux linkage is fixed to the nominal value of 0.26 Wb. The estimated and FEA-calculated HF inductances and estimated torque are shown in Figure 11.26. Obviously, the online identified HF inductances show a good agreement with the FE-predicted values, and the estimation error of machine torque is kept within ±5%.

11.5 Thermal Condition Monitoring

As one of the most important applications, parameter estimation techniques are widely used for thermal condition monitoring of PMSMs based on the temperature dependencies of the stator winding resistance and the rotor PM flux linkage [44]. Meanwhile, the windings and PMs are also the most vulnerable components for PMSMs. More specifically, the lifetime of the winding insulation is significantly affected by the stator winding temperature [45]. Furthermore, the irreversible partial demagnetization could be another critical issue when the PM temperature exceeds its maximum allowable temperature [46].

Nevertheless, compared to the measurements of rotor position and machine torque, the measurement of PM temperature is even more challenging during the rotation, i.e., additional measurement

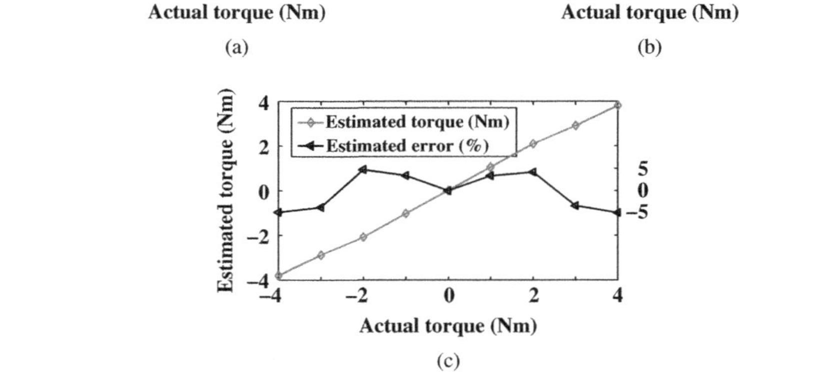

Figure 11.26 Torque estimation performance based on estimated HF inductances under different load conditions [39], IEEE: (a) FEA-calculated HF inductances, (b) HF signal injection estimated HF inductances, and (c) estimated torque and error.

Table 11.2 Temperature coefficients of copper and PM materials (%/°C).

	Copper	NdFeB	AlNiCo	SmCo	Ferrite
Temperature coefficients	0.30–0.40	−0.15 to −0.10	−0.03	−0.06 to −0.02	−0.03

and/or data transmission devices are usually required [47–49]. Therefore, the parameter estimation techniques have been widely used for thermal condition monitoring [42], [47–51], and the temperature dependencies of the stator resistance and the rotor flux linkage are expressed as:

$$R_{s,s} = R_{s,0}[1 + \alpha_{cu}(\vartheta_s - \vartheta_r)] \tag{11.37}$$

$$\psi_{pm,s} = \psi_{pm,0}\left[1 + \alpha_{pm}\left(\vartheta_{pm} - \vartheta_r\right)\right] \tag{11.38}$$

where $R_{s,s}$, $R_{s,0}$ and $\psi_{pm,s}$, $\psi_{pm,0}$ are the stator resistance and the rotor PM flux linkage at the actual temperatures, i.e., ϑ_s and ϑ_{pm}, and at the reference temperature ϑ_r. α_{cu} and α_{pm} are the temperature coefficients of copper and PM, respectively, as listed in Table 11.2. In addition, it is worth noting that ϑ_s and ϑ_{pm} are the average temperatures.

Most of the online estimation methods described in this book, including Chapters 3–7, have capabilities to track the changes in the thermal states of the stator winding and/or the PM. By way of example, when the d-axis current injection–based method described in Section 4.3 is employed, the variations of the estimated stator resistance and rotor PM flux linkage are presented in Figure 11.27.

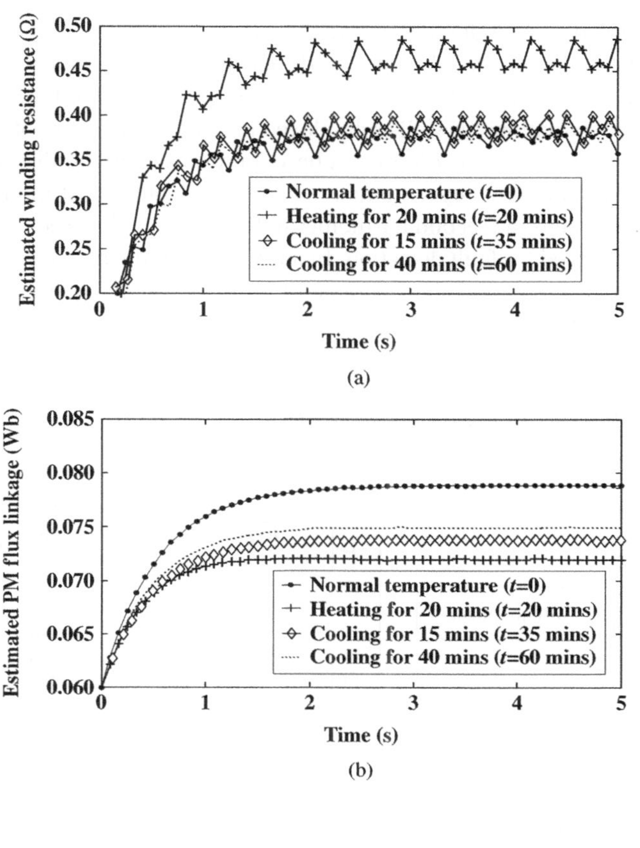

Figure 11.27 Estimated stator winding resistance and rotor PM flux linkage of prototype machine 1 by using injection of i_d =2.5 A under constant torque/speed control [42], IEEE: (a) estimated stator winding resistance and (b) estimated rotor PM flux linkage.

As can be observed, after 20 min of heating, the estimated stator winding resistance increases from 0.388 to 0.473 Ω, while the estimated rotor PM flux linkage decreases from 78.8 to 72.0 mWb. Furthermore, the corresponding average temperature changes can be calculated by Eqs. (11.37) and (11.38) together with the temperature coefficients listed in Table 11.2.

11.6 Fault Diagnosis

The online parameter estimation techniques are widely used in detecting fault-related signals and do not rely on additional measurement devices [52, 53], including inter-turn short-circuit (ITSC) faults, demagnetization faults, and eccentricity faults, etc.

11.6.1 Inter-turn Short-circuit Fault

ITSC fault is the short circuit between one or a few turns of winding, which is the most common electrical fault. ITSC fault is usually caused by mechanical stress, moisture, and partial discharge. When the ITSC fault occurs, a high circulating current is generated in the shorted circuit, which may propagate and lead to phase-to-phase, phase-to-ground, and demagnetization faults [54].

The residual-based detection method is extensively used for detecting ITSC fault, as illustrated in Figure 11.28, where the residual between the measured (fault) and estimated (healthy) quantities, e.g., current [54–56], back-EMF [57], voltage disturbance [58, 59], as well as torque [60], are used to reflect the deviations caused by the ITSC faults. Therefore, accurate modelling of the reference model in healthy mode is critical.

The healthy PMSM model in Ref. [54] is modelled considering the VSI non-linearity and the unbalanced inductances, and the current residual is used to detect the ITSC fault. Moreover, it has been demonstrated in Ref. [55] that the negative sequence current component is independent of machine parameter errors, and thus, it is used as a fault indicator to reduce the influence of parameter variations, as well as employed in Ref. [56]. In terms of EMF-based estimation in Ref. [57], it is established in the *abc* three-phase frame based on the FEA model, where the temperature effect is also considered by a thermal network. The voltage disturbances can also be indicators for the ITSC fault, and the second harmonics in the *dq*-axis voltage equations [58] and in the back-EMF [59] are observed for fault detection. In addition, with the aid of a torque transducer, the error of the measured torque and the calculated torque by torque equation at healthy condition is detected. It is shown that when the ITSC fault occurs, there is a direct current (dc) component in the residual, which can be used as the fault indicator.

By way of example, the waveforms and spectra of three phase currents of a 24-slot/8-pole SPMSM at healthy and ITSC fault conditions are shown in Figure 11.29. Obviously when ITSC

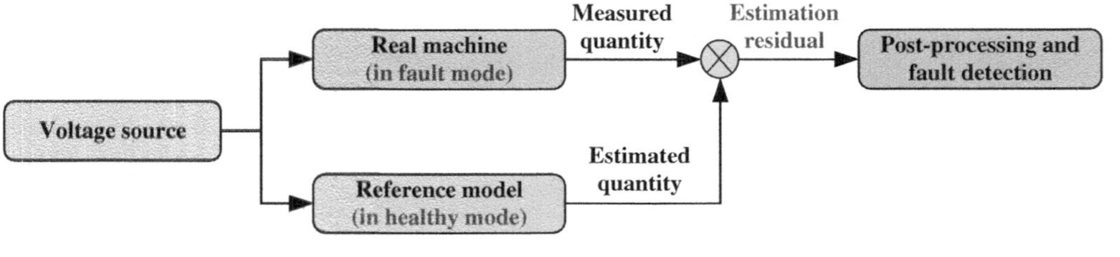

Figure 11.28 Illustration of residual-based ITSC detection methods.

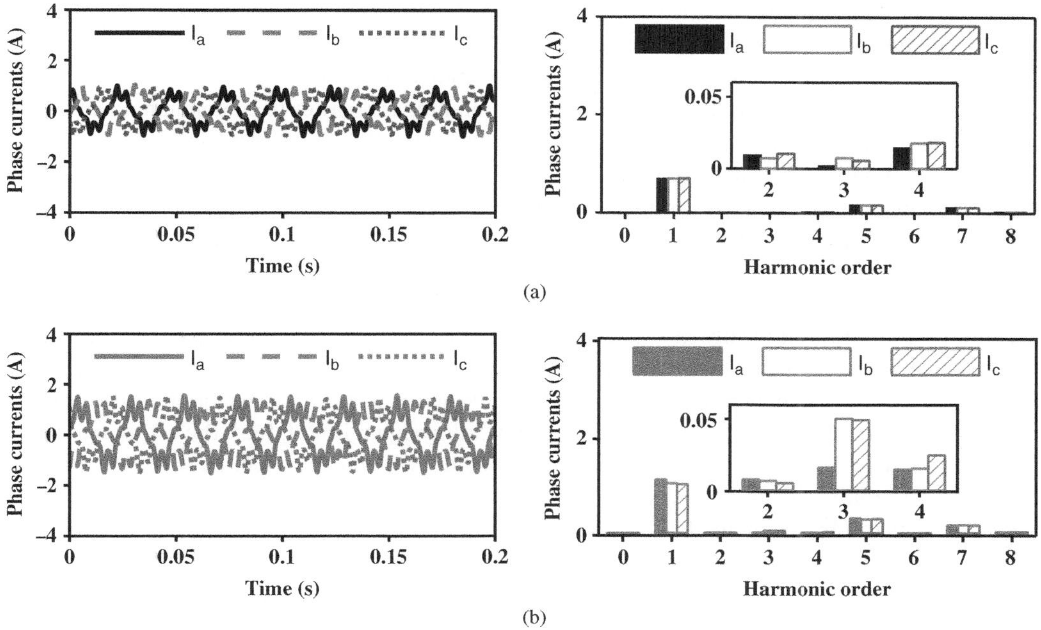

Figure 11.29 Waveforms and spectra of three-phase currents at healthy and ITSC conditions (600 r/min, SC ratio 11.25% at phase A): (a) healthy condition and (b) ITSC condition.

fault occurs, the third harmonics in phase currents increase significantly. Hence, there exist differences between the healthy phase currents and the faulty phase currents, which can be utilized for fault detection.

11.6.2 Demagnetization Fault

PMs are the most important and expensive components of PMSMs, and the demagnetization issue severely threatens the service life and performance of PMSMs [61], e.g., torque/power density, efficiency, torque ripple, etc. The demagnetization issue can be illustrated in Figure 11.30. As can be seen from Figure 11.30a, when the PM operating point is in the linear region above the keen point, i.e., P_1, only reversible demagnetization occurs, i.e., once the external magnetic field is removed, the PM operating point will be recovered along with the original demagnetization curve.

Figure 11.30 Illustration of irreversible demagnetization: (a) irreversible demagnetization due to large armature field and (b) irreversible demagnetization due to high temperature ($\vartheta_2 > \vartheta_1$).

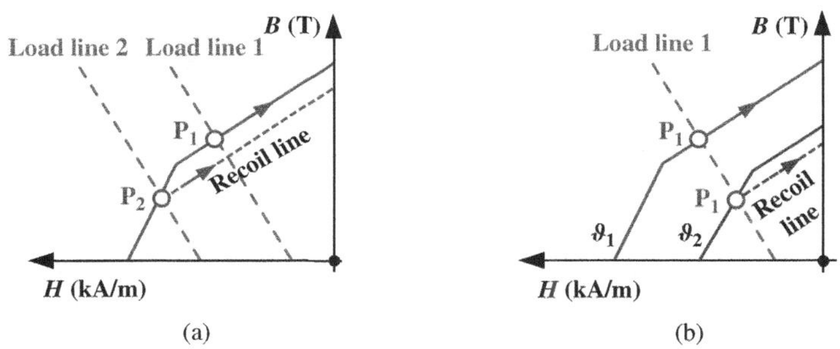

However, once the applied negative field strength exceeds the knee point of the B–H curve, i.e., $\boldsymbol{P_2}$, the remanence flux density will decrease dramatically. Consequently, when the external field is removed, the PM cannot recover its original magnetization curve but along the recoil line. Besides, the high-temperature rise will reduce the coercivity, making PM easier to be demagnetized by the armature field, as shown in Figure 11.30b. Moreover, when the PM temperature exceeds the Curie's temperature, it will be fully demagnetized.

Since the rotor flux linkage is proportional to the magnetic strength, the online parameter estimation of the rotor PM flux linkage or *d*-axis flux linkage is widely used to detect demagnetization faults and to evaluate the degree of demagnetization [62–64], where the parameter variation due to magnetic saturation and temperature changes should be considered [63–65]. More specifically, the variations of the analytically calculated *dq*-axis inductances are used in Ref. [63] to indicate the flux linkages in the demagnetized PMSMs. The sliding mode disturbance observed is employed in Ref. [64] to diagnose demagnetization fault, where the stator parameter mismatch impact can be eliminated. In addition, by injecting a *d*-axis HF pulsating signal in Ref. [65], the estimated *d*-axis inductance varies due to the change of saturation degree, which can reflect both the PM demagnetization and rotor with eccentricity.

By way of example, when one PM pole in a 12-slot/10-pole SPMSM is demagnetized, there will be some fractional harmonics emerging in the back-EMFs, i.e., the 3/5th and 7/5th harmonics, as shown in Figure 11.31. This feature enables the detection of partial demagnetization, and with an online back-EMF estimator, it is easy to detect the fault in real-time. It is worth noting that the fault-related harmonic order may vary according to the pole-slot combination and winding-configuration.

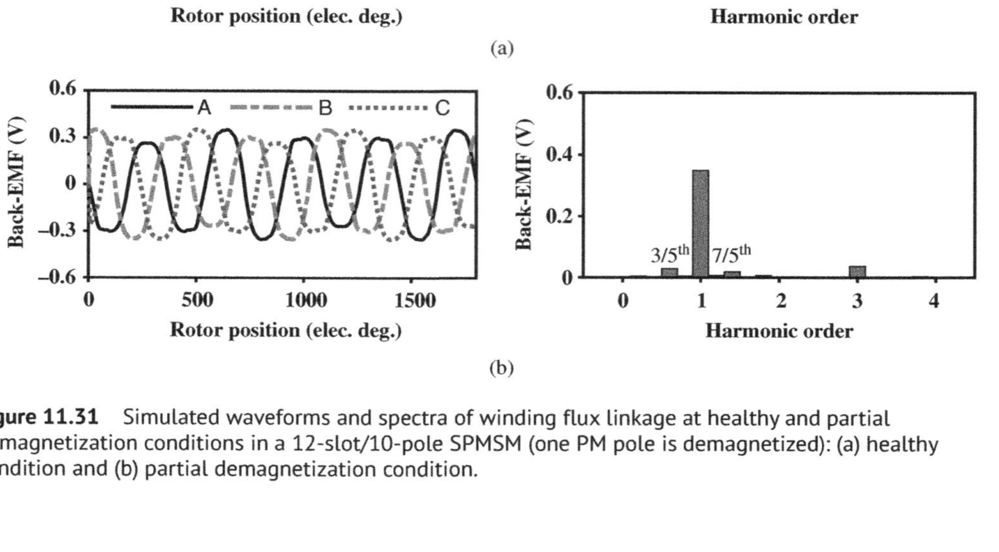

Figure 11.31 Simulated waveforms and spectra of winding flux linkage at healthy and partial demagnetization conditions in a 12-slot/10-pole SPMSM (one PM pole is demagnetized): (a) healthy condition and (b) partial demagnetization condition.

11.7 Summary

This chapter has described several important applications of online parameter estimation techniques, including improvements in control performance under sensored and sensorless control, precise torque estimation, thermal condition monitoring, as well as fault diagnosis, which are presented as follows and summarized in Table 11.3.

1. *Improvement in control performance:*
   - Both electrical and mechanical parameters are required for the design of PI regulators, which can achieve fast dynamic performance and small overshoot once the parameters are estimated delicately.
   - When the optimal MTPA current trajectory is determined by accurate electrical parameters instead of fixing to nominal values, the maximum output torque capability can be improved by more than 3%.
2. *Improvement in sensorless control and precise torque estimation:*
   The mutual inductances, representing the cross-coupling effect, have a great impact on the estimation performance of rotor position under sensorless control, whether it is implemented by the extended back-EMF-based method or the HF signal injection–based method. The same effect is also in the precise torque estimation.
3. *In terms of applications under sensorless control:*
   When the parameter estimation method is applied under sensorless control, the position error and parameter errors are interactive, which deteriorates both the sensorless control performance

Table 11.3 Summary of applications of online parameter estimation techniques.

Application (method)	Identified parameter	Magnetic saturation	VSI non-linearity	Impact/remark
Design of PI regulators (*multi-signal injection*)	Electrical: R_s, L_d, L_q, and ψ_{pm}	Self-	✓	1. Avoid trail-and-error 2. Good dynamic performance
	Mechanical: F and J	■	■	
MTPA current trajectory (*POPE method*)	R_s, L_d, and L_q,	Self-	✓	Enhance maximum torque production
Improvement in sensorless control (*FEA and HF signal injection*)	R_s, L_d, L_q, L_{qd}	Self- and mutual	✓	Improve control performance
Applications in sensorless control (*SWRPO*)	R_s, L_s, ψ_{pm}	■	✓	Overcome error transfer issue
Precise torque estimation (*HF signal injection*)	R_s, L_d, L_q, L_{qd}	Self- and mutual	■ (HF signal)	Improve estimation accuracy
Condition monitoring (*d-axis current injection*)	R_s and ψ_{pm}	■	✓	Average temperature estimation
ITSC fault (*residual-based method*)	R_s, L_d, L_q, L_{dq}, L_{qd}, and ψ_{pm}	Self- and mutual	✓	Establishment for accurate PMSM mode in healthy mode
Demagnetization fault (*HF injection, FEA, etc.*)	L_d, L_{dq}, and ψ_{pm}		✓	

"✓": considered; "■": not involved.

and the parameter estimation accuracy. The SWRPO injection can be employed to overcome this issue, and the estimated electrical parameters are immune to the position error under sensorless control.

4. *Thermal condition monitoring:*
   By utilizing the temperature dependency of stator winding resistance and rotor PM flux linkage, their parameter variations can effectively reflect the average temperature changes.
5. *Fault diagnosis:*
   The accurate machine parameters are critical in fault diagnosis, which can be used as the baseline to detect any deviation caused by either ITSC or demagnetization faults.

References

1 S.-K. Sul, *Control of Electric Machine Drive Systems*. New Jersey: Wiley, 2011.

2 S. Li and Z. Liu, "Adaptive speed control for permanent-magnet synchronous motor system with variations of load inertia," *IEEE Trans. Ind. Electron.*, vol. 56, no. 8, pp. 3050–3059, Aug. 2009.

3 K. Liu and Z. Q. Zhu, "Parameter estimation of PMSM for aiding PI regulator design of field oriented control," *2014 17th Int. Conf. Electr. Mach. Syst. (ICEMS)*, Hangzhou, China, 2014.

4 J. Ziegler and N. Nichols, "Optimum settings for automatic controllers," *J. Fluids Eng.*, vol. 64, no. 8, pp. 759–765, 1942.

5 F. Tinazzi, P. G. Carlet, S. Bolognani, and M. Zigliotto, "Motor parameter free predictive current control of synchronous motors by recursive least square self-commissioning model," *IEEE Trans. Ind. Electron.*, vol. 67, no. 11, pp. 9093–9100, Nov. 2020.

6 M. Khalilzadeh, S. Vaez-Zadeh, and M. S. Eslahi, "Parameter-free predictive control of IPM motor drives with direct selection of optimum inverter voltage vectors," *IEEE J. Emerg. Sel. Topics Power Electron.*, vol. 9, no. 1, pp. 327–334, Feb. 2021.

7 B. Bequette, *Process Control: Modeling, Design, and Simulation*. Upper Saddle River: Prentice Hall, 2003.

8 O. Jacobs, "The damping ratio of an optimal control system," *IEEE Trans. Automatic Control*, vol. 10, no. 4, pp. 473–476, Oct. 1965.

9 S. Morimoto, Y. Takeda, T. Hirasa, and K. Taniguchi, "Expansion of operating limits for permanent magnet motor by current vector control considering inverter capacity," *IEEE Trans. Ind. Appl.*, vol. 26, no. 5, pp. 866–871, Sep./Oct. 1990.

10 B. Wu and M. Narimani, "Control of synchronous motor drives," *High-Power Converters and AC Drives*, IEEE, 2017, pp. 353–391.

11 A. Consoli, G. Scarcella, G. Scelba, and A. Testa, "Steady-state and transient operation of IPMSMs under maximum-torque-per-ampere control," *IEEE Trans. Ind. Appl.*, vol. 46, no. 1, pp. 121–129, Jan./Feb. 2010.

12 S. Kim, Y. Yoon, S. Sul, and K. Ide, "Maximum torque per ampere (MTPA) control of an IPM machine based on signal injection considering inductance saturation," *IEEE Trans. Power Electron.*, vol. 28, no. 1, pp. 488–497, Jan. 2013.

13 A. Shinohara, Y. Inoue, S. Morimoto, and M. Sanada, "Direct calculation method of reference flux linkage for maximum torque per ampere control in DTC-Based IPMSM Drives," *IEEE Trans. Power Electron.*, vol. 32, no. 3, pp. 2114–2122, March 2017.

14 G. Liu, C. Song, and Q. Chen, "FCS-MPC-based fault-tolerant control of five-phase IPMSM for MTPA operation," *IEEE Trans. Power Electron.*, vol. 35, no. 3, pp. 2882–2894, March 2020.

15 P. Wang, Z. Q. Zhu, and D. Liang, "Improved position-offset based online parameter estimation of PMSMs under constant and variable speed operations," *IEEE Trans. Energy Convers.*, vol. 39, no. 2, pp. 1325–1340, June 2024.

16 K. Liu, J. Feng, S. Guo, L. Xiao, and Z. Q. Zhu, "Identification of flux linkage map of permanent magnet synchronous machines under uncertain circuit resistance and inverter nonlinearity," *IEEE Trans. Ind. Inform.*, vol. 14, no. 2, pp. 556–568, Feb. 2018.

17 Z. Q. Zhu and X. M. Wu, *Sensorless Control of Permanent Magnet Synchronous Machine Drives.* New York: Wiley and IEEE Press, 2023.

18 E. Armando, R. Bojoi, P. Guglielmi, G. Pellegrino, and M. Pastorelli, "Experimental identification of the magnetic model of synchronous machines," *IEEE Trans. Ind. Appl.*, vol. 49, no. 5, pp. 2116–2125, Sept./Oct. 2013.

19 Y. Inoue, Y. Kawaguchi, S. Morimoto, and M. Sanada, "Performance improvement of sensorless IPMSM drives in a low-speed region using online parameter identification," *IEEE Trans. Ind. Appl.*, vol. 47, no. 2, pp. 798–804, Mar./Apr. 2011.

20 S. J. Underwood and I. Husain, "Online parameter estimation and adaptive control of permanent-magnet synchronous machines," *IEEE Trans. Ind. Electron*, vol. 57, no. 7, pp. 2435–2443, July 2010.

21 Y. Inoue, K. Yamada, S. Morimoto, and M. Sanada, "Effectiveness of voltage error compensation and parameter identification for model-based sensorless control of IPMSM," *IEEE Trans. Ind. Appl.*, vol. 45, no. 1, pp. 213–221, Jan./Feb. 2009.

22 M. Rashed, P. Macconnell, A. Stronach, and P. Acarnley, "Sensorless indirect-rotor-field-orientation speed control of a permanent-magnet synchronous motor with stator-resistance estimation," *IEEE Trans. Ind. Electron.*, vol. 54, no. 3, pp. 1664–1675, Jun. 2007.

23 Y. Li, Z. Q. Zhu, D. Howe, and C. M. Bingham, "Improved rotor position estimation in extended back-EMF based sensorless PM brushless AC drives with magnetic saliency," *2007 IEEE Int. Electr. Mach. Drives Conf.*, Antalya, Turkey, 2007, pp. 214–219.

24 Y. Li, "Sensorless control of permanent magnet brushless AC motors accounting for cross-coupling magnetic saturation," PhD dissertation, University of Sheffield, Sheffield, U.K., June 2009.

25 L. M. Gong and Z.Q. Zhu, "A novel method for compensating inverter nonlinearity effects in carrier signal injection-based sensorless control from positive-sequence carrier current distortion," *IEEE Trans. Ind. Appl.*, vol. 47, no. 3, pp. 1283–1292, May/Jun. 2011.

26 P. Guglielmi, M. Pastorelli, and A. Vagati, "Cross saturation effects in IPM motors and related impact on zero-speed sensorless control," In *Conf. Rec. IEEE IAS Annu. Meeting*, 2005, vol. 4, pp. 2546–2552.

27 Y. Li, Z.Q. Zhu, D. Howe, C. M. Bingham, and D. A. Stone, "Improved rotor-position estimation by signal injection in brushless AC motors, accounting for cross-coupling magnetic saturation," *IEEE Trans. Ind. Appl.*, vol. 45, no. 5, pp. 1843–1850, Sept./Oct. 2009.

28 W. Xu and R. D. Lorenz, "High-frequency injection-based stator flux linkage and torque estimation for DB-DTFC implementation on IPMSMs considering cross-saturation effects," *IEEE Trans. Ind. Appl.*, vol. 50, no. 6, pp. 3805–3815, Nov./Dec. 2014.

29 J. Jang, S. Sul, J. Ha, K. Ide and M. Sawamura, "Sensorless drive of surface-mounted permanent-magnet motor by high-frequency signal injection based on magnetic saliency," *IEEE Trans.Ind. Appl.*, vol. 39, no. 4, pp. 1031–1039, July–Aug. 2003.

30 J. Jang, J. Ha, M. Ohto, K. Ide, and S. Sul, "Analysis of permanent-magnet machine for sensorless control based on high-frequency signal injection," *IEEE Trans. Ind. Appl.*, vol. 40, no. 6, pp. 1595–1604, Nov./Dec. 2004.

31 Z. Q. Zhu, P. Wang, N. M. A. Freire, Ziad Azar, and X. M. Wu, "A novel rotor position-offset injection-based online parameter estimation of sensorless controlled surface-mounted PMSMs," *IEEE Trans. Energy Convers.*, vol. 39, no. 3, pp. 1930–1946, Sep. 2024.

32 Z.Q. Zhu and D. Howe, "Electrical machines and drives for electric, hybrid, and fuel cell vehicles," *Proc. IEEE*, vol. 95, no. 4, pp. 746–765, April 2007.

33 T. Ito and K. Akatsu, "Electromagnetic force acquisition distributed in electric motor to reduce vibration," *IEEE Trans. Ind. Appl.*, vol. 53, no. 2, pp. 1001–1008, Mar./Apr. 2017.

34 Z. Q. Chen, M. Tomita, S. Doki, and S. Okuma, "An extended electromotive force model for sensorless control of interior permanent-magnet synchronous motors," *IEEE Trans. Ind. Electron.*, vol. 50, no. 2, pp. 288–295, Apr 2003.

35 R. Hejny and R. Lorenz, "Evaluating the practical low-speed limits for back-EMF tracking-based sensorless speed control using drive stiffness as a key metric," *IEEE Trans. Ind. Appl.*, vol. 47, no. 3, pp. 1337–1343, May 2011.

36 G. Chen and S. Yang, "Phase voltage measurement for permanent magnet machine sensorless drive using controller capture modulator," *IEEE Trans. Ind. Electron*, vol. 67, no. 1, pp. 49–58, Jan. 2020.

37 M. Martinez, D. Reigosa, D. Fernández, J. M. Guerrero, and F. Briz, "Enhancement of permanent-magnet synchronous machines torque estimation using pulsating high-frequency current injection," *IEEE Trans. Ind. Appl.*, vol. 56, no. 1, pp. 358–366, Jan./Feb. 2020.

38 Y. G. Kang, D. Reigosa, B. Sarlioglu, and R. D. Lorenz, "D- and Q-axis inductance estimation and self-sensing condition monitoring using 45° angle high-frequency injection," *IEEE Trans. Ind. Appl.*, vol. 57, no. 1, pp. 506–515, Jan./Feb. 2021.

39 B. Shuang and Z. -Q. Zhu, "Simultaneous sensorless rotor position and torque estimation for IPMSM at standstill and low speed based on high-frequency square wave voltage injection," *IEEE Trans. Ind. Electron.*, vol. 69, no. 9, pp. 8791–8802, Sept. 2022.

40 B. Shuang, "Sensorless torque and rotor position control for interior permanent magnet machines at standstill and low speed," PhD dissertation, University of Sheffield, Sheffield, U.K., March 2021.

41 S. Ichikawa, M. Tomita, S. Doki, and S. Okuma, "Sensorless control of permanent-magnet synchronous motors using online parameter identification based on system identification theory," *IEEE Trans. Ind. Electron.*, vol. 53, no. 2, pp. 363–372, Apr. 2006.

42 K. Liu, Z. Q. Zhu, and D. A. Stone, "Parameter estimation for condition monitoring of PMSM stator winding and rotor permanent magnets," *IEEE Trans. Ind. Electron.*, vol. 60, no. 12, pp. 5902–5913, Dec. 2013.

43 H. Lee, M. L. Woldesemayat, and K. Nam, "Zero torque control for EV coasting considering cross-coupling inductance," *IEEE Trans. Ind. Electron.*, vol. 64, no. 8, pp. 6096–6104, Aug. 2017.

44 Z. Q. Zhu and D. Liang, "Perspective of thermal analysis and management for permanent magnet machines, with particular reference to hotspot temperatures," *Energies*, vol. 15, no. 21, pp. 1–53, Nov. 2022.

45 "IEEE standard for the preparation of test procedures for the thermal evaluation of solid electrical insulating materials," *IEEE Std 98-2002*, pp. 1–16, 2002.

46 T. Sebastian, "Temperature effects on torque production and efficiency of PM motors using NdFeB magnets," *IEEE Trans. Ind. Appl.*, vol. 31, no. 2, pp. 353–357, March/April 1995.

47 D. D. Reigosa, F. Briz, P. García, J. M. Guerrero, and M. W. Degner, "Magnet temperature estimation in surface PM machines using high-frequency signal injection," *IEEE Trans. Ind. Appl.*, vol. 46, no. 4, pp. 1468–1475, July/Aug. 2010.

48 M. Ganchev, H. Umschaden, and H. Kapeller, "Rotor temperature distribution measuring system," *Proc. 2011 - 37th Annu. Conf. IEEE Ind. Electron. Soc.*, Melbourne, Australia, Nov. 2011, pp. 2006- 2011.

49 D. Liang, "Critical temperature estimation for permanent magnet synchronous machines," PhD dissertation, University of Sheffield, Sheffield, U.K., Feb. 2022.

50 S. Wilson, P. Stewart, and B. Taylor, "Methods of resistance estimation in permanent magnet synchronous motors for real-time thermal management," *IEEE Trans. Energy Convers.*, vol. 25, no. 3, pp. 698–707, Sept. 2010.

51 O. Wallscheid, A. Specht, and J. Böcker, "Observing the permanent-magnet temperature of synchronous motors based on electrical fundamental wave model quantities," *IEEE Trans. Ind. Electron.*, vol. 64, no. 5, pp. 3921–3929, May 2017.

52 Z. Gao, C. Cecati, and S. Ding, "A survey of fault diagnosis and fault-tolerant techniques—Part I: fault diagnosis with model-based and signal-based approaches," *IEEE Trans. Ind. Electron.*, vol. 62, no. 6, pp. 3757–3767, Jun. 2015.

53 Z. Gao, C. Cecati, and S. Ding, "A survey of fault diagnosis and fault-tolerant techniques—part II: fault diagnosis with knowledge-based and hybrid/active approaches," *IEEE Trans. Ind. Electron.*, vol. 62, no. 6, pp. 3757–3767, Jun. 2015.

54 N. Leboeuf, T. Boileau, B. Nahid-Mobarakeh, G. Clerc, and F. Meibody-Tabar, "Real-time detection of interturn faults in PM drives using back-EMF estimation and residual analysis," *IEEE Trans. Ind. Appl.*, vol. 47, no. 6, pp. 2402–2412, Nov./Dec. 2011.

55 M. A. Mazzoletti, G. R. Bossio, C. H. De Angelo, and D. R. Espinoza-Trejo, "A Model-based strategy for interturn short-circuit fault diagnosis in PMSM," *IEEE Trans. Ind. Electron.*, vol. 64, no. 9, pp. 7218–7228, Sept. 2017.

56 A. Mahmoudi, I. Jlassi, A. J. M. Cardoso, K. Yahia, and M. Sahraoui, "Inter-turn short-circuit faults diagnosis in synchronous reluctance machines, using the Luenberger state observer and current's second-order harmonic," *IEEE Trans. Ind. Electron.*, vol. 69, no. 8, pp. 8420–8429, August 2022.

57 A. Sarikhani and O. A. Mohammed, "Inter-turn fault detection in PM synchronous machines by physics-based back electromotive force estimation," *IEEE Trans. Ind. Electron.*, vol. 60, no. 8, pp. 3472–3484, Aug. 2013.

58 J. Hang, Q. Hu, W. Sun, X. Ren, S. Ding, Y. Huang, and W. Hua, "A voltage-distortion-based method for robust detection and location of interturn fault in permanent magnet synchronous machine," *IEEE Trans. Power Electron.*, vol. 37, no. 9, pp. 11174–11186, Sept. 2022.

59 B. Du, S. Wu, S. Han, and S. Cui, "Interturn fault diagnosis strategy for interior permanent-magnet synchronous motor of electric vehicles based on digital signal processor," *IEEE Trans. Ind. Electron.*, vol. 63, no. 3, pp. 1694–1706, March 2016.

60 R. Cui, Y. Fan, and C. Li, "On-line inter-turn short-circuit fault diagnosis and torque ripple minimization control strategy based on OW five-phase BFTHE-IPM," *IEEE Trans. Energy Convers.*, vol. 33, no. 4, pp. 2200–2209, Dec. 2018.

61 G. Li, B. Ren, Z. Q. Zhu, M. P. Foster, and D. A. Stone, "Demagnetization withstand capability enhancement of surface mounted PM machines using stator modularity," *IEEE Trans. Ind. Appl.*, vol. 54, no. 2, pp. 1302–1311, 2018.

62 W. le Roux, R. G. Harley, and T. G. Habetler, "Detecting rotor faults in low power permanent magnet synchronous machines," *IEEE Trans. Power Electron.*, vol. 22, no. 1, pp. 322–328, Jan. 2007.

63 S. Moon, J. Lee, H. Jeong, and S. W. Kim, "Demagnetization fault diagnosis of a PMSM based on structure analysis of motor inductance," *IEEE Trans. Ind. Electron.*, vol. 63, no. 6, pp. 3795–3803, June 2016.

64 Y. Han, S. Chen, C. Gong, X. Zhao, F. Zhang, and Y. Li, "Accurate SM disturbance observer-based demagnetization fault diagnosis with parameter mismatch impacts eliminated for IPM motors," *IEEE Trans. Power Electron.*, pp. 1–5, 2023.
65 Z. Liu, J. Huang, and B. Li, "Diagnosing and distinguishing rotor eccentricity from partial demagnetisation of interior PMSM based on fluctuation of high-frequency d-axis inductance and rotor flux," *IET Electric Power Appl.*, vol. 11, no. 7, pp. 1265–1275, 2017.

A

Finite Element Calculation of Winding Inductances

This appendix describes the finite element (FE) method for calculating apparent/incremental self- and mutual-inductances in the *dq*-axes, which is used to validate the high-frequency-based inductance estimation methods in Chapter 7 and to illustrate the importance of accurate inductance estimation in sensorless control in Chapter 11.

First, the apparent and incremental inductances, i.e., L_d^i, L_q^i, L_{dq}^i, and L_{qd}^i with {i = "ap" and "inc"}, are defined as follows:

Apparent inductances

$$\begin{cases} L_d^{\text{ap}} = \dfrac{\psi_d(i_d, 0) - \psi_{pm}(i_d, i_q)}{i_d}, & L_{dq}^{\text{ap}} = \dfrac{\psi_d(i_d, i_q) - \psi_d(i_d, 0)}{i_q}, \\ L_{qd}^{\text{ap}} = \dfrac{\psi_q(i_d, i_q) - \psi_q(0, i_q)}{i_d}, & L_q^{\text{ap}} = \dfrac{\psi_q(0, i_q)}{i_q} \end{cases} \tag{A.1}$$

Incremental inductances

$$\begin{cases} L_d^{\text{inc}} = \dfrac{\partial \psi_d}{\partial i_d}, & L_{dq}^{\text{inc}} = \dfrac{\partial \psi_d}{\partial i_q}, \\ L_{qd}^{\text{inc}} = \dfrac{\partial \psi_q}{\partial i_d}, & L_q^{\text{inc}} = \dfrac{\partial \psi_q}{\partial i_q} \end{cases} \tag{A.2}$$

where ψ_d and ψ_q are the *d*-axis and *q*-axis flux linkages. ψ_{pm} is the rotor permanent magnet (PM) flux linkage, of which the working point is affected by *dq*-axis armature reaction fields (see Chapter 7). However, for highly saturated permanent magnet synchronous machines (PMSMs), due to the existence of PM flux linkage, the *d*-axis flux linkages, $\psi_d(i_d, 0)$ and $\psi_{pm}(i_d, i_q)$, are usually at different working points with different magnetic saturation levels, and it is necessary to employ a frozen permeability method in finite element analysis (FEA) [1–3] to calculate $\psi_d(i_d, 0)$ and $\psi_{pm}(i_d, i_q)$. Furthermore, due to high magnetic saturation, $\psi_d(i_d, i_q)$, $\psi_d(i_d, 0)$, $\psi_q(i_d, i_q)$, and $\psi_q(0, i_q)$ may be operating at different working points. L_{dq}^{inc} may be not equal to L_{qd}^{inc}, i.e., $L_{dq}^{\text{inc}} \neq L_{qd}^{\text{inc}}$. However, due to complexity in calculating the winding inductances and their difference is usually small, and hence, for simplicity, it is usually assumed that $L_{dq}^{\text{inc}} = L_{qd}^{\text{inc}}$.

In the FE calculation, the incremental self- and mutual-inductances are calculated as follows [4–6]:

$$L_{dh} = \frac{\left[\psi_d(i_d + \Delta i_d, i_q, \psi_{pm}) - \psi_d(i_d, i_q, \psi_{pm})\right]}{\Delta i_d} \tag{A.3a}$$

Parameter Estimation of Permanent Magnet Synchronous Machines, First Edition. Zi Qiang Zhu, Kan Liu, and Dawei Liang.

$$L_{qh} = \frac{[\psi_q(i_d, i_q + \Delta i_q, \psi_{pm}) - \psi_q(i_d, i_q, \psi_{pm})]}{\Delta i_q} \tag{A.3b}$$

$$L_{dqh} = \frac{[\psi_d(i_d, i_q + \Delta i_q, \psi_{pm}) - \psi_d(i_d, i_q, \psi_{pm})]}{\Delta i_q} \tag{A.3c}$$

$$L_{qdh} = \frac{[\psi_q(i_d + \Delta i_d, i_q, \psi_{pm}) - \psi_q(i_d, i_q, \psi_{pm})]}{\Delta i_d} \tag{A.3d}$$

where Δi_d and Δi_q are the dq-axis current increments. For simplicity, ψ_{pm}, the rotor PM flux linkage, is often fixed to the no-load value ($i_d = 0\,\text{A}$ and $i_q = 0\,\text{A}$). However, the influence of the magnetic saturation on rotor PM flux linkage can also be considered by using the frozen permeability method in FEA [1–3] or by subtracting contributions of dq-axis inductances from the dq-axis flux linkages [7]. By way of example, Prototype Machine 2 with 18 slots/6 poles (see Appendix B) is employed for illustration, of which the cross-section is shown in Figure A.1. The flowchart of FE calculation is presented in Figure A.2.

The two-dimensional (2D) electromagnetic FEA software ANSYS [8] is employed for calculation. After the setup of the geometric model, material properties, and current excitations, the calculated three-phase flux linkages, i.e., ψ_a, ψ_b, and ψ_c, at the current point $i_d = 0\,\text{A}$ and $i_q = 4\,\text{A}$ are shown in Figure A.3a. By using Park transformation $\boldsymbol{P}$, they can also be transferred into dq-axes in Figure A.3b, which is expressed as follows:

$$\begin{bmatrix} \psi_d \\ \psi_q \\ \psi_N \end{bmatrix} = \boldsymbol{P} \begin{bmatrix} \psi_a \\ \psi_b \\ \psi_c \end{bmatrix} \tag{A.4a}$$

$$\boldsymbol{P} = \frac{2}{3} \begin{bmatrix} \cos(\theta) & \cos(\theta - 2/3\pi) & \cos(\theta + 2/3\pi) \\ -\sin(\theta) & -\sin(\theta - 2/3\pi) & -\sin(\theta + 2/3\pi) \\ 1/2 & 1/2 & 1/2 \end{bmatrix} \tag{A.4b}$$

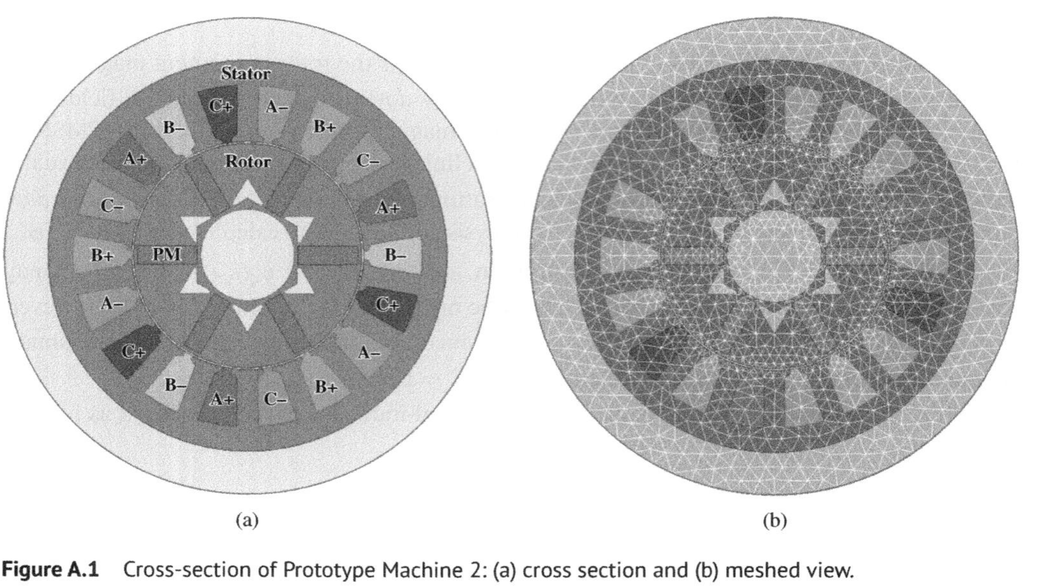

Figure A.1 Cross-section of Prototype Machine 2: (a) cross section and (b) meshed view.

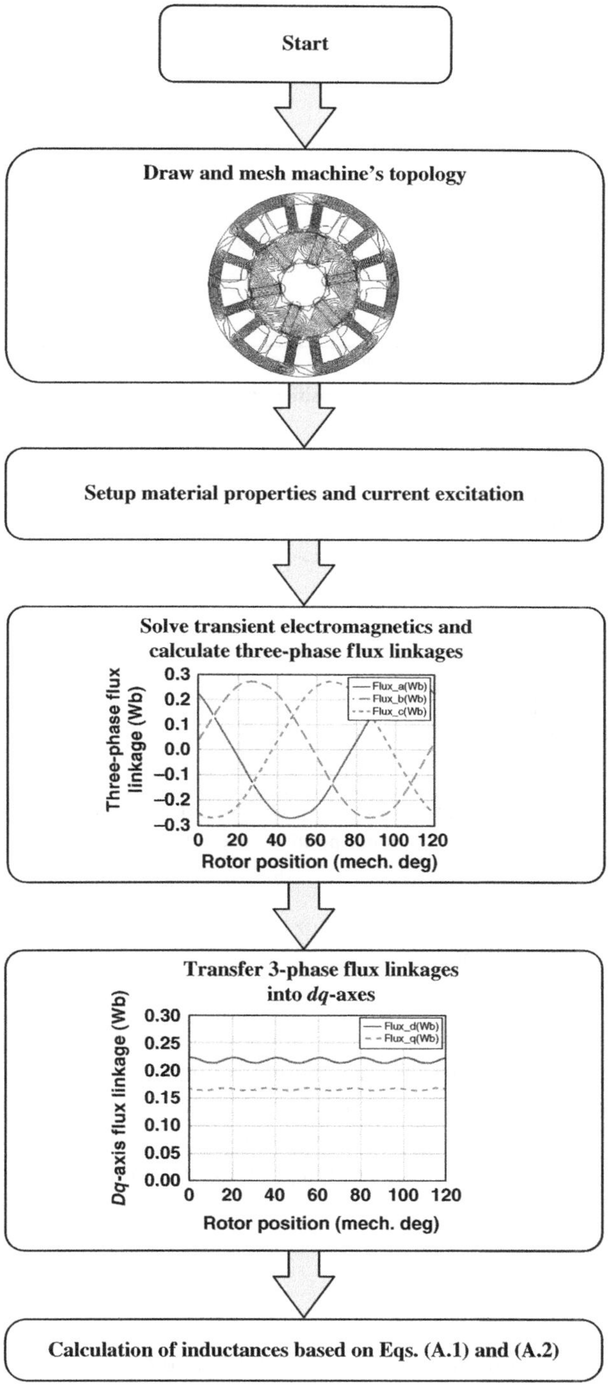

Figure A.2 Flowchart of FE calculation of inductances.

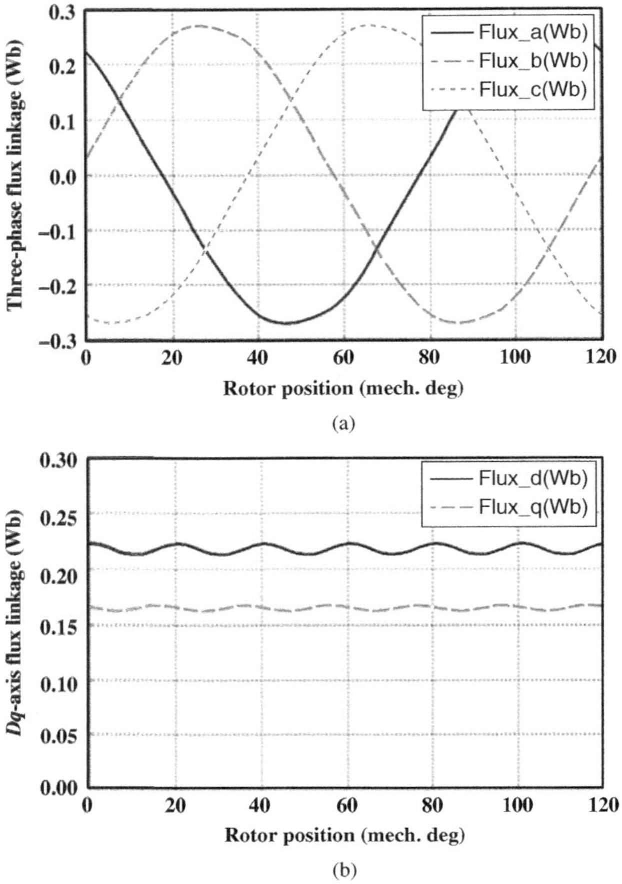

Figure A.3 FE-calculated three-phase and dq-axis flux linkages at $i_d = 0$ A and $i_q = 4$ A: (a) three-phase flux linkages and (b) dq-axis flux linkages.

As can be seen, the dq-axis flux linkages, i.e., ψ_d and ψ_q, contain a ripple component as a function of rotor position due to the harmonics in three-phase flux linkages, which is usually neglected during the parameter estimation. Therefore, the rotor position terms are eliminated from dq-axis flux linkages by averaging their values over one ripple period, i.e., $\overline{\psi_d}$ and $\overline{\psi}_q$, as calculated by Eq. (A.5). Afterwards, the variations in the dq-axis flux linkages with currents are shown in Figure A.4.

$$\begin{cases} \overline{\psi_d} = \dfrac{1}{2\pi}\int_0^{2\pi} \psi_d(i_d, i_q, \theta_r) d\theta_r \\ \overline{\psi_q} = \dfrac{1}{2\pi}\int_0^{2\pi} \psi_q(i_d, i_q, \theta_r) d\theta_r \end{cases} \tag{A.5}$$

The apparent self- and mutual-inductances, i.e., L_d, L_q, L_{dq}, and L_{qd}, can be calculated by Eq. (A.6), and the FE-calculated apparent inductances at various dq-axis currents are shown in Figure A.5.

$$L_d(i_d) = \frac{\overline{\psi_d}(i_d, 0) - \psi_{pm}}{i_d} \tag{A.6a}$$

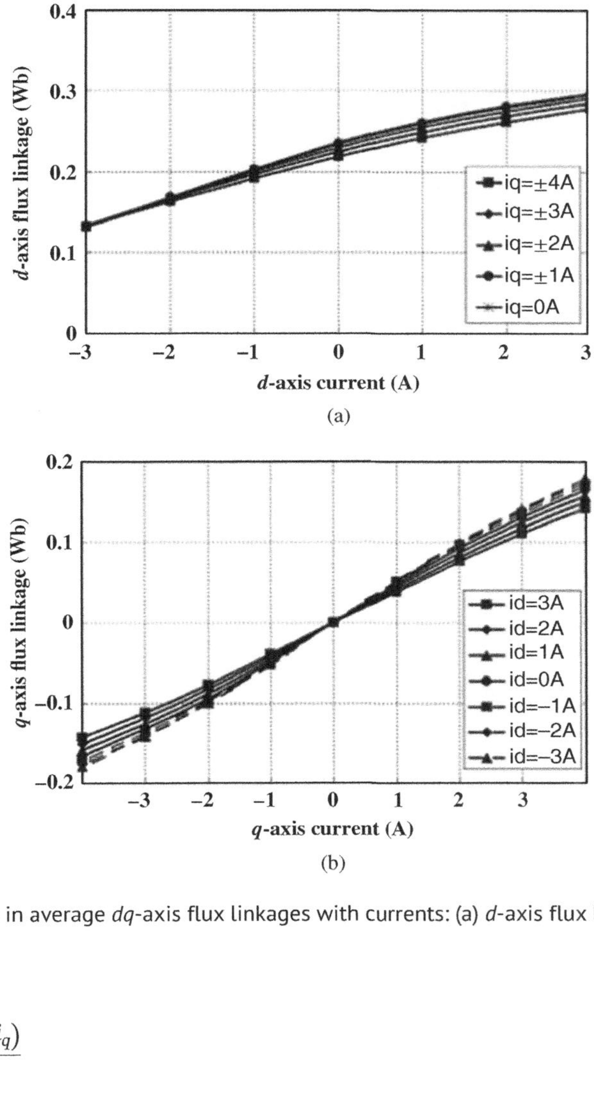

Figure A.4 Variations in average *dq*-axis flux linkages with currents: (a) *d*-axis flux linkage and (b) *q*-axis flux linkage.

$$L_q(i_q) = \frac{\overline{\psi_q}(0, i_q)}{i_q} \tag{A.6b}$$

$$L_{dq}(i_d, i_q) = \frac{\overline{\psi_d}(i_d, i_q) - \overline{\psi_d}(i_d, 0)}{i_q} \tag{A.6c}$$

$$L_{qd}(i_d, i_q) = \frac{\overline{\psi_q}(i_d, i_q) - \overline{\psi_q}(0, i_q)}{i_d} \tag{A.6d}$$

Furthermore, based on the definitions of incremental self- and mutual-inductances in Eq. (A.3), the FE-calculated results are shown in Figure A.6. The rated current of Prototype Machine 2 is 4 A, and the current increments, i.e., Δi_d and Δi_q, are selected as 0.1 A. As can be observed, the FE results exhibit a good agreement with the measured results in Chapter 7. Meanwhile, the improvement in sensorless control performance by considering cross-coupling magnetic saturation, described in Chapter 7, is also based on the above FE results.

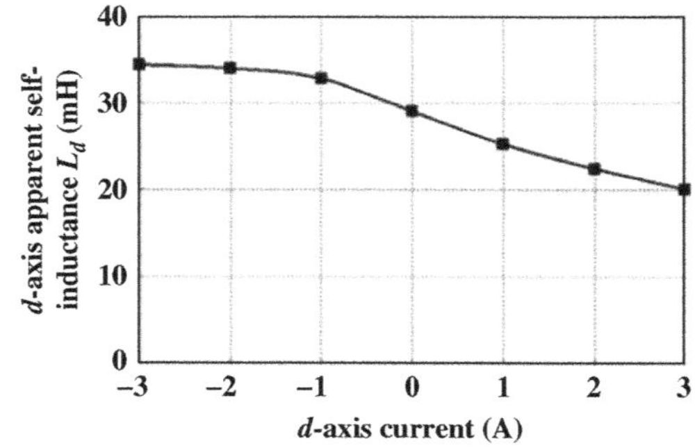

(a)

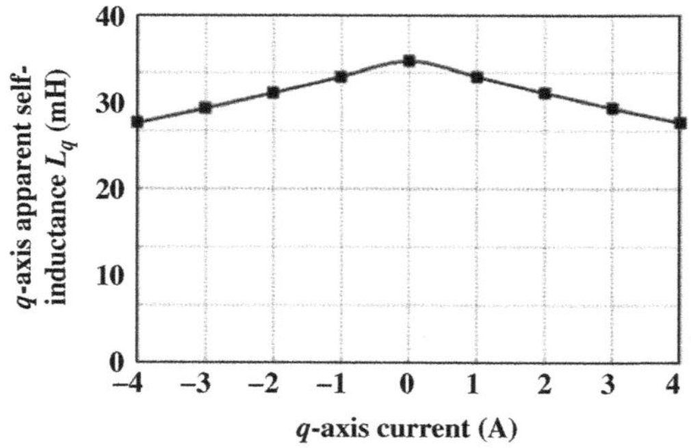

(b)

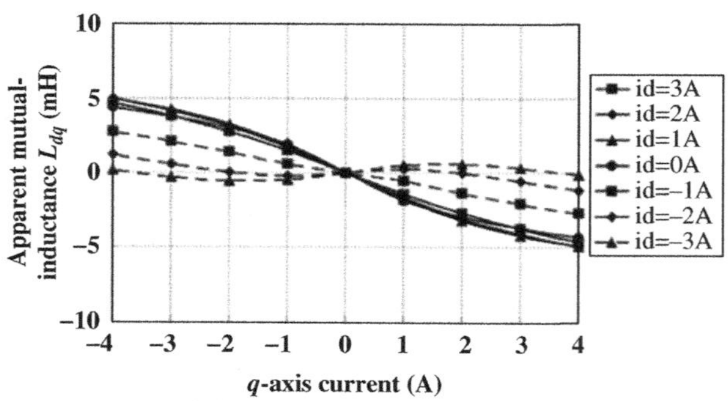

(c)

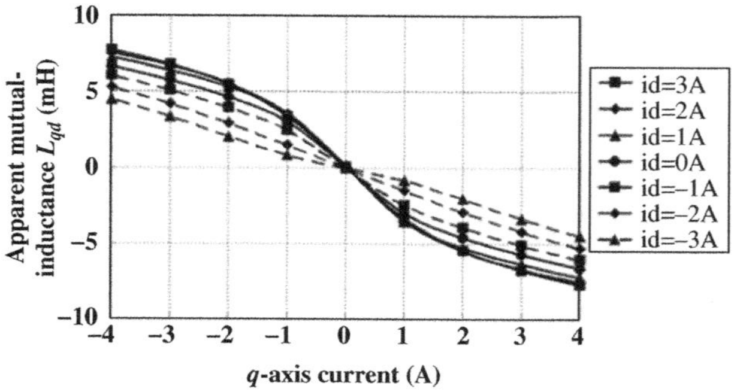

(d)

Figure A.5 Variations in dq-axis apparent self- and mutual-inductances with currents: (a) d-axis apparent self-inductance L_d, (b) q-axis apparent self-inductance L_q, (c) apparent mutual-inductance L_{dq}, and (d) apparent mutual-inductance L_{qd}.

Figure A.6 Variations in dq-axis incremental self- and mutual-inductances with currents: (a) d-axis incremental self-inductance L_d, (b) q-axis incremental self-inductance L_q, and (c) incremental mutual-inductance L_{dq}.

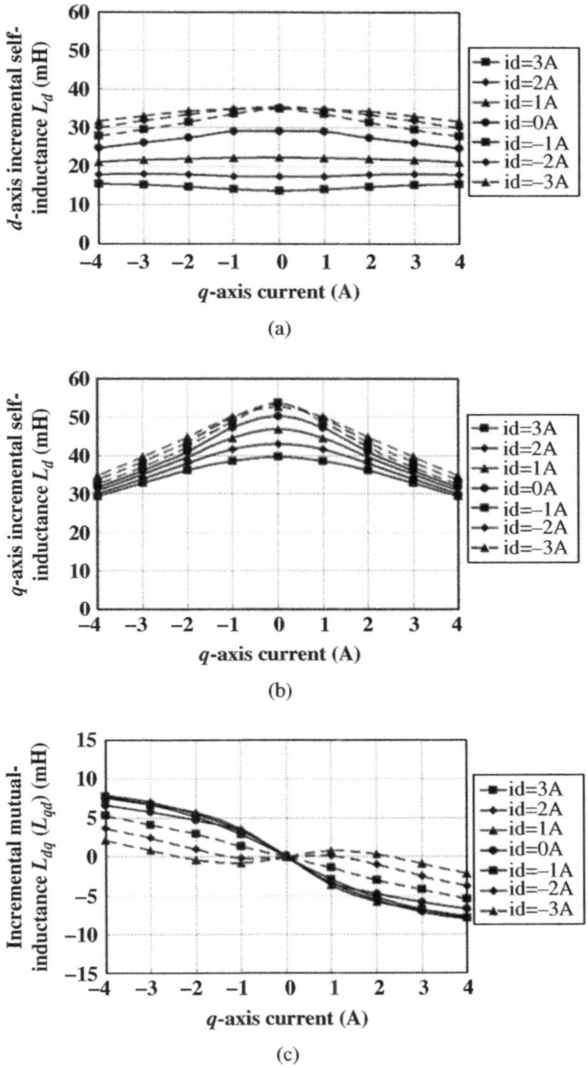

References

1 N. Bianchi and S. Bolognani, "Magnetic models of saturated interior permanent magnet motors based on finite element analysis," *Conf. Rec. IEEE IAS Annu. Meeting*, Oct. 12–15, 1998, pp. 27–34.

2 Z. Q. Zhu and W. Q. Chu, "Advanced frozen permeability technique and applications in developing high performance electrical machines," *Trans. China Electrotech. Society*, vol.31, no. 20, pp. 13–29, Oct. 2016.

3 W. Q. Chu and Z. Q. Zhu, "Average torque separation in permanent magnet synchronous machines using frozen permeability," *IEEE Trans. Magn.*, vol. 49, no. 3, pp. 1202–1210, March 2013.

4 Y. S. Chen "Motor topologies and control strategies for permanent magnet brushless AC drives," PhD dissertation, University of Sheffield, Sheffield, U.K., Sept. 1999.

5 Y. Li, "Sensorless control of permanent magnet brushless AC motors accounting for cross-coupling magnetic saturation," PhD dissertation, University of Sheffield, Sheffield, U.K., June 2009.

6 B. Shuang, "Sensorless torque and rotor position control for interior permanent magnet machines at standstill and low speed," PhD dissertation, University of Sheffield, Sheffield, U.K., March 2021.

7 K. Yamazaki and M. Kumagai, "Torque analysis of interior permanent-magnet synchronous motors by considering cross-magnetization: variation in torque components with permanent-magnet configurations," *IEEE Trans. Ind. Electron.*, vol. 61, no. 7, pp. 3192–3201, July 2014.

8 *ANSYS 17.1*. Ansys, inc. Accessed: 2018 [Online]. Available: https://www.ansys.com.

B

Specifications of Prototype Machines and Experimental Platforms

There are three prototype permanent magnet synchronous machines (PMSMs), Figure B.1, are used in this book and their specifications are listed in Table B.1. A typical laboratory experimental platform is shown in Figure B.2, which consists of a Prototype Machine 3, a dSPACE DS1006 control processer, a direct current (dc) power supply, a two-level voltage source inverter, and a load machine. The pulse width modulation (PWM) is generated with space vector pulse width modulation

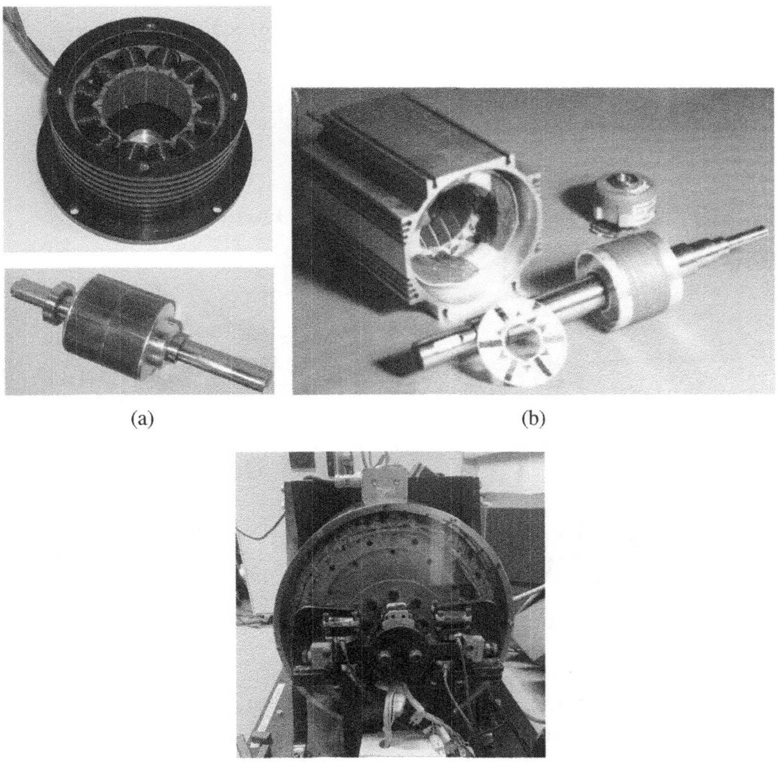

(a) (b)

(c)

Figure B.1 Prototype PMSMs: (a) Prototype Machine 1, (b) Prototype Machine 2, and (c) Prototype Machine 3.

Parameter Estimation of Permanent Magnet Synchronous Machines, First Edition. Zi Qiang Zhu, Kan Liu, and Dawei Liang.

Table B.1 Specifications of prototype PMSMs.

Parameters	Prototype Machine 1 (SPMSM)	Prototype Machine 2 (IPMSM)	Prototype Machine 3 (SPMSM)
Applications in Chapters	2, 3, 4, 5, 8, 9, 10	5, 6, 7, 8, 9, 10	11
Pole number	10	6	32
Slot number	12	18	96
Rated current (A)	10	4	5
Rated power (W)	300	600	3000
Rated speed (r/min)	400	1000	170
dc-link voltage (V)	36	158	600
Nominal phase resistance (Ω)	0.30	6.0	4.1
Terminal wire resistance (Ω)	0.2	0.2	0.125
Nominal d-axis inductances (mH)	3.24	38.10	21.0
Nominal q-axis inductances(mH)	3.24	55.00	21.0
Nominal rotor PM flux linkage (mWb)	70.7	236.0	1000
Nominal moment of inertia (kg m^2)	3.41×10^{-4}	3.61×10^{-4}	–

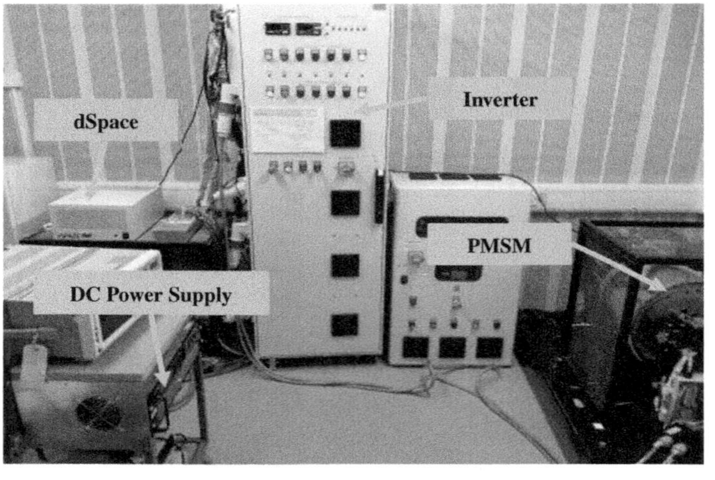

Figure B.2 Typical experimental platform.

(SVPWM) strategy. The prototype PMSMs, three-phase stator currents, and dc-link voltage are measured by Hall-effect sensors, and the rotor position and speed are measured by an incremental encoder, which are recorded by the ControlDesk software. In addition, the shaft torque can be measured by a torque sensor, Magtrol TM308/011, which is installed between the test machine and the load machine. The typical time constants of dSPACE-based drive system and the main characteristics of voltage source inverter (VSI) (Semikron SEMiX71GD12E4s) are tabulated in Table B.2.

Table B.2 Typical characteristics of drive system and VSI (at 25 °C).

Parameter	Value	Parameter	Value
Sampling time of drive system (μs)	400	Dead time in PWM (μs)	1.3
Turn on/off delay (μs)	0.16/0.433	Hardware dead time (μs)	4
Voltage drop of the active switch (V)	1.85	Switch control dead time (μs)	2
Voltage drop of the freewheeling diode (V)	2.20	Supply voltage range (V)	0–650

Index

Note: Page numbers in *italics* and **bold** refers to figures and tables respectively.

a

abc stationary reference frame, modelling in 30–32
acceleration/deceleration-based method 166–167, 170
Adaline estimator 52
Adaline neural network (ANN) 17, 24, 59, 153, 183, 192
  application to PMSM parameter estimation 193–195, *194*
  basic principle 192
  dq-axis inductance, stator winding resistance, and rotor PM flux linkage, estimated *195*
  structure *193*
  subnet structures of ANN estimators *194*
adaptive algorithm-based methods 183
adaptive law 184, 191–192
AI. *see* artificial intelligence (AI)
alternating current (ac) 12. *see also* direct current (dc)
  signal excitation 132
  three-phase 12
ANN. *see* Adaline neural network (ANN)
ANSYS software 244
apparent inductance(s) 7–8, *7*, 16, 82, 131, *131*, 132, 150, 243
  dq-axis 125, 133
  equivalent 9, 133
  estimation 17
  FE-calculated 246
  stator 16
apparent reluctivity 9, *9*
artificial intelligence (AI) 1

c

circumferentially magnetized IPM rotors 3
closed-loop transfer function 210, 212
constant torque/speed control 60, 64, *83*. *see also* variable speed control
  d-axis inductance, estimation of 82–84
  designed estimation method 78–79, 94–95, 99–101
  designed parameter estimation scheme 68–69
  dq-axis inductance map estimation errors *105*
  dq-axis inductances, estimation of *88*, *89*, 95–96, *96*
  dq-axis voltages and currents, measured during estimation *85*
  error analysis and experimental validation 69–72, *71*
  estimated parameters under constant/variable speed operations *103*
  estimation of R_s and ψ_{pm} based on measured inputs/outputs under *69*
  experimental and FEA results 84–89, 96–98, 101–103
  experimental platform and Prototype Machine 2, *102*
  external resistors, estimation results after addition of *87*
  FEA and POPE estimated *d*-axis inductance maps *104*
  FEA and POPE estimated *q*-axis inductance maps *104*
  gain functions *100*
  $i_d = 0$ and $i_d \neq 0$ under 99–103, *99*
  $i_d \neq 0$ under 94–98
  measured rotor speeds and reference *dq*-axis voltages *102*
  parameter estimation error variations *106*
  position offset influence on *dq*-axis back-EMFs **80**
  position offset-based estimation with $i_d = 0$ under 77
  principle of POPE with associated FOC system *78*
  Prototype Machines 1 and 2, main parameters of **84**
  q-axis inductance, estimation of 81–82

Parameter Estimation of Permanent Magnet Synchronous Machines, First Edition. Zi Qiang Zhu, Kan Liu, and Dawei Liang.

constant torque/speed control. *see also* variable speed control (*Continued*)
- q-axis inductances, comparison of *89*
- rotor PM flux linkage and inductances' saliency, estimation of 95
- rotor PM flux linkage and stator winding resistance, estimation of 79–81
- rotor PM flux linkage and winding resistance estimation *86*
- rotor PM flux linkages and variations, measured and estimated *86*, *87*, *98*
- stator winding resistance and rotor PM linkage *72*
- tracking temperature variations, estimated parameters for *72*
- winding resistance and rotor flux linkage estimation based on current injection 68–72

conventional estimation method 153
conventional Kalman filter 187–189
cost function(s) 91–92, 101, 113–115, 121
- flux linkage identification modeling 119–121
- minimization of 121
- rank-deficient 123

cross-coupling effect 1, 8, *8*, 121, 132, 135, 144, 154, 209, 223–224, 229
- HF square wave voltage injection 230–231
- magnetic 8
- in PMSMs 17

current injected-based parameter estimation scheme 184
current injections 59
- current injection–based parameter estimation *60*
- HF 153, 158
- multi-parameter estimation based error analysis and 60–68
- under constant torque/speed control 68–72

current variations
- dq-axis inductances, measured quantities for estimation of *156*, *157*
- electrical quantities measured for estimation of R_s and V_{dead} *154*
- estimation based on 153
- estimation results *155*
- winding resistance estimation without compensating for VSI non-linearity *156*

d

data recording and FE-predicted flux linkage maps 122, *123*
d-axis current injection–based estimation methods 60
d-axis inductance 3–4, 63, *67*, 75, 88, *88*
- estimation of 82–84, *118*
- maps predicted by FE and ICQGA-based method *127*

delay compensation angle 143, *143*
delay effect 143
- measured results 140, 142
- time 132–133, *134*, 136–137, 139, 144

demagnetization 236
- effect in d-axis 97
- faults 234–236, *236*
- irreversible 1, 11, *235*
- PM 209, 236
- reversible 235

digital signal processor (DSP) 25, *25*
direct current (dc) 30, 149, 234
- bias point 132
- link voltage 30, 60
- machines 1, 54, 170
- power supply 251
- voltage excitations 149

direct torque control (DTC) 1
distorted voltage 30, 32, 34–35, 43, 46, 52, 54, 64, 76, 89, 91, 100, 149, 153, 161, 213
- in *abc* stationary reference frame 32
- d-axis 35
- gain matrix of 33
- VSI-caused 16

dq-axis apparent inductances 125
dq-axis currents 23, 39, *39*, 43, 75, 78, 80, 90–91, 94, 96, 114–115, 120, 123, 125, 144, 183, 246
- comparison of rotor speeds and *180*
- at constant rotor speed 16
- HF 154, 157
- optimal 127
- sixth harmonics in 37
- step response of *215*
- and voltage with/without VSI compensation *45*

dq-axis equations 112
dq-axis flux linkage maps estimation 111, 119
- complete data recording process at different load points *123*
- cost function minimization 121
- cost function modeling for flux linkage identification 119–121
- data recording and FE-predicted flux linkage maps 122
- d-axis inductance maps *127*
- derivation of dq-axis inductances 125–127
- at different speed regions *125*
- dq-axis flux linkage maps and estimation errors, identified *124*
- experimental results 122–128
- FE-predicted dq-axis flux linkage maps *123*
- ICQGA-based identification fitness *124*
- identification 122–123
- measured data during variable speed control at load point *122*

MTPA current trajectory based on 127–128, *128*
Q-axis inductance maps *126*
uncertain circuit resistance, influence of 123–125
under uncertain circuit resistance *126*
dq-axis flux linkages 16, 96, 111, 119, 124–125, *126*, 131–132, 150, 153, 231, 244
estimation errors of 97
FE-calculated three-phase and *246*
FE-predicted 122
partial derivatives 16
surface fitting of 121
variations in average *247*
dq-axis incremental inductances 16, 149, 153, 156–158
dq-axis inductances 23–24, 27, 46, 51, 75, 77, 94, 108, 112, 119, 153, 166, 212, 216
derivation 125–127
estimation accuracy of 78
estimation 88, *88*, *89*, 95–96, *96*
incremental/apparent 14
matrix 137
measured quantities for estimation of *156*
nominal 154
saliency of 83
self-and mutual-magnetic saturation on 218
in SPM machines 3
subtracting contributions of 244
dq-axis voltage equation 8, 52, 76, 78, 94, 190, 218, 220
with positive and negative position offsets 99
second harmonics in 234
dq rotating reference frame, VSI modelling in 32–34
drive system of PMSM 11–12, *12*
DSP. *see* digital signal processor (DSP)
DTC. *see* direct torque control (DTC)

e

EA. *see* evolutionary algorithms (EA)
EKF. *see* extended Kalman filter (EKF)
electrical equations of PMSM 166
electrical machines 1
magnetic field distributions of 131
parameter identification of 121
parameter variations in 2
electric vehicles (EVs) 1–2, 209, 229
electromagnetic torque 7, 82, 165, 170
constant 114
prediction of 169
electromotive force (EMF) 4, 44, 54, 149
EMF. *see* electromotive force (EMF)
energy inductance 33
equivalent apparent inductance 9, 133
error analysis 47, 61
and experimental validation 69–72
multi-parameter estimation based on current injection and 60–68
error transfer issue 224, *225*, 228
estimation errors 103, 107, 122, 171, 220, 224
aggregate variance of 186
of dq-axis inductance maps *105*
of estimated HF inductances 146
of extend back-EMF-based sensorless control *221*
identified dq-axis flux linkage maps and *124*
variations of *108*
evolutionary algorithms (EA) 101, 111, 183–184
Euler's method 23
EVs. *see* electric vehicles (EVs)
extended back-EMF-based method 218–220, *219*
dq-axis apparent self and mutual inductances variations with currents *220*
estimation errors *221*
extended Kalman filter (EKF) 24, 59, 166, 183, 189
external rotor SPM machines 3

f

fast Fourier transform analysis (FFT analysis) 15, 35, *36*, 54
fault diagnosis 1, 17, 209, 234
demagnetization fault 235–236
ITSC faults 234–235
FE. *see* finite element (FE)
FEA. *see* finite element analysis (FEA)
FFT analysis. *see* fast Fourier transform analysis (FFT analysis)
field-oriented control (FOC) 1, 37, 209
design of PI regulators for 210–215
determination of MTPA current trajectory 215–216
dq-axis current loop and speed loop complete transfer function *211*
dq-axis currents, step response of *215*
electrical/mechanical parameters and optimized PI constants, estimated **214**
exemplary estimation procedure for design of PI regulators *213*
mechanical parameters, estimated *214*
parameter estimation 212–214
parameter-based PI regulator design 210–212
for PMSM *211*
rotor PM flux linkage estimation by fixing resistance *214*
rotor speed, step response of *215*
simplified transfer functions *212*
system 37, 52, 170
finite element (FE) 1
calculation of winding inductances 243
method 1, 17, 111, 131, 243

finite element analysis (FEA) 1, 14, 143, 152, 173, 243
FOC. *see* field-oriented control (FOC)
forgetting factor 186
fractional slot PM machines 4–5
frequency of perturbation signal 171
full-rank estimation model 100, 111
fundamental motion equation–based estimation method 168–169, *168*

g

GA. *see* genetic algorithm (GA)
gain factor 37
gain vector 186
Gauss–Newton method 183, 197–198
genetic algorithm (GA) 17, 91, 121, 183, 201
  application to parameter estimation for PMSMs 203
  basic principle 201–202
  dq-axis inductance, stator winding resistance, and rotor PM flux linkage, estimated *204*
  flowchart *202*
gradient-based methods 17, 183, 195–222, 204
  application to parameter estimation for PMSMs 198
  basic principle 196–198
  dq-axis inductance, stator winding resistance, and rotor PM flux linkage, estimated *199*

h

high-frequency (HF)
  components 35–36, *36*
  sinusoidal current injection–based offline estimation method 213, *213*
high-frequency current injection 153
  current variations, estimation based on 153–156
  dq-axis HF currents and voltages *159*
  measured rotor speed during *158*
  parameter estimation by 153–158
  zero-crossing detection, estimation based on 156–158
high-frequency pulsating voltage injection method 137–140
  position estimator, with aid of 137–139, *138*
  position estimator, without aid of 139–140, *139*
  sensorless position estimation based on *138*
high-frequency rotating voltage injection method 134–137, 140–142
  control diagram of HF inductance estimation *135*
  demodulation of amplitudes of positive and negative sequence HF currents *135*
  sensorless position estimation based on *136*
high-frequency signal injection method 132, 220
  calculated HF inductances by methods **144**
  combined HF rotating and pulsating voltage injection method 140–142, *140–141*
  d-axis pulsating signal injection-based sensorless control *222*
  dq-axis current sequences *144*
  estimation performance 142–144, *225*
  experimental results 142–146
  experimental setup *142*
  FE results, comparison with 144–145
  FEA calculated and estimated coupling factors, comparison of *223*
  HF inductances, comparison of *145*, **146**
  HF inductances, estimated variation of *143*
  high-frequency pulsating voltage injection method 137–140
  high-frequency rotating voltage injection method 134–137
  injected voltage, selection of 145–146
  injection sequences of, comparison of *132*
  magnetic saturation modelling and time delay effect in 133–134
  measured HF inductances, variation of *144*
  signal injection-based sensorless control waveforms of voltage and current *224*

i

ICA. *see* immune clonal algorithm (ICA)
ICQGA. *see* immune clonal-based quantum genetic algorithm (ICQGA)
IGBT. *see* insulated-gate bipolar transistor (IGBT)
ill-conditioned problem 46–47, *47*, 60, 63, 68–69, 119
immune clonal algorithm (ICA) 91, 121
immune clonal-based quantum genetic algorithm (ICQGA) 121
incremental *dq*-axis inductances 154
incremental inductance(s) 7, 9, 16, 131, *131*, 243
  dq-axis 8, 16, 125, 132, 149, 153, 156–158
  estimated *q*-axis 81
  HF 132
  integral 149
incremental reluctivity 9, *9*
inductance-capacitor-resistance (LCR) 160
inductances 1, 6–7, 51, 131
  apparent 7–8, *7*, 16, 82, 131, *131*, 132, 150, 243
  energy 9
  estimation of 112–119
  HF 136–137, *138*, 143
  incremental 7, 9, 16, 131, *131*, 243
  mutual 6, 133, 140, 243
  self 6, 133, 140, 243
  variations of 24
inductances' saliency, estimation of 95
injection angle 135, 143
insulated-gate bipolar transistor (IGBT) 29
interior permanent magnet synchronous machine (IPMSM) 5, 150

interior PM machines (IPM machines) 3–4, *4*
inter-turn short-circuit faults (ITSC faults) 209, 234–235, *235*
inverter non-linearity 2, 111
  dq-axis flux linkage maps estimation with 119–128
IPM machines. *see* interior PM machines (IPM machines)
IPMSM. *see* interior permanent magnet synchronous machine (IPMSM)
iron loss, influence of 11
irreversible demagnetization 1, 11, 235, *235*
ITSC faults. *see* inter-turn short-circuit faults (ITSC faults)
I-type IPM machines 4

k

Kalman filter (KF) 17, 187
  application to parameter estimation for PMSMs 190
  conventional Kalman filter 187–189
  dq-axis inductance, stator winding resistance, and rotor PM flux linkage, estimated *190*
  EKF 189
  parameter estimation based on *188*
KF. *see* Kalman filter (KF)

l

LCR. *see* inductance-capacitor-resistance (LCR)
least mean square (LMS) 192
least-square regression 185
linear recursive estimator 186
LMS. *see* least mean square (LMS)
low-pass filter (LPF) 26, 136, 139
  first-order Butterworth 26, 65
  time constant of measured *dq*-axis currents 177
LPF. *see* low-pass filter (LPF)
Lyapunov stability theory 190–191

m

magnetic saturation 131
  influence of 7–9, *7*
  modelling in high-frequency signal injection 133–134
magnetomotive forces (MMFs) 5, 131
maximum torque per ampere (MTPA) 1, 111, 209
  dq-axis flux linkage maps based current trajectory 127–128, *128*
  FOC-based MTPA control *216*
  MTPA current trajectory determination 215–216
  MTPA trajectories and output torques comparison *218*
  output torques and derived MTPA trajectories *217*
mechanical equations of PMSM 166
mechanical parameter estimation methods 2, 165–166
  acceleration/deceleration-based method 166–167
  comparison between schemes 171–173
  estimation moment of inertia of PMSM rotor *173*
  estimation results of scheme **171**, **172**
  experimental results 170–177
  fundamental motion equation–based estimation method 168–169, *168*
  MRAS observer–based method 167–168
  MRAS-based estimation *174*
  PI speed regulator, design of 177–180
  Prototype Machines 1 and 2, main design parameters of **171**
  rotor PM flux linkage, influence of variation in 173–175
  rotor speed and *q*-axis current measured and detected zero-crossing points *172*
  sinusoidal perturbation signal, sensitivity analysis of 175–177
  torque prediction, critical issue of 169–170
MMFs. *see* magnetomotive forces (MMFs)
model predictive control (MPC) 1, 210
model reference adaptive system (MRAS) 17, 24, 166, 183, 190
  application to parameter estimation for PMSMs 191–192
  basic principle 190–191
  dq-axis inductance, stator winding resistance, and rotor PM flux linkage, estimated *193*
  estimator *191*
  observer–based method 167–168
modern control and optimization theory–based parameter estimation algorithms 183, **205**
  ANN 192–195
  designed parameter estimation scheme 183–185
  GA 201–203
  gradient-based methods 195–198
  KF 187–190
  MRAS 190–192
  online parameter estimation *184*
  PSO 198–201
  RLS method 185–187
  simulated electrical quantities *185*
  three-step parameter estimation *184*
moment of inertia 165–166, 170–171
  combined 167, 210, 214
  of dc load machine and PMSM rotor 175
  estimation *173*
  rotor 7
  variation of estimated *176*
MPC. *see* model predictive control (MPC)
MRAS. *see* model reference adaptive system (MRAS)
MTPA. *see* maximum torque per ampere (MTPA)

multi-parameter estimation 24, 46, 60, *61*. *see also* parameter estimation
- *d*-axis current injection, estimated parameters based on *66*
- designed parameter estimation scheme 60–63
- error analysis 63–65
- experimental results 65–68
- iterative calculation of estimated parameters *63*
- measured variation with *d*-axis currents *65*
- Prototype Machine 1, estimated parameters for tracking temperature variations of *67*
- rank-deficient problem in 23
- sampled electrical angular speed and *dq*-axis currents and voltages *66*
- variations calculated and measured with different amplitudes *65*

multi-step estimation methods 150, 158
multi-step parameter estimation 158
- three-step estimation method 161–163
- two-step estimation method 159–161

mutual-magnetic saturation 131, 149–150, 218–220

n

negative sequence currents 135–136
Newton's method 196–197
non-linearity of VSI 29
- inverter 2
- modelling 30–34

non-overlapping stator windings 4
non-salient-pole permanent magnet synchronous machines 25, 51, 60, 70, 78, 84, 88, 108

o

offline parameter estimation 1–2, 14–16
- methods 149
- by signal injection 17

online compensation of VSI non-linearity 37–39
online estimation of rotor flux linkage 52–55, *53*
online parameter estimation 1–2, 16–17, *17*, 158, *184*
- application of 209
- error transfer issue 224, *225*
- ill-conditioned problem 46–47
- non-linearity of VSI 29–45
- for permanent magnet synchronous machines 149
- rank-deficient problem 23–29
- RLS algorithm in 185
- summary of applications of **237**
- SWRPO-based *226*

online tuning 37
on-state resistance
- of IGBT 30, 32–33
- of power devices, and equivalent core-loss resistance 111, 123
- of VSI and core-loss resistance 67

open-loop transfer function 105, 212
optimal current trajectory 210, 215–216

p

parameter estimation 132. *see also* mechanical parameter estimation methods
- applications 209
- control performance, improvement in 210–216
- designed parameter estimation scheme 183–185
- electrical 111
- fault diagnosis 234–236
- HF current injection 153–158
- of PMSMs 1
- precise torque estimation 229–232
- sensorless control, improvement in 216–229
- square voltage injection 150–153
- techniques 1, 14–17, *15*
- thermal condition monitoring 232–234
- variations of parameter estimation error *106*

particle swarm optimization (PSO) 17, 121, 183, 198
- application to parameter estimation for PMSMs 200–201
- basic principle 198–200
- *dq*-axis inductance, stator winding resistance, and rotor PM flux linkage, estimated *201*
- PSO-based parameter estimation *200*

permanent magnet machines (PM machines) 1, 2–5, 243
- basic equations and machine parameters 5–11
- mathematical model 7–11
- performance assessment 10
- physical properties **10**

permanent magnet synchronous machines (PMSMs) 1–2, *2*, *18*, 23, 51, 59, 111, 131, 149, 165, 209, 243
- application to parameter estimation for 186, 190–191, 193, 198, 200, 203
- correlation between voltages and space vectors **13**
- *dq*-axis equivalent circuits of *11*
- drives and control strategies 11–14
- electrical drive system for 2
- electrical parameters in 59
- fundamental mathematical model for 5–7
- highly saturated 243
- magnetic saturation in 131
- model-based sensorless control method 217–218
- multi-parameter estimation of 23
- non-salient-pole 108
- online parameter estimation under variable speed control *112*
- parameter estimation techniques 14–17
- prototype 251–252
- salient-pole 78, 94
- switching states and gate signals *14*

system control diagram for *41*
thermal conditions of 111, 232–234
phasor analysis of rotor position offset in PMSMs 75–77, *77*
PI speed regulator. *see* proportional–integral speed regulator (PI speed regulator)
PM machines. *see* permanent magnet machines (PM machines)
PMSMs. *see* permanent magnet synchronous machines (PMSMs)
POPE method. *see* position offset-based parameter estimation method (POPE method)
Popov's integral inequality 191–192
position estimator 132, 137, 143
with aid of 137–139
without aid of 139–140
position-offset-based online parameter estimation under sensorless control 224, *229*
phasor diagram of SPMSM with stator resistance error *226*
SWRPO-based online parameter estimation under sensorless control *226*
position offset-based parameter estimation method (POPE method) 75, 170, 173
data measurement process *170*
estimation errors, variations of *108*
features of POPE methods under different control strategies **76**
$i_d = 0$ and $i_d \neq 0$ under constant and variable speed control 99–103
$i_d = 0$ under constant torque/speed control 77–89
$i_d \neq 0$ under constant torque/speed control 94–98
$i_d = 0$ under variable speed control 89–94
iterative computation process *79*
phasor analysis of rotor position offset in 76–77, *77*
position offset injection 2, 17, 76, 111
analysis of amplitude of 103–107
current loop before and after *107*
I-type system *107*
machine state without 100
SPMSM on 224
upper and lower limits of 76
positive sequence currents 136
posterior distribution 189
precise torque estimation 229
based on estimated HF inductances 231–232, *232*
HF square wave voltage injection 230–231, *230*
predict phase in Kalman filter 188, 189
proportional–integral speed regulator (PI speed regulator) 12, 26, 64, 166
curve of friction coefficient *179*
design of 177–180
FOC-based mechanical parameter estimation system *177*
performance comparison between designed speed regulators *179*
rotor speeds and *dq*-axis currents, comparison of *180*
step response of rotor speed in forward and reverse directions *179*
transfer function of speed loop of field-oriented control *178*
prototype permanent magnet synchronous machines 251, *251*
characteristics of drive system and VSI **253**
experimental platform *252*
specifications of **252**
PSO. *see* particle swarm optimization (PSO)
pulse width modulation (PWM) 134, 251
PWM. *see* pulse width modulation (PWM)

q

q-axis flux linkage 120, 150–152
error *97*, *98*
estimated *125*, *153*, *247*
q-axis inductance 3–4, 62, 89, 219
estimation of *43*, 81–82, *82*, *83*, 88, *152*
incremental 82
maps predicted by FE and ICQGA-based method *126*
nominal 37
percentage error 81
QGA. *see* quantum genetic algorithm (QGA)
quantum genetic algorithm (QGA) 91, 115, 202
quasi Newton's method 183, 197

r

radial-flux rotor PM machines 3, *3*
radially magnetized IPM rotors 3
rank-deficient problem 16, 23, 59
estimated parameters of schemes *27*, *28*, *29*
experimental analysis and results 24–29, *26*, **29**
Prototype Machine 1, design parameters and specification of **25**
rank-deficient issue 23–24
recursive least squares (RLS) 17, 24, 59
algorithm 183
application to parameter estimation for PMSMs 186
basic principle 185–186
dq-axis inductance, stator winding resistance, and rotor PM flux linkage, estimated *187*
estimator 165
method 24, 185
residual-based ITSC detection methods 234, *234*
RLS. *see* recursive least squares (RLS)
rotor angular velocity error 168

rotor PM flux linkage 23, 35, 41, 51–52, 54, 59–62, 77, 87, 91, *97*, 119, 149, 159–160, 236
  complete estimation flowchart for *96*
  estimated and FE-predicted variations, comparison of *93*
  estimated and measured, comparison of *56*
  estimation of 79–81, 94–95, 112–119, *174*, *175*, 228
  FEA results, comparison of estimated and *175*
  influence of variation in 173–175
  line back-EMF and FFT results under no-load condition *54*
  at load condition 87
  magnetic saturation on 244
  at no-load condition 56, 87
  online estimation 52–55, *53*
  in PMSMs 97
  Prototype Machine 1 with buried thermocouples *54*
  stator winding resistance and estimated *233*
  thermal condition monitoring of rotor PM 55–57, *56*
  variations in 68, 85, *87*, *98*
  VSI non-linearity compensation, comparison of estimated with and without *55*
  winding resistance estimation and *160*
rotor position 75, 81, 136, 142, 154–155, 223, 246, 252
  error 219
  measurement error of 81
  offsets 16, 24, 88
  precision of 209
  sensor 2
  sensorless control 132
  variation of measured HF inductances with different *144*
rotor-saliency-based method 217–218
rotor speed 16, 26, 39, 55, 59, 62, 65, 68, 71, 78, 80, 82, 86, 95–96, 99, 165, 184, 210, 218
  harmonics of 35
  steady-state 61
  step response of 215
  variable 90
  waveforms and FFT results of measured *36*

S

salient-pole permanent magnet synchronous machines 25, 78, 84, 88, 94, 108
self-magnetic saturation 131
sensitivity analysis of sinusoidal perturbation signal 175–177
sensorless control 1–2, 16–17, 216, 237
  application of parameter estimation under 224–229
  experimental results 228
  extended back-EMF-based method 218–220
  high-frequency signal injection–based method 220–224
  improvement in sensorless control performance 218–224
  parameter estimation results *228*
  position-offset-based online parameter estimation under 224–228
  rotor position 132
signal-to-noise ratio analysis (SNR analysis) 76, 89, 156, 170
sinusoidal perturbation signal
  moment of inertia variation of estimated dc load machine and PMSM rotor *176*
  sensitivity analysis of 175–177
sinusoidal reference space vector 13
SNR analysis. *see* signal-to-noise ratio analysis (SNR analysis)
space vector pulse width modulation (SVPWM) 12–14, *13*, 30, 251–252
  strategy 252
  switch on/off signals of 30, *30*
SPM machines. *see* surface-mounted PM machines (SPM machines)
spoke-type IPM rotors 3–4, *3*
square voltage injection
  dq-axis flux linkage maps, estimated *153*
  parameter estimation by 150–153
  q-axis flux linkages, estimated without and with compensation *152*
  q-axis inductance, estimated against FEA results *152*
  q-axis voltage and *dq*-axis currents, measured *151*
square wave rotor position-offset (SWRPO) 224–225, *226*, 227
square-wave voltage injection 150, *151*
state-space model 183, 190
stator inductance 225
  dq-axis 190, 192, *194*
  estimation of 226–227, *226*, *228*
stator resistance 1, 23, 35, 52, 100, 111, 124, 133, 149–150, 213, 218, 224
  error 226
  estimation of 112–119, 227
  nominal 117
  temperature dependencies 11
  uncertainties 108
  variations of 16
stator winding resistance 6, 9–10, 24, 59–60, 62, 64, *66*, 68, 71, 75, 77, 149, 213
  correlation of measured *56*
  estimation of *44*, 79–81
  temperature-dependent 51, 232
  value 52
  voltage drop of 44
stator winding topologies 4, *5*

steady-state dq-axis equations 59
steady-state PMSM model 34, 52, 113
  with discrete sampling instant 119
steepest descent method 183, 196, 198
surface-fitting/curve-fitting techniques 120
surface-mounted PM machines (SPM machines) 3, *3*
  non-salient-pole 70
SVPWM. *see* space vector pulse width modulation (SVPWM)
SWRPO. *see* square wave rotor position-offset (SWRPO)

t

temperature 71, 117, 236
  coefficients 10
  effects 2
  housing 98
  influence of 5, 9–11, *10*
  PM 55, 94
  stator resistance dependencies and the rotor flux linkage 11
  stator winding 51
  stator winding resistance correlation and measured *56*
  tracking capability 67
temperature-dependent stator winding resistance 51
THD. *see* total harmonic distortion (THD)
thermal condition monitoring 1, 17, 232–233
  of rotor PM 51, 55–57, *56*
  stator winding resistance and estimated rotor PM flux linkage *233*
  temperature coefficients of copper and PM materials **233**
three-phase current waveforms 39, *41*
three-phase PMSM model 5, *5*, *7*
three-phase VSI 12
three-step estimation methods 149–150, 161
  flowchart *161*
  measured electrical quantities *162*
  rotor PM flux linkage, estimated *162*
  stator winding resistance, estimated *163*
time delay effect 132, 136, 139, 144
  in high-frequency signal injection 133–134, *134*
  influence of 137
torque prediction, critical issue of 169–170
torque transducer 51, 234
total harmonic distortion (THD) 4
transfer function *211*, *212*
  closed-loop 212
  of mechanical model of PMSM drive system 167, *167*
  open-loop 105, 167, 212
  of PI regulator 105
  of speed loop under FOC 177, *178*
tuning factor 37–38
two-dimensional FEA (2D FEA) 89, 244
two-step estimation methods 149, 159
  rotor PM flux linkage and stator winding resistance, estimated *160*
  winding resistance and rotor PM flux linkage, estimation of *160*, *161*
two-variable quadratic polynomial function 96, 120

u

uncertainties of circuit resistance
  dq-axis flux linkage maps estimation with 119–128
  influence of 123–125
update phase in Kalman filter 188, 189

v

variable speed control 16, 75, 89
  complete estimation method *91*
  d-axis inductance under constant load torque, estimation of *118*
  designed estimation method 90–92
  dq-axis flux linkage maps estimation with uncertainties 119–128
  experimental and FEA results 92–94
  experimental results 115–119
  $i_d = 0$ and $i_d \neq 0$ under 99–103
  identifiability analysis and influence of VSI non-linearity 113
  improved estimation scheme 113–115, *114*
  measured data under *116*
  measured electrical quantities *92*
  measurement process of POPE under *90*
  online parameter estimation under 111
  parameters based on POPE, estimated *93*
  parameters estimated under *117*, *118*
  position offset-based estimation with $i_d = 0$ under 89
  Prototype Machine 2, specification and measured parameters of **116**
  rotor PM flux linkage variations, comparison of estimated and FE-predicted *93*
  stator resistance, inductances, and rotor PM flux linkage estimation 112–119
viscous friction coefficient 1, 165–167, 169, 210
voltage fluctuation 106–107
  q-axis 224–225, 227–228
voltage source inverter (VSI) 11, 29, 51, 62, 252
  abc stationary reference frame, modelling in 30–32
  acquisition of high-frequency component *36*
  $\alpha\beta$-axis currents *40*
  designed experimental investigation *42*
  dq rotating reference frame, modelling in 32–34
  dq-axis currents *39*, *45*

voltage source inverter (VSI) (*Continued*)
electrical parameters **41**
estimated V_{dead} *37*, *39*
influences of VSI non-linearity 39–45
non-linearity estimation and compensation 34–39, *38*, *40*
q-axis inductance, estimated *43*
relationship between current directions and *dq*-axis voltages **34**
sixth harmonic components *35*
stator winding resistance and rotor flux linkage, estimated *44*, *45*
theoretical, equivalent, and real IGBT output voltages and drive signals *31*
VSI-based three-phase PMSM drive system *30*
waveforms *34*, *36*
VSI. *see* voltage source inverter (VSI)

W
winding inductances 17
average *dq*-axis flux linkages with currents, variations in *247*
cross-section of Prototype Machine 2, *244*
dq-axis apparent self-and mutual-inductances with currents, variations in *248*
dq-axis incremental self-and mutual-inductances with currents, variations in *249*
FE-calculated three-phase and *dq*-axis flux linkages *246*
finite element calculation of 243
flowchart of FE calculation of inductances *245*
winding resistance 11, 24–25, 33, 38, 41, 43, 51, 153, 159, 184, 193, 203
estimated rotor PM flux linkage and *161*
estimation of 46
values of 210

Z
zero-crossing detection, estimation based on 156–158
Ziegler–Nichols method 210

Printed and bound by CPI Group (UK) Ltd, Croydon, CR0 4YY
17/09/2026
14964335-0003